ASTRONOMY AND SATELLITE C(
PATHWAYS FORWA

IAU SYMPOSIUM

COVER ILLUSTRATION:

'Swamped Skies' by Joshua Rozells (image cropped with the photographer's permission).

IAU SYMPOSIUM PROCEEDINGS SERIES

INTERNATIONAL ASTRONOMICAL UNION
UNION ASTRONOMIQUE INTERNATIONALE

International Astronomical Union

ASTRONOMY AND SATELLITE CONSTELLATIONS: PATHWAYS FORWARD

PROCEEDINGS OF THE 385th SYMPOSIUM OF THE INTERNATIONAL ASTRONOMICAL UNION LA PALMA, CANARY ISLANDS, SPAIN 2–6 OCTOBER, 2023

Edited by

PETER GRIMLEY
AsterEd Consulting, UK

JOHN BARENTINE
Dark Sky Consulting, LLC, USA

DAVID GALADÍ-ENRÍQUEZ
Universidad de Córdoba, Spain

and

CONSTANCE WALKER
NSF NOIRLab/IAU CPS, USA

CAMBRIDGE
UNIVERSITY PRESS

CAMBRIDGE UNIVERSITY PRESS
University Printing House, Cambridge CB2 8BS, United Kingdom
1 Liberty Plaza, Floor 20, New York, NY 10006, USA
10 Stamford Road, Oakleigh, Melbourne 3166, Australia

First published 2026

Printed in Great Britain by Henry Ling Limited, The Dorset Press, Dorchester, DT1 1HQ

Typeset in System LaTeX 2ε

A catalogue record for this book is available from the British Library of Congress Cataloguing in Publication data

This journal issue has been printed on FSC™-certified paper and cover board. FSC is an independent, non-governmental, not-for-profit organization established to promote the responsible management of the world's forests. Please see www.fsc.org for information.

ISBN 9781009399227 hardback
ISSN 1743-9213

Table of Contents

Preface

Introduction

Piero Benvenuti

Since the launch of the first 60 Starlink satellites in May 2019, the astronomical community has been concerned with the impact of the large satellite constellations on the appearance of the night sky and on astronomy. By now this issue has become a hot topic of discussion not only within the scientific community, but also on social media and amongst decision makers. Undoubtedly, connecting to the Internet directly from space represents an important technological advancement, but the necessity of mitigating the interference that the large constellations have on the dark and radio-quiet sky is equally important for the future of astronomy.

Indeed, "Dark and Quiet Skies for Science and Society" was the theme of four online conferences jointly organised in 2020 and 2021 by the International Astronomical Union (IAU) and the UN Office for Outer Space Affairs (UNOOSA) with the collaboration of the Government of Spain. Given the importance of the theme and the interest of the astronomical community in the matter, in 2021 it was decided to submit a proposal for a dedicated IAU Symposium with the title "Astronomy and Satellite Constellations: Pathways Forward" to be held in La Palma (Canary Islands, Spain) in October 2023.

The Symposium, whose results are outlined here, was a great success and represents an important step forward in the definition of the specific impact of the constellations on diverse types of astronomical observations, in both the optical/infrared and the radio domains, as well as that on various cultural heritages. The success of the Symposium stems from the previous conferences and from the work of the IAU Centre for the Protection of Dark and Quiet Sky from Satellite Constellation Interference (IAU CPS), and had the additional benefit of an in-person component. The pleasant and friendly atmosphere of La Palma — "la Isla bonita" — helped to stimulate intense and animated discussions amongst more than 100 experts, joined remotely by an equal number of colleagues, producing the notable results that are outlined here.

A great thanks goes to the local organisers for their impeccable efforts and to all the participants for opening the path to effective and viable solutions to mitigate the interference of the satellite constellations.

Goals of the symposium

Connie Walker

The Symposium was planned as a dedicated space where astronomers, industry, space lawyers and other interested communities and stakeholders could share their perspectives or the current status of their work with respect to large satellite constellations and their impact on astronomy and the night sky. Through presentations and panel discussions, we explored the current status of mitigation strategies in terms of satellite observations, software development, satellite tracking, space situational awareness, telescope mitigation measures, brightness testing and modelling, spectrum management, policy developments, industry perspectives, and community engagement, to name just a few. Participants shared their current or future lines of research in these areas and discussions that followed identified gaps that should be addressed. (Individual slide sets and videos of the presentations are available at `https://bit.ly/IAUS385programme`.) By networking and exchanging information, headway was made towards the ultimate goal of mitigating the impacts of satellite constellations on astronomy, although there remains a long road ahead.

Session summaries and recommendations for actions going forward

The Symposium sessions were defined thematically by the functions of the IAU CPS "hubs": the Community Engagement Hub, SatHub, the Policy Hub and the Industry & Technology Hub.

Community Engagement

Jessica Heim and John Barentine

The Community Engagement sessions were fortunate to have presentations from Elder Wilfred Buck, a member of the Opaskwayak Cree Nation, and other traditional knowledge keepers. The speakers shared perspectives on their relationship with the sky and cosmos, and discussed ways to co-create collaborative endeavours related to the night sky and space. We heard from speakers with a background in history and science and technology studies, who helped us to examine past and present space and astronomy issues from different lenses, and to explore environmental and cultural aspects of these topics.

SatHub

Siegfried Eggl, Mike Peel and Meredith Rawls

SatHub focuses on observations, data analysis, software, services, and training materials to improve our understanding of the impact of satellite constellations on astronomy and observers worldwide. The session organised by SatHub leaders at the Symposium demonstrated that satellite constellation interference occurs across astronomical disciplines. Thirty-three domain experts provided unique insights into the technical challenges astronomers face as a result of the ever-increasing number and brightness of commercial constellation satellites in low Earth orbit (LEO).

Ultra-bright beams of satellites communicating with their ground receivers, as one example, can render radio observatories unable to observe the cosmos for extended periods of time. Unintended electromagnetic radiation from satellites, too, can interfere with radio observations. National borders and policies can put limits on radio protection zones; however, these only extend to ground-based transmitters, not satellite transmissions. In addition, large satellites that reflect sunlight back to the ground can be some of the brightest objects in the night sky. The sunlight reflected off such satellites would cause "whiteouts" of entire CCDs of the Vera C. Rubin Observatory LSST Camera. While there is only presently one such prototype satellite, BlueWalker 3, it is a clear case for satellite operators to proactively avoid in the future. Modelling the non-uniform distribution and brightness of LEO satellites and predicting when they reflect the most sunlight is one possible mitigation strategy for ground-based observatories, but there are significant challenges. Brightness modelling requires either partly privileged satellite operator information on satellites and laboratory resources, or substantial observation time on already oversubscribed telescopes.

The crowding of LEO further hampers active satellite avoidance strategies for telescopes because of the ever-increasing rate of satellite manoeuvres. A large number of collision avoidance manoeuvres limits the accuracy of long-term satellite orbit predictions. Moreover, crowded near-Earth space comes with additional challenges for adaptive optics as artificial guide stars crucial for wavefront sensing have to be turned off by default when certain satellites pass overhead to avoid damaging them. Last but not least, a large number of constellation satellites are on their way to changing the appearance of the night sky forever, which should not be taken lightly. However, both astronomers and industry partners of the CPS are working on strategies for co-existence, and this was a key topic of discussion during the Symposium's SatHub panel.

Satellite beam steering, conscious choices of the positioning of ground stations, and clear ITU regulation can help protect radio astronomy. Reducing backscattered light towards Earth's surface, predictive models for satellite spectra and brightness, studies of how detectors respond to bright streaks, and satellite monitoring capabilities are all crucial to minimising interference for optical astronomy (both imaging and spectroscopy). The key to mitigating satellite constellation interference for all sky observers is to engage with all stakeholders in an open and constructive manner. This is one of the principles of the IAU CPS.

Policy

Andrew Williams and Richard Green

The IAU CPS Policy Hub is engaging a growing network of over 150 people in research and policy work and regular interactions with government affairs representatives from several national astronomical societies, observatories and spectrum managers. The Policy Hub aims to raise awareness of astronomy requirements in space policy-making circles, coordinate policy work performed by national societies and observatories, facilitate proposals for new regulation in coordination with national points of contact, coordinate spectrum management processes with emerging optical-related issues, and identify future threats.

In the policy session, 28 experts on space law, regulation, policy and government affairs provided a range of talks on the dark and quiet skies policy issues. The overall theme was one of progress. In only a few years, the work of policy advocates and the broader astronomy community has resulted in increasing awareness of dark and quiet skies concerns for space policy. Presenters noted the increasing recognition of the topic at high levels, as evidenced by various proclamations and initiatives including the European Union Council Conclusions on Fair and Sustainable Space, the European Space Agency-led Clean Space Charter and Zero Debris initiatives, the Science Ministerial declaration from the G7 summit in May 2023, the UK Earth Space Sustainability Initiative, and endeavours at UN COPUOS.

Yet despite the growing political awareness, regulation and space policies that address dark and quiet sky concerns during licensing and on-orbit operations remain nascent. A review of national space laws and regulations showed that few countries consider dark skies in their national regulatory frameworks, whereas quiet skies remain managed via the national spectrum licensing processes. Nevertheless, some promising changes are emerging.

In the United States, the Federal Communications Commission (FCC) now mandates that space operators enter into a coordination agreement with the National Science Foundation (NSF) for the issuance of operational licenses. The European Union has authorised spending on, and is now actively planning, an independent satellite constellation — the IRIS2 — the design requirements of which include provisions to minimise optical brightness. The European Space Agency has created a "Clean Space" programme, which includes dark and quiet skies protection, and the German space strategy aligns with this objective. In Chile, a robust effort is underway to make environmental assessments a prerequisite for licensing. The UK's Astra-Carta and the Earth Space Sustainability Initiative (ESSI) were identified as having elements dedicated to dark and quiet skies protection. In Asia, space powers such as India and China are making strides in their national space programmes, but have yet to broach dark and quiet skies concerns.

The Symposium highlighted several significant work outputs. Firstly, an IAU CPS Policy Position Paper is in the final stages of preparation, and valuable insights were incorporated from the Symposium. Secondly, the Space Policy and Law Research Team is actively conducting a detailed analysis of legal and policy questions, with plans to release a comprehensive report to the CPS in 2024. The group recommended declaring astronomy as a form of space exploration, in order to support interactions with space policy making.

Additionally, a noteworthy partnership with the Space Sustainability Rating is progressing, resulting in the development of an initial framework for a dedicated module on dark and quiet skies. The Policy Hub also disclosed the creation of a consolidated requirements document, harmonising the diverse requirements arising from the SATCON1, 2, and D&QS1, 2 conferences. Lastly, a lunar policy study has been completed, with the establishment of a new lunar working group in progress.

A recurring theme throughout the session was the consensus on the necessity for continuous interaction with policymakers and industry stakeholders to raise awareness and advocate for protecting dark and quiet skies. The session underscored the importance of ongoing efforts to integrate dark and quiet skies considerations into various policies and initiatives, fostering a global commitment to this critical cause.

Industry

Tim Stevenson, Chris Hofer and Patricia Cooper

The Industry and Technology Hub session consisted of a panel including a representative from SpaceX, who provided an update on Starlink satellites and their mitigation measures. The Q&A highlighted a number of concerns regarding relationships with regulators, the state of the market and limits to it, and the flow of information from operators to support independent researchers.

Session on UN COPUOS and the Group of Friends

Mila Francisco

Spain and Chile are spearheading the efforts to protect the dark and quiet skies at the United Nations Committee on the Peaceful Uses of Outer Space (UN COPUOS). The Symposium had the honour of having Mila Francisco Ferrada (in person) and Jose Manuel Ramirez Arrazola (remotely) from the Chilean and Spanish delegations to UN COPUOS respectively and also Nathalie Richard (in person) representing the UN Office for Outer Space Affairs (UN OOSA). Together with the IAU CPS Director Piero Benvenuti and Ashley Zauder from the U.S. National Science Foundation, Federico Di Vruno from SKAO/CPS conducted an informative session on dark and quiet skies at the UN COPUOS with several presentations and a Q&A panel.

During the session the work of UNOOSA and its role in space capacity building and as secretariat of UN COPUOS were explained, along with the history of UN COPUOS, its functioning and committees, its treaties and how stakeholders can get involved. Also covered was the story of how the IAU engaged with UN COPUOS since 2018 on the topic of dark and quiet skies and how it reached a point where an agenda item was proposed in 2023 (unfortunately not accepted owing to lack of consensus).

In this open session, Jose Manuel Ramirez Arrazola and Mila Francisco Ferrera described the proposal for a Group of Friends on the Dark and Quiet Skies for Science and Society (GOF). In a following closed session, the Group of Friends was officially created with the support of the COPUOS delegations from Bulgaria, Chile, Germany, Italy, Luxemburg, New Zealand, Romania, Slovakia, South Africa, Spain, Switzerland, United Kingdom, USA and the Permanent Observers Committee on Space Research (COSPAR),

European Southern Observatory (ESO), International Academy of Astronautics (IAA), International Astronomical Union (IAU), and Square Kilometre Array Observatory (SKAO).

The session finished with a panel of experts who answered questions from the audience on topics like satellite licences, future activities on the Moon, and ongoing discussions at UN COPUOS on dark and quiet skies.

ITU

Veronique Glaude

Veronique Glaude from the International Telecommunication Union (ITU) gave a presentation about the ITU in general and in particular about the rules of satellite constellations. She highlighted the role that ITU has in maintaining the Radio Regulations, the international treaty which defines the international use of the radio spectrum, and how the ITU applies the regulations and also updates them every four years.

In particular for the satellite constellations she detailed the "satellite filling process" and the specific rules that apply in the case of non-geostationary satellites, like power limits, coordination requirements, and protection of the geostationary satellites among others.

In a separate presentation, Federico Di Vruno described the ITU-R regulatory framework and its procedures. He focused on how spectrum managers work at the ITU, how the ITU study groups work and what is the preparation process towards the World Radiocommunication Conference (WRC23). It was also mentioned that for the WRC23 there are several proposals for agenda items for WRC27 (as one conference also defines the agenda for the following one).

Concluding remarks

Federico di Vruno

IAU Symposium 385 was the first hybrid meeting of experts focused solely on the topic of the impact of satellite constellations on astronomy. The participants included astronomers, Indigenous Peoples, spectrum managers, and space policy experts, all highlighting the importance that this topic currently holds. The discussions were non-stop, not only during the presentations but also in the coffee breaks and during the many excursions that the local organising committee planned for us.

We highlight the quality of the research from the astronomical community, where several researchers have volunteered time away from their normal field of study to better understand the optical and radio frequency characteristics of satellite constellations and to assess their impact on astronomy observations.

The general conclusion of the Symposium is that we now understand much more about the constellations, and now it is time for concrete actions from both industry and academia. As a conclusion activity, the participants worked in a proposed list of actions, which are summarised here:

- Develop an orchestrated strategy: to find regulations, controls, moratoriums, etc
- Develop engaging media campaigns like "seatbelts for satellites"
- Increase the efforts in outreach to the general public, like influencers, secondary and primary school teachers, planetariums, etc.
- Seek broader coalitions with other scientific areas like the environmental sciences
- Make research more widely available to the public (simulations, observations, etc)
- Work on having a consistent message from the astronomical community and other interested actors in the protection of dark and quiet skies.

- Consider the trade-off between strong actions such as lawsuits and consensual policy actions
- Engage the youth to protect dark and quiet skies
- Make the protection of space a human-interest story, where dark and quiet skies are an important part of it but not all
- Consider answering the question "how many satellites is too many for astronomy"
- Suggest standards such as "sustainability by design"
- Consider making available agreement templates between countries and companies
- Keep using scientific research as a driver for change

Editors

SOC

LOC

List of Participants

LAST NAME	FIRST NAME	COUNTRY
Adam	Christian	Chile
Aguiar Álvarez	Amanda	Spain
Ahmadi	Hanane	Germany
Albrecht	Rudolf	Austria
Alemu	Etsegenet	Ethiopia
Alqassab	Myriam	Bahrain
Alvarez	Brian	United States
Andreasyan	Hasmik	Armenia
Angulo Rodríguez	María Isabel	Spain
Argo	Megan	United Kingdom
Arjó Francés	Sara	Spain
Artigas	Carmen	Uruguay
Aubé	Martin	Canada
Bailey	Judith	Australia
Bannister	Michele	Australia
Barbosa	Domingos	Portugal
Baroch Lopez	David	Spain
Barstow	Martin	United Kingdom
Bassa	Cees	Netherlands
Bautista Durán	Marta	Spain
Benbekhti-Winkel	Nadya	Germany
Benkhaldoun	Zouhair	Morocco
Benvenuti	Piero	Italy
Benz	Willy	Switzerland
Betmann	Maximilian	Germany
Bialek	Spencer	Canada
Billot	Nicholas	France
Blagojevich	Tamara	Serbia, Republic of
Boley	Aaron	Canada
Bonilla	Eliud	United States
Boskri	Abdelkarim	Morocco
Bradshaw	Andrew	United States
Breslin	Peter	Netherlands
Brooks	Jacob	United Kingdom
Bruno	Sarah Marie	United States
Buck	Wilfred	United States
Cabrera Lavers	Antonio Luis	Spain
Campion	Nicholas	United Kingdom
Caraveo	Patrizia	Italy
Cavallaro	Cassandra	United Kingdom
Chavez	Juan Carlos	United States
Christou	Julian	United States
Clark	Stuart	United Kingdom
Cooper	Patricia	United States
Corradi	Romano Luigi Maria	Spain
Corral Escobedo	Luis José	Mexico
Dadighat	Michelle	United States
Dalal	Roohi	United States
Damé	Luc	France

Daniels	Paul	United Kingdom
Davis	Christopher	United States
De Pree	Chris	United States
Decker	Shelley	United Kingdom
DeMarchi	Lindsay	United States
Despois	Didier	France
Devarakonda	Yaswant	United States
Di Vruno	Federico	United Kingdom
Diaz Castro	Francisco Javier	Spain
Domínguez	Mariano	Argentina
Dominici	Tania	Brazil
Douchin	Dimitri	Australia
Drago-González	Álex	Finland
Duelli	Markus	United States
Eaves	Megan	United Kingdom
Eggl	Siegfried	United States
Engelbrecht	Brandon	South Africa
Etemadi Tajbakhsh	Shahriar	United Kingdom
Farina	Cecilia	Spain
Francisco	Mila	Chile
Galadí-Enríquez	David	Spain
Gális	Rudolf	Slovakia
Giovanardi	Fabio	Italy
Glaude	Veronique	Switzerland
Goldstein	David	United States
Goodman	Matthew	United States
Green	Richard	United States
Grigg	Dylan	Australia
Grimley	Peter	United Kingdom
Grob	Thomas	Germany
Grosse	Doris	Australia
Gupta	Neeraj	India
Hainaut	Olivier	Germany
Hnel	Andreas	Germany
Hartley	Ruskin	United States
Harvey	Alvin	United States
Hasan	Priya	India
Hellbourg	Gregory	United States
Herrero Casas	Enrique	Spain
Hertz	Paul	United States
Hiramatsu	Masaaki	Japan
Hu	Tianzhu Hu	China
Hurtado	Norma	France
Hwang	Narae	Korea, Rep of
Idelbacha	Hamid	Morocco
Indermuehle	Balthasar	Australia
Isidro	Mathieu	United Kingdom
Ivantsov	Anatoliy	Belgium
Jilete Calleja	Beatriz	Spain
Johnson	Christopher	Netherlands
Johnson	Jourdan	United States
Johnson	Nathan	United States
Jozsa	Gyula	Germany
Kaab	Mohamed	Jordan
Kadyrova	Daiana	Kazakhstan

Kaley-Isley	Jonathan	Netherlands
Koplow	David Alexander	United States
Krafton	Kelsie	United States
Kragie	Eileen	United States
Krantz	Harrison	United States
Langlois	Pierre	Canada
Lawler	Samantha	Canada
Lawrence	Andrew	Scotland
Le Pellec	Marie	France
Linde	Peter	Sweden
Lindqvist	Gunnar Jerry Michael	Sweden
Longa	Penelope	Chile
López-Pérez	José Antonio	Spain
Lourenco	Liroy	Australia
Lucatello	Sara	Italy
Luis Valle	Andrea María	Germany
Machon	Miloslav	Czech Republic
MacLeod	Kenneth	United States
Madkour	Waleed	Netherlands
Malkin	Zinovy	Russian Federation
Manasoiu	Otilia	United Kingdom
Marcoux	Laura-Alexe	Canada
Markiewicz	Monika	United Kingdom
Masi	Gianluca	Italy
Massey	Robert	United Kingdom
McCoy	Michael	United States
Mejuto	Javier	Honduros
Micheva	Genoveva	Germany
Molinari	Emilio Carlo	Italy
Moro	Rafael	United States
Most Farjana	Sharmin	India
Muñoz Tuñón	Casiana	Spain
Navarro Jiménez	Silvana Guadalupe	Mexico
Neilson	Hilding	Canada
Nichols	Tiffany	United States
Nicotera	Lorena	Italy
Nolasco	Juan	Portugal
Núñez de Murga	Jorge	Spain
Oliver	Amy	United States
Ong	Isabella	Singapore
Otarola	Ángel	Chile
Pashkovskii	Vladimir	Germany
Peel	Mike	United Kingdom
Persaud	Donny	United States
Pietu	Vincent	France
Plavchan	Peter	United States
Polin	Daniel	United States
Pratiwi	Nindhita	Indonesia
Pugh	Teznie	United States
Python	Vincent Paul	Switzerland
Rajagopalan	Rajeswari Pillai	India
Rand	Lisa Ruth	United States
Redman	Matt	Ireland
Reinert	Thomas	United States

Renchin	Tsolmon	Mongolia
Ricard	Nathalie	Switzerland
Riddick	Fiona	Spain
Riolobos	Javier	Spain
Risacher	Christophe	France
Rotola	Giuliana	Italy
Saada	Adrien	Switzerland
Sabolová	Katarína	Slovakia
Seaman	Robert	United States
Sefako	Ramotholo	South Africa
Seiser	Samuel David	Austria
Seitzer	Patrick	United States
Semkov	Evgeni	Bulgaria
Serebryanskiy	Aleksander	Kazakhstan
Šilha	Jiří	Slovakia
Siminski	Jan	Germany
Singh	Sanjay	India
Siringo	Giorgio	Chile
Snyder	Adam	United States
Sokol	Joshua	United States
Solter-Hunt	Sierra	United States
Sorokin	Boris	Russian Federation
Spencer	Samuel Timothy	Germany
Stark	David	United States
Stevenson	TJ	United Kingdom
Stocker	Greg	United Kingdom
Stoner	Jeff	United States
Strasser	Jonas	Germany
Sun	Weijia	Germany
Sundin	Maria	Sweden
Tahar	Mohammad Redzuan	Malaysia
Teimoorinia	Hossen	Canada
Thiele	Sarah	United States
Thomson	Fionagh	United Kingdom
Tornatore	Vincenza	Italy
Tregloan-Reed	Jeremy	Chile
Tsakiri	Maria	Greece
Turnshek	Diane	United States
Tyson	Tony	United States
Unda-Sanzana	Eduardo	Chile
Ussia	Giuseppe	Italy
Vaduvescu	Ovidiu	Spain
van den Berg	Maureen	United States
Vanderley	Bevin	United States
Vanotti	Maurizio	United Kingdom
Varela	Antonia	Spain
Venkatesan	Aparna	United States
Wahid	Kamaruzzaman	Malaysia
Walker	Connie	United States
Wallner	Stefan	Austria
Watson	Fred	Australia
Watson	Robert	Australia
Webster	Rachel	Australia

Wesselius	Paul	Netherlands
Williams	Andrew	Germany
Winkel	Benjamin	Germany
Woods	Paul	United Kingdom
Woollands	Robyn	United States
Wooten	Michelle	United States
Yakushina	Yana	Belgium
Yap	Xiao-Shan	Netherlands
Zamora Sánchez	Olga María	Spain
Zancanaro	Yannis	France
Zhumagulov	Aslan	Kazakhstan
Zilkova	Danica	Slovakia
Zúñiga Fernández	Sebastián	Belgium

Credit: Antonio Gonzáles (`www.cielos-lapalma.es`)

Credit: Antonio Gonzáles (www.cielos-lapalma.es)

Credit: Antonio Gonzáles (www.cielos-lapalma.es)

Astronomy and Satellite Constellations: Pathways Forward
Proceedings IAU Symposium No. 385, 2026
P. Grimley, J. Barentine, D. Galadí-Enríquez & C. Walker
doi:10.1017/S1743921324000899

Streaks, signals, and space junk: facing the challenges of satellite constellations

Karen L. Howard

PhD; Director of Science and Technology Assessment, U.S. Government Accountability Office (GAO)
email: HowardK@gao.gov

Abstract. Satellites provide important data and services, such as communications, internet access, Earth observation, and technologies like GPS that provide positioning, navigation, and timing. However, the launch, operation, and disposal of an increasing number of satellites could cause or increase several potential effects, such as hindering optical and radio astronomy, creating additional space debris, and affecting the upper atmosphere through launch and reentry emissions. To help policymakers address the challenges presented by the ongoing increase of satellites in orbit, the U.S. Government Accountability Office (GAO) issued a report: Large Constellations of Satellites: Mitigating Environmental and Other Effects (GAO-22-105166). This keynote address discussed the report findings, including policy options directed to a range of policymakers that may include legislative bodies, government agencies, researchers, standards-setting organizations, industry, and other groups.

Keywords. Astronomy, space policy, satellite constellations, space junk, orbital debris, policy options, U.S. Government Accountability Office, GAO

1. Introduction

Sputnik 1, the first artificial satellite in space, launched in 1957. The number of active satellites increased steadily for the next 50 years, then surged from around 1400 in 2015 to almost 5500 by spring of 2022. By January 2023, there were more than 6700 active satellites. These statistics have been obtained following the practice used by the Defense Intelligence Agency in DIA (2023), analysing the public satellite database maintained from UCS (2023). This trend is expected to continue to accelerate, with multiple experts reporting that around 58 000 additional satellites could be launched by the end of the decade. Ongoing and proposed launches of a series of large constellations of satellites are the primary drivers of this increase. Private companies plan to use these satellite constellations to provide services such as broadband internet access in underserved rural communities. Communications satellites comprised 66 percent of active satellites in 2022, while 21 percent were for remote sensing, the next largest category. Of the total satellites in orbit in 2022, around 63 percent were operated by U.S.-based entities, including government and commercial operators (UCS 2023). However, space is a shared environment, and several dozen other countries are currently operating satellites. This is an issue that requires international attention and better collaboration.

2. Potential effects of large satellite constellations

Satellites provide important data and services, such as communications, internet access, Earth observation, and technologies like GPS that provide positioning, navigation, and timing. However, the launch, operation, and disposal of a rapidly increasing number of satellites could cause or heighten several potential effects. For example, sunlight reflections and radio transmissions from satellites can affect telescopes, which could make it more difficult for astronomers to assess risks associated with near-Earth asteroids or to observe other celestial objects. Other potential effects include emissions from the rockets that carry satellites into space and from reentry of satellites at the end of life. Such emissions can cause temperature changes in the atmosphere and affect the ozone layer. Finally, orbital debris from the increasing number of satellites complicates satellite operations and poses other risks.

In light of the broad congressional interest in commercial satellites and to assist the United States Congress with its oversight responsibilities, we prepared the report GAO (2022). This keynote address provided a summary of the report findings, including: (1) the potential effects from projected increases in large constellations of satellites; (2) current or emerging technologies and approaches to evaluate and mitigate these effects, along with challenges to developing or implementing these technologies and approaches; and (3) policy options that might help address the challenges. These policy options are directed to a range of policymakers that may include legislative bodies, government agencies, researchers, standards setting organizations, industry, and other groups.

GAO (2022) assessed technologies and approaches to evaluate and mitigate the following potential effects of satellite constellations:

- **Reflected sunlight.** Satellites can reflect sunlight that affects ground-based optical astronomy, especially for wide-field imaging telescopes. This is most problematic near dusk and dawn, and when satellites are at higher altitudes in low Earth orbit. These reflections also affect cultural and religious use of the night sky, interfere with amateur astronomy and astrophotography, and increase the night sky's overall brightness. These reflections can also affect data collection from other satellites, including space-based telescopes such as Hubble. Experts and stakeholders suggested several possible approaches to mitigate these light reflections, including satellite design choices to darken satellites, enhanced models that could allow for establishing brightness standards, collaboration on orbit assignments, shared data to enable better predictions of timing for avoidance, and improved image processing technology to remove satellite light streaks.
- **Radio transmissions.** Satellites can also transmit radio signals that obstruct observations of faint, naturally occurring radiofrequency signals. Satellite operators and astronomers are beginning to explore ways of mitigating these effects through consideration of satellite design and operation (e.g., radio beam size, timing of transmissions). Other potential mitigations include strengthening protections for spectrum and radio astronomy observations, improving data sharing between satellite operators and the astronomy community, and improving hardware and data collection for radio telescopes.
- **Emissions into the upper atmosphere.** Rocket launches and satellite reentries produce particles and gases that can affect atmospheric temperatures and deplete the ozone layer. The dramatic increase in the number of satellites will lead to a similar increase in end-of-life satellite reentry. Limiting the use of rocket engines that produce certain harmful emissions could help mitigate the effects of launches. However, the size and significance of these effects for both launch and reentry emissions are poorly understood due to a lack of observational data, and it is not yet clear whether mitigation will be warranted. Potential approaches include increasing collection of observational data, establishing emissions databases, setting metrics for comparison, and sharing data to help assess whether mitigations are warranted.
- **Increase in orbital debris.** Orbital debris is highly varied in size and comes from many sources, including collisions between large pieces of debris or between active satellites and orbital debris, destruction of satellites by antisatellite weapons, and loss of

structural integrity for aging satellites. Debris in space can damage or destroy satellites, which affects commercial services, scientific observation, and national security. Debris also increases diffuse brightness of the night sky and risks of human injury or casualty and property damage when pieces fail to completely disintegrate upon reentry. Possible mitigations include better characterization of debris, improved space situational awareness and adherence to operational guidelines, designing for complete disintegration upon reentry (although this has potential tradeoffs with increased atmospheric emissions), better control of satellite reentry timing and location, and removal or repurposing of debris. However, achieving these mitigations is challenging.

3. Policy options

The global nature of these effects gives rise to some common challenges in the U.S. and internationally, including: insufficient knowledge of the scale of these effects; insufficient technology development to help address them; lack of shared data; absence of standards, regulations, and agreements; and insufficient organization and leadership.

GAO developed the following policy options to help address challenges with evaluating and mitigating the effects of large constellations of satellites. We presented such options to help policymakers consider and choose those appropriate to the goals they hope to achieve. Policymakers may include legislative bodies, government agencies, researchers, standards-setting organizations, industry, and other groups.

Policy options:

- **Build knowledge and develop technologies** (GAO 2022, p. 57)
  - Policymakers could support research on the extent of potential effects, as well as development of technologies to address them.
- **Improve data sharing** (GAO 2022, p. 57)
  - Policymakers could facilitate improved sharing of relevant data about satellite constellations.
- **Establish standards, regulations, and agreements** (GAO 2022, p. 58)
  - Policymakers could establish appropriate standards, regulations, and agreements to help mitigate potential effects of satellite constellations.
- **Improve organization and leadership** (GAO 2022, p. 59)
  - Policymakers could build national and international organizational and leadership structures that facilitate effective mitigation of the potential effects of satellite constellations.

Ideally these policy options would be considered concurrently given the lengthy timeframes that may be needed to implement some of them and the rapidly growing number of satellites.

Policymakers may be better positioned to act on this complex issue if they consider interrelationships among these policy options. For example, implementing the fourth option (improving organization and leadership) may improve policymakers' ability to implement the first and second options (building knowledge, developing technologies, and improving data sharing). Similarly, implementing the first option may help with the third option (establishing standards, regulations, and agreements). More generally, trade-offs between mitigations may emerge, the ongoing increase in new constellations may introduce unexpected changes, and a large and diverse set of interests from the global community may shift over time, all of which present persistent uncertainties. To address these complexities and uncertainties, the full report presents the policy options in a framework, which may help policymakers strategically choose options to both realize the benefits and mitigate the potential effects of large constellations of satellites.

International challenges call for international cooperation and action, and for parallel efforts within a framework that acknowledges the interrelated nature of these policy options and the tradeoffs involved.

4. Scope and methodology

We focused this technology assessment on large commercial constellations of satellites, considering direct environmental and other effects that the constellations may introduce or exacerbate. We reviewed literature; interviewed agency officials, industry representatives, and experts in academia and at a federally funded research and development center; and conducted a 2-day meeting of 15 experts. We selected the experts based on their technical, legal, economic, or policy expertise, with a goal of obtaining a balanced and diverse set of views from government scientists, nongovernmental experts, industry representatives, and academic researchers. For more information on the objectives, scope, and methodology of this technology assessment, see Appendix I of GAO (2022).

References

Defense Intelligence Agency (DIA) 2022, *Challenges to Security in Space: Reliance in an Era of Competition and Expansion*, ISBN 978-0-16-0955-66-2, available at https://www.dia.mil/Portals/110/Documents/News/Military_Power_Publications/Challenges_Security_Space_2022.pdf

Government Accountability Office (GAO) 2022, *Large Constellations of Satellites: Mitigating Environmental and Other Effects* (GAO-22-105166), available at (https://www.gao.gov/products/GAO-22-105166)

Union of Concerned Scientists (UCS) 2023, *UCS Satellite Database*, Cambridge, Mass: Jan. 1, 2023, accessed Sept. 29, 2023, available at https://www.ucsusa.org/resources/satellite-database

Astronomy and Satellite Constellations: Pathways Forward
Proceedings IAU Symposium No. 385, 2026
P. Grimley, J. Barentine, D. Galadí-Enríquez & C. Walker
doi:10.1017/S1743921324000851

From Sputnik to Starlink: historical perspectives on astronomy and satellite communications

Lisa Ruth Rand

Division of the Humanities and Social Sciences, California Institute of Technology
email: rand@caltech.edu

Abstract. As the number and variety of communications satellites proliferate in the rapidly accelerating New Space Age, specialists and lay observers alike have weighed the implications of a crowded orbital environment for astronomical research. Although so-called "megaconstellations" may be a relatively new development, conflict over the anticipated effects of space technology on astronomy has a much longer history, going back to the very first experimental communications satellites. Astronomers have long acted as environmental watchdogs for outer space even as the power landscape in the international aerospace industry shifted during the half century since Sputnik. What lessons might current astronomers take from past conflicts over the integrity of the night sky to address contemporary and future risk?

Keywords. Space history, project West Ford, satellite constellations

---

The age of commercial satellite constellations presents distinct challenges to astronomers. However, these challenges are not without historical precedent. The international astronomical community, and the IAU as an institution, addressed remarkably similar technological and political dilemmas in the early years of the satellite age when faced with uncertainty about interference from novel communications satellite systems. Understanding the historical landscape of these dilemmas may yield relevant context for current conversation and controversy about the shifting landscape of satellite communications and the politics and power that undergirds those shifts. In particular, the history of the first communications satellite system, eventually known as Project West Ford, has become increasingly relevant 60 years after its second, successful launch.

The designers of the first satellites constructed them in a variety of shapes, forms, and purposes. Some bore scientific instruments, and some did little more than announce their presence in a strange new environment. The first artificial satellite, Sputnik 1, reached orbit in October 1957. While its famous "bleep"—a taunt or a cheer depending on the political inclinations of the listener—fostered unprecedented if tense connections across a rapidly shrinking world, they were just that—signals. The first Sputniks, and the earliest American satellites that joined them, were not true communications satellites in that they could not relay a signal sent from the ground to another point on the ground.

Satellites were not the only technical forays into orbit during this time. Shortly after Sputnik 1 ascended into the October sky both the United States and the Soviet Union began conducting nuclear weapons tests at the edges of outer space. Project Argus and Operation Hardtack detonated several devices at altitudes ranging from 43 to 540 kilometers in 1958, some of them alarming observers in Hawaii with unannounced fireballs (Honolulu Star-Bulletin 1958a,b,c; Hoerlin 1976). These practices continued into the early 1960s with tests like Operation Star Fish Prime whose 1.4 megaton yield detonated at an altitude of 400 kilometers disabled several operating satellites (Bing 1962). After each test the land area and airspace below experienced extended periods of communications blackout due to ionospheric disruption. In this period of tense nuclear conflict tenuously held at bay by the threat of mutually assured destruction, the potential use of nuclear weaponry in outer space to isolate combatants without causing material harm on the ground below drew attention from American defense officials as both a tool and a threat (Department of Defense 1961; Fleming 2011).

In the midst of this geopolitical instability and technological flexibility, the United States Army Signal Corps convened a study in Cape Cod, Massachusetts. The assembled group of technical and scientific experts were charged with the main goal of considering ways to implement satellite technology to increase communications reliability and coverage for the United States armed services in the face of such novel threats (Terrill 1999). One of the promising ideas to come out of this study suggested a passive approach to satellite communications built upon World War II radar counteroffensive techniques.†

This proposed experimental system did not resemble the substantial, electronics-bearing, individually operating satellites commonly designed today. Rather, each satellite in the system that came to be known initially as Project Needles consisted of an ethereal copper filament, only 1.77 centimeters long and thinner than a human hair. The test version of the system included 400 million of these dipoles deployed into a polar orbit of about 3500 kilometers above the Earth's surface. Once diffused into a ring the dipoles would act as a passive reflection medium for microwave signals sent from the ground, a mechanism that mimicked the properties of the natural ionosphere whose vulnerability to nuclear attack had inspired the summer study (Morrow and MacLellan 1961).

Like the ionosphere, its designers predicted that Project Needles could be used for long-range radio communications by scattering microwave transmissions tuned to the dipoles. The system held promise to even improve upon the natural ionosphere, given its relative resilience, lower susceptibility to seasonal variation, jamming, and physical attack. All expensive equipment requiring maintenance would be constructed on the ground, and all space components made of cheap, readily available industrial materials that could be installed piggybacked on another defense payload. If two belts were to be established, one with a polar inclination and another equatorial, such a dipole scatter system could provide worldwide coverage with opportunities for nearly infinite channels (Richmond 1960).

The MIT Lincoln Laboratory sought to build and test this system, and secured funding from the US Air Force to develop and launch a preliminary version. (Terrill 1999) However, Lincoln Laboratory researchers grew concerned that the reflective dipoles might have detrimental effects on astronomical research. They sought the guidance of the newly formed Space Science Board (SSB) of the National Academy of Sciences to vet the safety of the classified project. The Board convened a committee of astronomers and engineers to consider how the dipoles might interact with the outer space environment and anticipate any potential impacts on scientific research. The committee was also charged with predicting how the project might be received by scientists and lay citizens at home and abroad (Goldberg 1961a).

In April 1960 the committee delivered a preliminary report urging the Air Force not to conduct the test over concerns about potential long-term detrimental outcomes that

† For the World War II countermeasure ancestor to Project Needles, see "Project Window" and "Project Chaff," in Buderi (1996), pp. 207–208.

a successful system might yield for optical and radio astronomy. The committee's recommendation concerned less the test belt as proposed but rather the precedent that its success might set. Should dipole belts become a common form for communications satellites, multiple, denser fields of reflective material could dramatically intensify background noise and the occurrence of false signals. The committee also predicted an adverse reaction from astronomers around the world to what they might see as a threat to their current and future observational practices. Rather than risk this outcome, the committee suggested the project be shelved (Department of State 1960).

In June 1960, a second draft of the report changed the overall position to one of cautious recommendation that the test be carried out with proper precautions. Members of the committee clashed with SSB chair Lloyd Berkner, whose familiarity with federal agencies led him to suggest that working with the military rather than in direct opposition could be a better approach to protect astronomers' interest given that the project was likely to proceed regardless of dissent (Berkner 1961). Compromise came in an agreement to declassify parts of the project under the name Project West Ford to solicit feedback on its safety from a broader quorum of experts (Goldberg 1961b).

Members of the ad hoc committee and researchers at the MIT Lincoln Laboratory published a series of articles about the dipole system in April 1961 edition of *The Astronomical Journal.* The articles emphasized the safety of the test project to radio and optical astronomy and outlined predictions for a short orbital lifetime for all dipoles in the proposed single belt.† Rather than assure astronomers of the project's innocuousness, the publication unleashed a torrent of dissent that came to a head the following summer at the 11th General Assembly meeting of the IAU in Berkeley, California.

At this meeting the assembly put forward a resolution against Project West Ford. The resolution issued a warning that the integrity of orbital space for scientific research could be threatened by such satellite systems. Through the resolution the IAU membership positioned astronomers as canaries in the coal mine of outer space, stating that no group has the right to change Earth's environment (including outer space) in any significant way without full international study and agreement. The resolution requested that governments with space industries consult with the IAU before launching potentially dangerous space projects so that safety may be agreed upon by a quorum of experts (IAU 1961). In the coming months similar resolutions issued forth from organizations including the American Astronomical Society, the International Union of Radio Science, the French and Belgian National Academies of Science, the Royal Astronomical Society, and the Committee on Space Research of the International Council of Scientific Unions.

These collective expressions of resistance to an American military project reflected many astronomers' concerns about longstanding observational practices. Many also feared negative effects on anticipated novel methods expected to change astronomy as a discipline in the near future. The same IAU meeting hosted an exhibit dedicated to explaining the design and functioning of the first orbiting astronomical observatories scheduled to be operational within two to three years. Fifteen Space observatories promised to open new parts of the spectrum to observation from above the obstructive effects of the atmosphere. Extending some of the same effects of the industrial atmosphere to new heights might force space telescopes into higher, riskier, and more expensive altitudes. The American Astronomical Society's anti-West Ford resolution specifically condemned the project as a threat to space observatories, suggesting that a future of dipole belts in space might end this revolution before it got off the ground (AAS 1961).

In spite of this dissent the Air Force launched an initial West Ford package a few months later in October 1961. A failure of the dipole's dispenser mechanism prevented the copper filaments from deploying, and the Air Force proceeded apace with a second attempt.

† See *The Astronomical Journal*, Vol. 66, No. 3 (April 1961).

Concurrently astronomers ramped up their protests, engaging with public audiences to amplify the case against West Ford. Many who sought to put an end to West Ford held a deep conviction in the cultural and even moral authority enjoyed by astronomers at home and abroad and expected it would bolster their cause in the court of public opinion. They provided interviews to journalists writing for popular presses, many of whom took up the case against West Ford as an issue of environmental morality.

Many articles in mainstream presses portrayed West Ford as akin to celestial littering. Littering represented an ongoing locus of conflict in cities around the world, and it was an act increasingly bound to criminal consequences since the early 1950s in America. If urban-dwellers could be fined for dumping unwanted waste into an empty lot, why should governments be allowed to get away with the same crime in outer space? Such articles highlighted the quantity of dipoles and planetary scale of their dispersal, unleashed by a national military without the consent of citizens on the ground below.

At the same time, Rachel Carson's book *Silent Spring* had rapidly become an international best-seller. The book introduced to mainstream readers the world over the unintended, often invisible environmental consequences wrought by pesticide programs and nuclear tests—also large-scale government efforts whose compounding consequences could neither be fully predicted nor controlled. Carson suggested that coalitions of unbiased experts and an educated populace provided the best bulwark against such large-scale environmental threats imposed by governments upon non-consenting societies (Carson 2002). Whether knowingly or not, astronomers' efforts to publicize their fears about West Ford yielded just this kind of coalition. And it eventually bore fruit.

The MIT Lincoln Laboratory heeded requests by astronomers to alter the dispenser payload to ensure that the dipoles would only deploy once they reached a resonant orbit. At the planned 3500-kilometer altitude, West Ford's designers expected sunlight pressure to push the high mass-to-area ratio dipoles back into the atmosphere over the course of a few short years. This would render temporary any impediments to present and future observing practices (Tukey 1963). Satisfied that their concerns were being heard, most astronomers stood down and waited on the results of the experiment.

Following these changes the second launch successfully deployed the dipoles in May of 1963. The belt transmitted a signal between California and Massachusetts, and pieces in newspapers that had previously run articles about the environmental dangers of the test cheekily began to describe the dipole belt as "talking back" (Associated Press 1963a,b,c). Fortunately, the second Project West Ford test did not yield any of the feared effects: The dipoles were barely detectable, with no measurable impact on the practices of professional astronomers (IEEE 1964). As planned, solar energy, friction, and gravity brought down most of the dipoles within a few years, though not all—some dipoles succumbed to surface tension and clumped together, reconfiguring their mass-to-area ratio and reducing the efficacy of solar pressure in driving their orbital decay (Shapiro 1966).

The Project West Ford test was successful on nearly all metrics. It completed the intended mission, with no adverse effects on astronomy. But the experiment was never repeated, and currently no dipole belts provide microwave scattering services from orbit. Like most histories of technology a number of technical, social, cultural, and legal reasons may have confirmed the demise of this form of communications satellite. Higher fidelity active satellites with ever-improving electronics on board quickly became the state of the art and status quo, though the Starfish Prime test demonstrated their susceptibility to nuclear attack. The Partial Nuclear Test Ban Treaty banning all atmospheric testing of nuclear weapons entered into force the same year as the successful West Ford test, rendering its services less urgent than before. West Ford also suffered a poor reputation following the public outcry—a reputation that endures in periodic popular news articles that decry it as a "crazy" Cold-War era geoengineering scheme or inaccurately characterize it as a Saturn-like ring around Earth (Jackson, Symon and Farkas 2013; Hanson 2013).

While a major news item at the time, Project West Ford has largely faded from popular memory in the 60 years since its second launch. Yet its legacy endures in the language of the Outer Space Treaty—the preeminent document of international space governance which, like the IAU resolution, calls for international consultation before the launch of any potentially harmful space experiments. Echoes of the project and its surrounding controversy emerged anew in recent debates over the risks posed by the latest novel communications satellite system.

Following the first Starlink launches in 2019 astronomers' public responses in notable cases revived rhetorically identical critiques raised during the Project West Ford controversy. Much like West Ford, Starlink's creators have touted the sheer number of proposed satellites in each system—hundreds of millions in West Ford and tens of thousands for Starlink, though obviously quite divergent in mass, size, altitude, magnitude, and longevity. The two projects have notable differences beyond the shape and number of satellites aloft: West Ford was a publicly funded military project, Starlink a privately funded commercial project; the regulatory landscape has shifted immensely in favor of protecting free-market autonomy since the early 1960s; and the user base of Starlink is far broader than that of West Ford. In 2019 and 2022 SpaceX signed agreements with the U.S. National Science Foundation meant to facilitate discussion of potential methods to mitigate the constellation's impact on ground-based astronomy (National Science Foundation 2023).

While a lot has changed in the intervening half century between Project West Ford and the rise of active commercial satellite constellations, much of the rhetoric and discourse surrounding each technological inflection point bear remarkable similarities. By discussing these similarities in more detail, and by understanding the long history of astronomers' attention and responses to the potential harms of novel communications satellites, this history may offer practical lessons for the current moment. Rather than starting from scratch, this allows us to examine what has worked, what hasn't, and what lessons we might take from the rhymes of history to apply to the current moment.

References

American Astronomical Society (AAS) 1961, "Resolution Adopted by the Society at Its Annual Business Meeting, Nantucket, Mass., June 20, 1961," June 20, 1961, 1 of 2 14.C Satellites 14.C.7m Project West Ford (Needles) 1961–62, National Archives at College Park, College Park, MD

Associated Press 1963a, Bath Times, May 25

Associated Press 1963b, The Call, May 25

Associated Press 1963c, The Montreal Star, May 27

Berkner, L. V. 1961, Letter to Leo Goldberg, January 1, 1961, NRAO - Director's Office - Professional Organizations/Committees - SSB West Ford, Box 1, NAS-SSB - Project West Ford - File 1, National Radio Astronomy Observatory

Bing, G. F. 1962, Assistant Director Nuclear Test Detection Office, "Radiation Damage to Solar Cells Attributed to Trapped Electrons from STARFISH PRIME," Memorandum, August 17, 1962, National Security File, Files of Charles E. Johnson, Box 30, Atmospheric Tests, LBJ Library

Buderi, R. 1996, *The Invention That Changed the World: How a Small Group of Radar Pioneers Won the Second World War and Launched a Technological Revolution*, New York: Simon & Schuster

Carson, R. 2002, *Silent Spring* 40th anniversary ed., Boston: Houghton Mifflin

U.S. Department of Defense. 1961, "Department of Defense White Paper on Project West Ford," White Paper, September 1961, National Radio Astronomy Observatory - Director's Office - Professional Organizations/Committees - NAS Space Science Board - Project West Ford, Box 1, NAS-SSB - Project West Ford - File 1, National Radio Astronomy Observatory

Fleming, J. R. 2011, Annals of Iowa, 70(4), 314

Goldberg, L. 1961a, AJ, 66, 105

Goldberg, L. 1961b, Letter to L. V. Berkner, January 3, 1961, NRAO - Director's Office - Professional Organizations/Committees - SSB West Ford, Box 1, NAS-SSB - Project West Ford - File 1, National Radio Astronomy Observatory

Hanson, J. 2013, Wired Science, August 13, `http://www.wired.com/wiredscience/2013/08/project-west-ford/`

Hoerlin, H. 1976, Technical report LA-6405, Los Alamos National Laboratory, Los Alamos, New Mexico

Honolulu Star-Bulletin 1958a, "Nuclear Rocket Exploded at Johnston; Failure to Warn Hawaii Is Protested", August 1

Honolulu Star-Bulletin 1958b, "'Should I Leave?' Californian Asks After Explosion", August 1

Honolulu Star-Bulletin 1958c, "Thousands in Honolulu View Blast", August 1

International Astronomical Union (IAU) 1961, "IAU News Bulletin 8 - Thursday, August 24, 1961, Berkeley, California," August 24, 1961, Series HUG FP 83.17 Leo Goldberg International Astronomical Union (IAU) Correspondence, 1948-1979; Box 6; IAU Berkeley 1961, Harvard University Archives

Institute of Electrical and Electronics Engineers (IEEE) 1964, Proc IEEE, 52, 449

Jackson, X., Symon, E. V., & Farkas, I. 2013, Cracked.com, November 16, `http://www.cracked.com/article_20661_6-insane-schemes-attempted-at-dawn-space-travel.html`

Morrow, W. E, & MacLellan, D.C., 1961, AJ, 66(3), 107

U.S. National Science Foundation. 2023, "NSF Statement on NSF and SpaceX Astronomy Coordination Agreement," January 10, `https://new.nsf.gov/news/statement-nsf-astronomy-coordination-agreement`

Richmond, A. S. 1960, "Orbital Scatter Communication," September 9, 1960, Satellites: West Ford 1960-1986 Folder 11429, NASA Headquarters Historical Reference Collection, Washington, DC

Shapiro, I. I. 1961, *Science*, 154(3755), 1445

U.S. Department of State. 1960, "Memo on the April 15th 1960 Meeting of the Space Science Board Ad Hoc Committee on Project NEEDLES," April 15, 1960, 14.B Outer Space 14.B.19 Project Needles 1960; General Records of the Department of State, 1763–2002, Record Group 59 Box 347, National Archives at College Park, College Park, MD

Terrill, D. R. 1999, *Iowa Enters the Space Age: James Van Allen, Earth's Radiation Belts, and Experiments to Disrupt Them*, Maxwell Air Force Base, Alabama: Air University Press.

Tukey, J. W., et al. 1963, "Steps Taken by U.S. Government and Lincoln Laboratory with Regard to Consultation and Information Exchange with Foreign and Domestic Scientists on Project WEST FORD", Washington, DC: Office of Science and Technology

Astronomy and Satellite Constellations: Pathways Forward
Proceedings IAU Symposium No. 385, 2026
P. Grimley, J. Barentine, D. Galadí-Enríquez & C. Walker
doi:10.1017/S1743921324001017

Overview of Indigenous rights and outer space for the IAU-CPS Policy Hub

Hilding Neilson

Department of Physics & Physical Oceanography,
Memorial University of Newfoundland & Labrador, St. John's, NL A1B 3X7, Canada
email: hneilson@mun.ca

Abstract. As part of the mission of the International Astronomical Union Centre for the Protection of the Dark and Quiet Sky from Satellite Constellation Interference (IAU-CPS) Policy Hub to consider national and international regulations about the usage and sustainability in outer space, we also included discussion specific to the rights of Indigenous peoples with respect to outer space under the context of the United Nations Declaration for the Rights of Indigenous Peoples (UNDRIP). In this work, we review how some of the articles of UNDRIP require various actors in the use and exploitation of outer space including satellite companies, nation states, and professional/academic astronomy to consult and support Indigenous peoples/nations and respect Indigenous sovereignties. This work is concluded with recommendations for consulting and collaborating with Indigenous peoples and recommendations for moving from the traditional colonial exploitation of outer space and building an anti-colonial future in relationship with outer space.

Keywords. Indigenous Rights, UNDRIP, Space Ethics, outer space Policy

---

1. Introduction

The purpose of this section is to explicitly highlight Indigenous rights and perspectives with respect to outer space and to consider the importance of Indigenous methods, consultations, and sovereignties. These issues are crucial for any advocacy and discussion around any policy document related to space sustainability and the impacts of satellite constellations.

The contributor Hilding Neilson (he/him) acknowledges that he is Mi'kmaq and a member of the Qalipu First Nation in Turtle Island (North America). His contribution is based on that perspective and based on numerous works of interdisciplinary research in astronomy, Indigenous studies, and research ethics.

I further acknowledge that any policy work relating to the field of astronomy benefits from historic and contemporary colonization. More specifically, we acknowledge that astronomers from around the globe and the International Astronomical Union have benefited from the taking of Indigenous lands for ground-based telescopes without free and informed consent or with manufactured consent. This issue is particularly important today as Kanaka Maoli have spent decades protesting the Thirty Meter Telescope planned for construction on Mauna Kea in Hawai'i. But, astronomers have benefited from the construction and planned construction of optical and radio facilities on Indigenous Lands the world over. There is some irony to the astronomy committee advocating for Indigenous voices in the discussion of building a framework for sharing outer space while preferring new observatories over respecting Indigenous rights (Prescod-Weinstein et al. 2020; Neilson & Lawler, 2019; Kahanamoku et al. 2020; Salazar, 2014; and more).

Even in the light of these issues of colonization that support Astronomy communities, the goals of this section are:

(1) to highlight Indigenous rights with respect to outer space activities through the United Nations Declaration of the Rights of Indigenous Peoples (UNDRIP) †,
(2) to highlight the need for anti-colonial methods when parties advocate for new rules and regulations with respect to outer space activities,
(3) to discuss how Indigenous methodologies can allow for novel regulations and perspectives on how humanity and private and public interests interact with outer space.

Indigenous peoples have lived in regions around the world since time immemorial and continue to live in their territories or elsewhere today. Many peoples have been removed from their territories because of various forms of genocides, but their rights and knowledges exist today and will continue to do so into the future. As such, it is necessary for any policy analysis that wishes to be inclusive and equitable to be respectful of Indigenous knowledges. It is also necessary to note that there is no one Indigenous knowledge; every Nation or group has its own knowledge and its own methodologies. Therefore, to be inclusive of Indigenous communities there must be consultations with as many different communities, and nations as possible.

2. Indigenous rights to outer space via UNDRIP

UNDRIP was adopted by the UN General Assembly in 2007 and has become a pathway for reconciliation in many nations. In particular, for any discussion about outer space, we cite a number of articles that establish the rights of Indigenous peoples.

The noted articles include:

Article 1: Indigenous peoples have the right to the full enjoyment, as a collective or as individuals, of all human rights and fundamental freedoms as recognized in the Charter of the United Nations, the Universal Declaration of Human Rights and international human rights law.

Article 3: Indigenous peoples have the right to self-determination. By virtue of that right they freely determine their political status and freely pursue their economic, social and cultural development.

Article 5: Indigenous peoples have the right to maintain and strengthen their distinct political, legal, economic, social and cultural institutions, while retaining their right to participate fully, if they so choose, in the political, economic, social and cultural life of the State.

Article 7.1: Indigenous individuals have the rights to life, physical and mental integrity, liberty and security of person.

Article 8.1: Indigenous peoples and individuals have the right not to be subjected to forced assimilation or destruction of their culture.

Article 11.1: Indigenous peoples have the right to practice and revitalize their cultural traditions and customs. This includes the right to maintain, protect and develop the past, present and future manifestations of their cultures, such as archaeological and historical sites, artefacts, designs, ceremonies, technologies and visual and performing arts and literature.

Article 11.2: States shall provide redress through effective mechanisms, which may include restitution, developed in conjunction with indigenous peoples, with respect to their cultural, intellectual, religious and spiritual property taken without

† https://www.un.org/esa/socdev/unpfii/documents/DRIPS_en.pdf

their free, prior and informed consent or in violation of their laws, traditions and customs.

Article 12.1: Indigenous peoples have the right to manifest, practice, develop and teach their spiritual and religious traditions, customs and ceremonies; the right to maintain, protect, and have access in privacy to their religious and cultural sites; the right to the use and control of their ceremonial objects; and the right to the repatriation of their human remains.

Article 13.1: Indigenous peoples have the right to revitalize, use, develop and transmit to future generations their histories, languages, oral traditions, philosophies, writing systems and literatures, and to designate and retain their own names for communities, places and persons.

Article 18: Indigenous peoples have the right to participate in decision-making in matters which would affect their rights, through representatives chosen by themselves in accordance with their own procedures, as well as to maintain and develop their own indigenous decision-making institutions.

Article 20.1: Indigenous peoples have the right to maintain and develop their political, economic and social systems or institutions, to be secure in the enjoyment of their own means of subsistence and development, and to engage freely in all their traditional and other economic activities.

Article 20.2: Indigenous peoples deprived of their means of subsistence and development are entitled to just and fair redress.

Article 25: Indigenous peoples have the right to maintain and strengthen their distinctive spiritual relationship with their traditionally owned or otherwise occupied and used lands, territories, waters and coastal seas and other resources and to uphold their responsibilities to future generations in this regard.

Article 26.1: Indigenous peoples have the right to the lands, territories and resources which they have traditionally owned, occupied or otherwise used or acquired.

Article 27: States shall establish and implement, in conjunction with indigenous peoples concerned, a fair, independent, impartial, open and transparent process, giving due recognition to indigenous peoples' laws, traditions, customs and land tenure systems, to recognize and adjudicate the rights of indigenous peoples pertaining to their lands, territories and resources, including those which were traditionally owned or otherwise occupied or used. Indigenous peoples shall have the right to participate in this process.

Article 29.1: Indigenous peoples have the right to the conservation and protection of the environment and the productive capacity of their lands or territories and resources. States shall establish and implement assistance programmes for indigenous peoples for such conservation and protection, without discrimination.

Article 29.2: States shall take effective measures to ensure that no storage or disposal of hazardous materials shall take place in the lands or territories of indigenous peoples without their free, prior and informed consent.

Article 31.1: Indigenous peoples have the right to maintain, control, protect and develop their cultural heritage, traditional knowledge and traditional cultural expressions, as well as the manifestations of their sciences, technologies and cultures, including human and genetic resources, seeds, medicines, knowledge of the properties of fauna and flora, oral traditions, literatures, designs, sports and traditional games and visual and performing arts. They also have the right to maintain, control, protect and develop their intellectual property over such cultural heritage, traditional knowledge, and traditional cultural expressions.

Article 32.1: Indigenous peoples have the right to determine and develop priorities and strategies for the development or use of their lands or territories and other resources.
Article 32.2: States shall consult and cooperate in good faith with the indigenous peoples concerned through their own representative institutions in order to obtain their free and informed consent prior to the approval of any project affecting their lands or territories and other resources, particularly in connection with the development, utilization or exploitation of mineral, water or other resources.
Article 32.3: States shall provide effective mechanisms for just and fair redress for any such activities, and appropriate measures shall be taken to mitigate adverse environmental, economic, social, cultural or spiritual impact.
Article 42: The United Nations, its bodies, including the Permanent Forum on Indigenous Issues, and specialized agencies, including at the country level, and States shall promote respect for and full application of the provisions of this Declaration and follow up the effectiveness of this Declaration.

All of these articles have some impact on Indigenous rights with respect to outer space even though the phrase "outer space" and this list, while detailed, is not meant to be exhaustive. We can consider these articles as addressing three different themes: 1) Cultural rights with respect to outer space; 2) Economic fights with respect to outer space; and 3) Responsibilities with respect to outer space. It is in these three categories where Indigenous rights can be elucidated in more detail.

3. Cultural rights and outer space

Indigenous Peoples have lived in their territories since time immemorial. As such Indigenous Peoples have lived with the night sky in various ways also since time immemorial and those relationships manifest in numerous ways. For instance, the Cree peoples on Turtle Island (North America) refers to themselves as Star People and as a people who come from the stars (Buck 2018). What it means to come from the stars is concept that only Cree Elders and Knowledge Keepers can truly explain, but the concept does illustrate an intimate and relational connection with the night sky. These relationships are apparent globally through Indigenous star stories and constructs such as Medicine Wheels, methods for navigating waters using the stars, calendars, and more. Regardless of how non-Indigenous persons might judges the values of these knowledges, i.e., wayfinding and navigation versus navigation with the Global Positioning System, the rights of Indigenous peoples to maintain, use, and live their cultural practices is part of rights from UNDRIP, in particular Article 31.1. As such, any actions that remove access to the night sky for ritual, culture, etc., is a violation of UNDRIP.

Two ways that Indigenous peoples can lose access the night sky are: light pollution and satellite pollution. Hamacher et al. (2020) have noted that light pollution is a form of cultural genocide; that is, if Indigenous stories are written in the stars, or in the case of some Nations, stories are written in the dark spaces between stars then removing visibility to those constellations or "dark constellations" is colonization and assimilation. This is particularly true as Indigenous star knowledge is transferred as part of oral traditions; then, light pollution is a direct assault on Indigenous knowledges. This is also true for satellite pollution. Bright satellites (in this case bright does not necessarily have to mean visible by the unaided eye) that can be seen in various ways also impact how Indigenous peoples interact with the night sky. It is unlikely that light pollution from satellite onstellations will ever have a significant impact on Indigenous methods of wayfinding or time-keeping, but a bright satellite constellations will impact cultural relationships with the night sky and require Indigenous peoples to adapt their knowledges to a view of the night sky changed by humans without their consultation or consent. That, too, is colonization.

The ongoing impacts of light pollution and satellite pollution are a clear violation of Indigenous rights if they negatively impact the ability of peoples to transmit their knowledges and continue religious and spiritual practices (Articles 12.1 and 13.1). A lack of consent can be construed as a violation of Article 8 that demands that Indigenous peoples not be forced to assimilate. As such, any regulations and policies with respect to human interaction with outer space and the night sky require consultation and consent of Indigenous peoples from the world over.

4. Economic rights and outer space

Many of the Articles of UNDRIP highlight the rights of Indigenous peoples to manage their economic well-being and the right to access their traditional territories for development. This includes: Articles 3; 8, and 32. The rights of Indigenous peoples to control economic development in outer space on Indigenous lands can be considered through two lenses: 1) from the usage of lands for launch capabilities, and 2) from the usage of outer space itself. There continues to be significant interest in creating more launch facilities to expedite human activity in space. Many of these launch facilities are being built on traditional Indigenous lands without free and prior consent and without economic benefit for the peoples living on those lands. As launch facilities can contribute significant pollution via noise, and blowing dust around, and in waste if launches fail, then there is an ethical, if not, legal, obligation, for space companies to negotiate and consider the rights of Indigenous peoples.

As for outer space itself, it may seem counterintuitive to consider that Indigenous peoples would have any more economic rights than any other group on Earth. However, the UNDRIP guarantees the rights of Indigenous peoples to use their traditional lands. Furthermore, traditional lands do not have to be defined solely by the colonial definitions of borders as defined by Articles 26 and 27. Those articles define lands in terms of "traditionally owned, occupied or otherwise used or acquired". Historically, nation states may have defined traditional lands in very strict terms of the written records of colonizing nations and by archaeological evidence; i.e., did Indigenous peoples leave physical evidence on specific spots of land. These methods were specifically designed to dispossess Indigenous peoples of traditional lands and do not reflect Indigenous claims and oral knowledges. As such, the rights and responsibilities of Indigenous peoples with respect to land should respect any Indigenous groups' definition of Land which would include access to all components that contribute to the ecosystem and culture such as water, air, the night sky, and yes, outer space.

These rights are enshrined in many oral traditions. In North America alone, many First Nations peoples refer to themselves as coming from the stars, for which it is not within the rights of anyone from outside those cultures to define what "coming from the stars" mean. Many stories include people coming and going from outer space, making Indigenous peoples "users" of and visitors to outer space. Denying rights to Indigenous peoples to be part of the regulation of how humans operate in outer space because there is no evidence that they use outer space or used technology to go there is to perpetuate colonization.

As such, it is reasonable to argue that celestial objects, such as the Moon, and outer space, itself, are part of the traditional lands of Indigenous peoples and in the framework of UNDRIP means that Indigenous peoples have the right to participate economically as well as to have a say in how the outer space environment is regulated.

5. The obligation of the United Nations and space actors

Because Indigenous peoples have rights to participate in the space economy and to have cultural rights to space protected as they choose, it is then incumbent on the United Nations and nation states to uphold UNDRIP and work with Indigenous peoples in developing the path for humanities present and future in outer space. There are a number of ways to approach this collaboration and consultation, which include:

- Organizations choosing to engage in dialogues about outer space regulation must consult with and equitably include Indigenous peoples in that work. It is not sufficient to include Indigenous peoples solely through simplistic lenses of 'Cultural Engagement' or 'Stakeholders'.
- Nation states must consult and work with Indigenous peoples in the framework of free and prior informed consent (FPIC) to develop an anti-colonial framework to outer space activities that moves beyond the simple colonial framework that has traditionally existed.
- Nation states must consult and work with Indigenous peoples to support their ability to benefit from the space economy in as equitable of ways as other actors.
- The United Nations, in developing new regulations for the use of outer space, must consult and work with Indigenous peoples under the framework of free and prior informed consent in ways that acknowledge Indigenous peoples as nations unto themselves and respects their sovereignties.

These recommendations do not have a clear basis in international law, but are anti-colonial methods to support the rights of Indigenous peoples. They are a starting point for developing a new anti-colonial framework for our interactions with outer space and celestial objects in ways that acknowledge and respect Indigenous sovereignties in their respective lands and respects both their cultural and economic rights.

6. Indigenous rights and sustainability of outer space

One of the ways we might consider supporting Indigenous cultural and economic rights in outer space would be moving to a paradigm that goes beyond traditional settler perspectives of sustainability that are arguably framed around the issue of exploitation of resources and how much impact on the environment is permissible before the damage to the environment is irreversible or human usage of that environment is unsafe. Historically, Indigenous peoples have suffered under this kind of discussion. The nuclear tests of the Manhattan Project impact Navajo peoples, while nuclear bomb testing in the Pacific continues to have negative health impacts on peoples there. Uranium mining has harmed both Navajo and Inuit communities. Despite all of that, many consider nuclear power to be a sustainable and environmentally friendly method of electricity generation.

There are many more examples of how the history and context of environmentalism and sustainability has not included Indigenous peoples nor respected their land rights or sovereignties. The current global context of the Outer Space Treaty and the Artemis Accords is based on how humans, or more specifically nation states, operate with respect to one another in outer space. These agreements have no context for how humans or nation states or private companies operate with respect to outer space. Effectively, these agreements, more so the Artemis Accords, focus on ways to peacefully exploit outer space and protect the "rights" of those doing so.

One way to consider our interaction in space is via a treaty: not a treaty about human activities in outer space, but a treaty about our responsibilities in outer space. Many Indigenous peoples traditionally lived in treaty and relationship with animals, plants, water, air, etc. These treaties can and have included outer space. In the Mi'kmaw cosmology of six worlds, the Land above Clouds, or outer space, is part of a treaty. Country et al. (2023) notes that Sky Country (Space) is to be cared for, not exploited. These treaties do not outline rights to outer space, but ethical obligations and responsibilities.

It is reasonable that any anti-colonial perspective on outer space not only include the rights of Indigenous peoples, but also the rights of celestial objects and outer space itself and thus humanity's responsibilities for engaging in activities in outer space. The ways this would work would not be obvious, especially in a global discussion that involves colonial nation states. However, many scholars have noted that responsibilities can include giving back in trade when using resources This option requires humanity to operate in a system that is not focused solely on exploitation but supporting an ecosystem at the same time. In outer space, the concept of an ecosystem would be difficult to understand for many because in the Western tradition, outer space and celestial objects are treated as dead/non-living, and in the Western system where humans are the apex of a hierarchy, humans would have a right to exploit however they see fit. For many Indigenous peoples these treaties with land, animals, and space mean that they are animate or living. As such any resource usage must be in trade and must benefit space or the Moon.

One might question how we can determine what would be of benefit for space or celestial objects and who would determine the answer to that question. It is not obvious how to answer this question because any international committee to address such a question can have dramatically different opinions and powers, but options might include if we take a resource on the Moon, then we have to return it. If we use water on the Moon for sustaining settlements or for propulsion then we might have an obligation to return an equal amount of water to the Moon. This concept is a way to maintain sustainability in a way that is not common on Earth but will also require changes to how we relate to life on Earth. If light pollution is treated this way, then we have an obligation to the animals that are impacted by light pollution and have to give back to them in ways that support them.

This concept is simply a different perspective on much of the current international discussion that is, by and large, ignoring the rights and voices of Indigenous peoples in favour of various colonial and capitalist interests. But, if we choose to follow UNDRIP and work with Indigenous nations the world over then there may be more options than the bimodal and colonial discussion of exploitation and sustainability.

References

Buck, W. 2018, Tipiskawi kisik: Night sky star stories. Manitoba First Nations Education Resource Centre Incorporated

Country, B., Ganambarr, R., Ganambarr-Stubbs, M. et al. 2023, The Routledge Handbook of Social Studies of outer space, 217–225

Hamacher, D. W., De Napoli, K. and Mott, B. 2020, arXiv e-print, arXiv:2001.11527

Kahanamoku, S., Anolani Alegado, R., Kagawa-Viviani, A. et al. 2020, arXiv e-prints, arXiv:2001.00970

Neilson, H. and Lawler, S. 2019, In Canadian Long Range Plan for Astronomy and Astrophysics White Papers, volume 2020, 2

Prescod-Weinstein, C., Walkowicz, L. M., Tuttle, S. et al. 2020, arXiv e-prints, page arXiv:2001.00674

Salazar, J. A. 2014, PhD thesis, University of Hawai'i at Manoa

Astronomy and Satellite Constellations: Pathways Forward
Proceedings IAU Symposium No. 385, 2026
P. Grimley, J. Barentine, D. Galadí-Enríquez & C. Walker
doi:10.1017/S1743921324001005

Indigenous knowledges and kinship as a model for our future in outer space

Hilding Neilson

Department of Physics & Physical Oceanography,
Memorial University of Newfoundland & Labrador, St. John's, NL A1B 3X7, Canada
email: hneilson@mun.ca

Abstract. Commercial endeavours have already compromised our relationship with space. The Artemis Accords are creating a framework that will commercialize the Moon and further impact that relation. To confront that impact, a number of organizations have begun to develop new principles of sustainability in space, many of which are borne out of the capitalist and colonial frameworks that have harmed water, nature, peoples and more on Earth. Indigenous methodologies and ways of knowing offer different paths for living in relationship with space and the Moon. While Indigenous knowledges are not homogeneous, there are lessons we can use from some of common methods. In this talk we will review some Indigenous methodologies, including the concept of kinship and discuss how kinship can inform our actions both on Earth and in space.

Keywords. Indigenous Methods, Indigenous Rights

1. Introduction

The Outer Space Treaty (†) notes that activities in outer space should be for the "benefit of all mankind". However, the current explosive growth of the private space industry has impacted the human relationship with the night sky, celestial objects such as the Moon, and outer space itself. In that sense, not all of humanity seems to benefit from these endeavours. In particular, a number of actors, including the International Astronomical Union, have been developing new reports and conferences with the aim of mitigating the impact of a minimally regulated space industry including the SATCON1 and SATCON2 reports led by the National Science Foundation's National Optical-Infrared Astronomy Research Laboratory (Walker et al. 2020b; Hall, Walker, Rawls et al. 2021) and the Dark & Quiet Skies for Science and Society conferences led by the International Astronomical Union (IAU) and the United Nations Office of Outer Space Affairs and others (Walker et al. 2020a). In these reports, Indigenous peoples are treated as a stakeholder along with planetaria and amateur astronomy groups and in some of the reports as victims of this growing light and space pollution. In the latter sense arguments are made that light pollution will negatively impact ancient practices and connections with the night sky, usually without input from Indigenous peoples and ignoring that Indigenous peoples have modern cultures.

In this presentation, I aim to consider the discussion of protecting the night sky and outer space through a lens of Indigenous methodologies as a third perspective from the current binary discussion that pits outer space as only an exploitable resource that benefits the growing space-industrial complex against the argument led by astronomers that outer space must be treated as a natural environment that must be protected and be

† https://www.unoosa.org/oosa/en/ourwork/spacelaw/treaties/outerspacetreaty.html

untouched (e.g., Lawrence, Rawls, Jah et al. 2022). This third method is motivated from the perspective that both sides of current discussion are colonial, and that both minimize the rights of Indigenous peoples with respect to outer space. To that end this presentation is designed to advocate for **policy changes** that centre and include Indigenous peoples, nations and methods and moves beyond community engagement (Barentine & Heim 2023).

Before continuing the discussion, it is necessary to acknowledge that the author is Mi'kmaq and Settler and can only speak to that experience. Furthermore, as a professional astronomer the author acknowledges that the field of professional astronomy has benefited from past, current, and ongoing colonization, particularly in Chile†, Hawai'i (Salazar 2014), and elsewhere. There is an irony that the professional astronomy community seeks to engage with Indigenous peoples with respect to activities in outer space while ignoring and dismissing Indigenous peoples when building observatories.

2. The coloniality of space exploitation and space environmentalism

Much of the current discourse in the current era of outer space exploration has become a narrative of space exploitation. This narrative has been growing with the Artemis Accords‡ that seek to provide a set of guidelines for resource extraction in outer space and on the Moon, with an eye to asteroid mining and Martian exploration. The Accords, while designed to not directly conflict with the Outer Space Treaty, allows for operators to remove material from bodies in outer space and not to be interfered with in locations where operators have equipment and satellites. The Accords claim that no private operator or nation state can claim territory in outer space, yet they allow for operators to remove material and effectively 'squat'. Even if this does not mean ownership in outer space, it follows a path of the 19th-century Gold Rush where prospectors staked claims and took whatever was considered valuable while often leaving nothing but waste. In that case, many of these gold rushes significantly harmed Indigenous peoples, stole Indigenous lands, and had significant negative impact on the local plants, animals, and waters that had been there since time immemorial. As such the Artemis Accords are colonial.

That colonialism is not new. Treviño (2020) noted how the history of space exploration is part of a colonial narrative that has existed for centuries. From the first manual on the colonization of Mars by Werner Von Braun, the development of outer space activities have been justified by the same narrative that colonists used to settle the Americas: *terra nullius*, and Manifest Destiny. In many ways the growth of light pollution around the world is also part of this colonialism and *terra nullius*. Hamacher et al. (2020) wrote that "Light Pollution is Cultural Genocide" on the principle that many Indigenous nations and peoples write stories in their constellations (both star-based and "dark" constellations). These stories carry significant cultural, scientific and holistic knowledges that continue to be a crucial part of those cultures. In the same way that light pollution is colonisation, so too is light pollution from satellites. Whiles light pollution acts to erases Indigenous knowledges and stories, the growth of bright satellites aim to rewrite them. This, too, is a significant colonial problem.

On the other hand, an increasingly commonly touted solution for regulation is to treat outer space as part of the environment. While this might appear to be a way to include Indigenous peoples and methods in the conversation about regulating outer space, much of the environmental movement has been colonial. One such example is the development of the National Park systems in Canada and the United States that stole land from Indigenous peoples in the name of conservationism. It is notable that the clause in the

† https://astrobites.org/2019/09/10/astronomical-observatories-and-indigenous-communities-in-chile/

‡ `https://www.nasa.gov/artemis-accords/`

Artemis Accords citing the need to preserve sites of cultural importance will potentially act in the same manner, creating "National Parks" on the Moon based on the interests of Eurocentric cultures.

Furthermore, how this outer space environmentalism will be implemented will arguably be centred on the interests of nation states, especially the richest states that can afford the investment necessary for outer space exploitation. Part of the space interests of these states is professional astronomy, and hence will include the need to support ground-based astronomy. Since most of the largest radio and optical telescopes that exist or are planned are on Indigenous lands, then any outer space environmental movement could easily have negative consequences for Indigenous peoples.

3. Outer space is part of the land

Indigenous peoples have been connected to their homelands since time immemorial. That connection includes a deep, holistic relationship with the night sky, the motions of the Moon and the planets and more. That connection includes oral traditions that use the night sky for rituals, calendars, ethics, and more. Many peoples tell stories of the North Star as a guide star, and many in the Pacific use the stars for way-finding (Ruggles 2015). In the far north, Inuit peoples use the Flickering Star as a guide for weather prediction (MacDonald & Schledermann 1998), while in the southern hemisphere, many nations use dark constellations for time-keeping (Urton 1981). The Mi'kmaw people use the constellation of Muin and the Seven Bird Hunters to share about circumpolar stars, hunting ethics, biology and more (Marshall et al. 2010). The night sky is an important part of Indigenous knowledges and cultures since time immemorial and for many peoples the night sky is a reflection of the land below.

Space exploration is also part of the oral traditions. For instance, many stories about the constellation that is called the Pleiades in the Eurocentric tradition is about travellers to the sky in a number of Indigenous traditions (Lankford 2007). In many cultures, the people come from the sky (Buck 2018). Indigenous oral traditions contain many stories and knowledges about people who explore the Moon and outer space. That exploration does not occur in a physical sense that we tend to consider space exploration, i.e., foot- and rover prints on celestial bodies, nor does it occur in the same sense that academic astronomy explores outer space through the interpretation of photons and gravitational waves. This space exploration occurs in ways centred on Indigenous methodologies that must also be part of any conversation about outer space exploration and exploitation if we are to act in anti-colonial ways.

These relationships denote deep connections with outer space for navigation, time-keeping, and more. For many Indigenous nations, these connections come with protocols, treaties, and responsibilities. Country et al. (2023) note that for Yol*n*u peoples in Australia, interactions with Sky Country requires people to follow protocols. In Mi'kmaw cosmology of six worlds, the sky above is its own world (Whitehead & Kaubach 1988). As such, interactions between L'nu† and the sky above is governed by treaties, similar to treaties between different Indigenous nations or with different animals, plants, water, and more. These treaties outline both rights and responsibilities for all sides. In that situation, honouring those responsibilities would preclude peoples from the pure exploitation of outer space, but would not necessarily require ending any form of space exploration in the Eurocentric sense.

The combination of interactions with outer space since time immemorial along with the protocols and treaties in outer space mean that Indigenous peoples and nations have sovereignties and responsibilities with respect to the night sky and how we interact with

† L'nu is the Mi'kmaq word for people.

outer space. **Therefore, however nation states choose to agree to exploit outer space, there must be consideration for Indigenous rights and methodologies.**

4. Indigenous methodologies in outer space

Much of how we view our place with respect to exploring and exploiting outer space derives from the Western perspective of sciences and the methods used. However, Indigenous methodologies offer different perspectives on the human relationship with outer space that would be a third way between the current perspectives of outer space conservation or outer space exploitation.

The concept of Indigenous methods or sciences have been discussed by numerous authors in many different research fields (e.g., Cajete 2000; Battiste 2013; Lipe 2019). To summarize, one can list these Indigenous methods of knowing to be:

- What's above reflects below.

This concept is a comparison of how humans are related to what happens in the sky. For instance, many constellations reflect stories of nature and the land, be it the story of Muin and the Seven Bird Hunters in Mi'kmak'i (Marshall et al. 2010) or the Cree sky stories (Buck 2018). These stories speak to relationships between the land and the night sky and how perspectives depend on where on the land one is.

- Knowledge is relational.

For many Indigenous peoples, knowledge is not something universal but depends on the person learning and their perspectives. Furthermore, knowledge is relational because knowledge is something that is used; knowledge is active and to be shared ethically. Knowledge is not something that is simply obtained.

- Indigenous methods considers multiple variables that interact concurrently.

In traditional Western/Eurocentric sciences, the method to understand a phenomenon is to attempt to reduce it to the minimal number of variables. To reduce variables to the minimal number usually requires measuring average values and missing how some of those variables interact. By considering multiple variables concurrently, knowledge keepers and Elders develop and grow the ability to predict and manipulate phenomena.

- Knowledge is (w)holistic.

For many Indigenous peoples knowledges have many purposes. A star story is more than what might be classified solely as astronomy. Star stories such as Muin and the Seven Bird hunters reflect biology and nature, ethics and community along the motions of stars. As such Indigenous knowledges tend to not be placed into disciplines akin to how traditional Western/Eurocentric knowledges are.

- Nature is holy, sacred, and familial.

In traditional Western sciences, nature is usually treated as a hierarchy. In that sense, we tend to place humans and human rights above the rights of animals, plants, rocks, etc. One way this is seen is how elements of nature are treated as resources for the service of humans. For many Indigenous nations, nature is animate and has rights. Interactions with nature operate in reciprocity where people give back when they take from nature (Kimmerer 2013). Another way to view this is through the lens of treaty. In a treaty with nature, Indigenous peoples have responsibilities to support nature and the relationship is give and take. In the Eurocentric model the relationship is to take and control.

It must be noted that there is no such thing as a pan-Indigenous knowledge system: there are Mi'kmaw ways of knowing, Salish ways of knowing and so on. Every culture and nation will have its own methodologies and relationships with nature. But, there are commonalities that can be considered such as the methods listed. These methods are significantly different from Eurocentric/Western methods in how humans relate with nature, relate with the sky, and, most importantly, our responsibilities we owe nature and the sky. As such, many Indigenous peoples consider the stars, planets and the Moon to be a relation, or an ancestor. This perspective is different from the Western perspective where the Moon is a dead rock, the stars are not alive and so on.

As such, these methods suggest a different way to interact and live with the night sky, the planets, and the Moon, and how we would operate in outer space. Since the planets and the Moon are relations or kin, then these bodies have rights and are alive or animate in a sense that is antithetical to Western/Eurocentric methods. Furthermore, this also suggests that the outer space is a land unto itself that requires relationship building and can be claimed or abused by humans. This is completely different from the two current leading narratives about humanity's future in outer space.

One narrative is the traditional colonial narrative that it is humanity's destiny to travel into outer space and settle other worlds: a Manifest Destiny of outer space. That Manifest Destiny is being accelerated by the Artemis Accords that invite private interests to mine and use outer space, treating it as a resource to be exploited and to profit from. The second narrative is outer space as an environment that must be protected for sustainable usage by humans. This narrative seeks to create protections but environmentalism on Earth can be colonial and anti-Indigenous. The primary example of this the development of national parks in Canada and the United States that removed Indigenous peoples from their traditional lands to protect those lands for the common good. While there are additional examples that can be cited, the key issue regarding both of these narratives is that they both claim to seek to use outer space, the Moon and more for the benefit of humanity.

The Indigenous ways of knowing cited here tell us that there is a third way to proceed where we seek to act in treaty or relationship with outer space as an land where humans are guests when we operate there. This act of treaty does not mean humanity must cease any operations in outer space, but that those operations must be carried out as good guests. This may seem like a strange concept, but it is consistent with how Indigenous peoples lived on the lands of Turtle Island (North America) and beyond for millennia (Kimmerer 2013).

What would being a good guest in outer space look like? It is not obvious, but operators in outer space would not simply take and satellite operators would not simply fill the orbital space around the Earth with satellite constellations, and it would require minimising the existence of 'dead' satellites in orbit. Being a good guest requires operators and users in outer space to give back and support the nature of outer space, the Moon and celestial objects. How we give back to outer space is something that humanity and nations can work on developing beyond debating who decides how outer space will be used to serve humanity or, more likely, serve a privileged few.

5. Summary

The goal of this work is to illustrate a third path to how humans can approach our relationships with and actions in outer space based on Indigenous methods and ways of knowing. In this approach, I suggest that one way to be more consistent with these Indigenous methods to consider our treaties and relationships as seeking to not treat outer space as a resource for the benefit or profit of humanity, but, instead, that we live in treaty with outer space as its own land and place with its own rights of existence. While, this behaviour would be dramatically different than what humanity has done previously in outer space and is doing today, it is possible to change our systems and allow humans and Western societies to live in relationship with outer space. I argue that the first step in building this relationship is to develop protocols for what humans can give to outer space when we take from it, whether this taking is of water from the Moon, minerals from asteroids or even from the near-vacuum of outer space when we place satellites.

While this work illustrates how Indigenous methods can inform policy regarding outer space, but more than that it notes how Indigenous peoples and nations should be a part of any policy with respect to the usage of outer space and that participation should be equitable and respectful.

References

Barentine, J. C. & Heim, J. 2023, arXiv preprint arXiv:2311.02184

Battiste, M. 2013, Decolonizing Education: Nourishing the Learning Spirit. Purich Publishing Limited

Buck, W. 2018, Tipiskawi kisik: Night sky star stories. Manitoba First Nations Education Resource Centre Incorporated

Cajete, G. 2000, Native Science: Natural Laws of Interdependence, Clear Light Publishers

Country, B. Ganambarr, R. Ganambarr-Stubbs, M., et al. 2023, The Routledge Handbook of Social Studies of outer space, 217–225

Hall, J. Walker, C. Rawls, M., et al. SATCON2: Executive summary. Bulletin of the AAS, 2021

Hamacher, D. W., De Napoli, K., & Mott, B. arXiv preprint arXiv:2001.11527, 2020

Kimmerer, R. Braiding sweetgrass: Indigenous wisdom, scientific knowledge and the teachings of plants. Milkweed editions, 2013

Lankford, G. E. Reachable stars: Patterns in the ethnoastronomy of eastern North America. University of Alabama Press, 2007

Lawrence, A., Rawls, M. K., Jah, M., et al. Nat Astro, 6(4):428–435, 2022

Lipe, D. 2019, Indigenous education: New directions in theory and practice, 453

MacDonald, J. & Schledermann, P., 1998, Arctic, 51(4), 394

Marshall, L., Marshall, M., Harris, P., & Bartlett, C. 2010, Muin and the Seven Bird Hunters: A Mi'Kmaw Night Sky Story, Cape Breton University Press

Ruggles, C. L. N. 2015, Archaeoastronomy in Polynesia, Springer New York, 2231

Treviño, N. B. 2020, PhD thesis, The University of Western Ontario

Urton, G. 1981, Proc Amer Phill Soc, 125(2), 110

Walker, C., Di Pippo, S., Aubé, M., et al. 2020, Report to UN Office for outer space Affairs. Report I: `https://noirlab.edu/public/products/techdocs/techdoc021`

Walker, C., Hall, J., Allen, L., et al. 2020, BAAS, 52(2)

Whitehead, R. H., & Kaulbach, K. 1988, Stories from the six worlds: Micmac legends. Halifax, Canada: Nimbus Publishing Ltd.

Astronomy and Satellite Constellations: Pathways Forward
Proceedings IAU Symposium No. 385, 2026
P. Grimley, J. Barentine, D. Galadí-Enríquez & C. Walker
doi:10.1017/S1743921324001443

Countering the threat of space pollution: awareness raising via sustainability narratives

Nicholas Campion

University of Wales Trinity Saint David
email: nickcampion12@gmail.com

Abstract. This paper engages with the conference themes of community engagement and awareness raising, through proposing narratives which provide alternative stories to the belief in unrestricted technological development. The plan to launch up to 400,000 satellites into Earth orbit is an example of unregulated corporate capitalism: many private corporations and governments are encouraging rapid growth of the space sector without sufficient consideration of broader and potentially consequences for Earth and space environments or the interests and welfare of the wider community. My question in this paper is how far sustainability narratives from the humanities and social sciences can assist in the attempts to limit pollution of the sky and near-Earth environment.

Keywords. satellite constellations, sustainability narratives, harmony

1. Introduction

The discussion I want to raise in this paper is how far sustainability narratives from the humanities and social sciences can assist in the attempts to limit pollution of the sky and near-Earth environment. At my institution, the University of Wales Trinity Saint David, we now host a UNESCO Bridges Hub,† the purpose of which is to employ the humanities and social sciences to broaden the appeal and success of sustainability science. In this respect I would like to begin with a quote from the so-called La Palma Declaration of 2007 (Starlight Initiative 2007):

> ... a view of the starlight has been and is an inspiration for all humankind, ... [and] throughout history, the contemplation of the firmament has sustained many of the scientific and technical developments that define progress.

The declaration was put together by a coalition including UNESCO, UNWTO and the IAU, but it is noticeable how, having established a broad remit ('an inspiration of mankind') this is then narrowed down to 'scientific thoughts'. My question here is why there is no appeal to a wider constituency, namely the humanities and social sciences in academia, and those in the wider public who may have a spiritual or religious conception of the universe. In part. the reason may be associated with anti-religious polemics amongst astronomers (Tyson 1999). The meeting point between astronomy and spirituality, though, is found in the concept of 'awe', a non-scientific word commonly used by astronomers to justify their interest in the sky (Campion 2011; De Witte 2022). My point here is that an over-emphasis on the word science may limit the scope

† https://www.uwtsd.ac.uk/unesco-bridges-hub/ [accessed 3 November 2023].

and appeal of the resistance to sky and near-Earth pollution. How many amateur astronomers, who spend their nights tracking dark sky objects, would agree with the Roman Emperor Marcus Aurelius (161-180 CE)? He was the author of the following inspirational passage: "Survey the circling stars as though yourself were in mid-course with them. Often picture the changing and re-changing dance of the elements. Visions of this kind purge away the dross of our Earth-bound life." (Aurelius 1964). These days one might say that dark skies are beneficial for one's well-being (Campion 2016a). As Michael Dahlstrom wrote, emphasising science may be less effective than employing narratives that make sense to an otherwise non-specialist public (Dahlstrom 2014)‡

> when the context moves from data collection to the communication of science to nonexpert audiences, stories, anecdotes, and narratives become not only more appropriate but potentially more important. Research suggests that narratives are easier to comprehend and audiences find them more engaging than traditional logical-scientific communication ... engagement into the world of a narrative ... uses enough emotional and cognitive resources that it is difficult for audiences to generate counter-arguments against the evaluations to which they are exposed.

SpaceX/Starlink's promotion emphasises sustainability, environmental awareness and the good of humanity (SpaceX 2022). Counter narratives emphasise the polluting effects of SpaceX activities (Whittaker 2018). The key is found in effective science communication (Besley & Dudo 2022).

2. Contexts

Light pollution became a major concern for astronomers in the 1980s: the International Dark Skies Association was founded in the 1980s. But recently, events took a more intense turn. As O'Callaghan (2022) wrote,

> The mega constellation era began in May 2019, when Elon Musk's firm SpaceX launched the first 60 satellites in its Starlink constellation. Today the constellation's numbers have swelled to more than 3,000 and account for fully half of all active satellites in space.

O'Callaghan is concerned with the environmental harm resulting from satellite constellations. Venkatesan, Lowenthal, Prem & Vidaurri (2020) adopt an apocalyptic narrative and raise the problem of disadvantaged communities:

> the rapid increase in satellite constellations is a simmering crisis that is silently approaching the point of no return ... The rush to claim near-Earth space is also leaving out the world's most minoritized communities, including Indigenous peoples, who need to be involved as stakeholders in decision-making.

At the extreme end of the problem is the weaponisation of space (Tyson & Lang 2018). The metaphor of space as a new and completely unregulated 'Wild West' has become common for writers on the topic (Cruddas 2020). Narratives tend to be doom-laden and emphasise negative messages.

In addition, opinions about whether the Outer Space Treaty of 1967 or the United Nations Office for Outer Space Affairs (UNOOSA), can be effective are divided. For example, commentators are divided as to the balance between Space Law on the one hand, and national governments or private corporations in relation to asteroid mining. UNOOSA has excellent objectives: space for climate action supports UN Sustainable

‡ See also Downs (2014); Jones & Anderson Crow (2017); Muindi, Ramachandran & Tsai (2020).

Development Goal 13: "Take urgent action to combat climate change and its impacts." (United Nations 2015). It is difficult to see how this has any relationship to the problems outlined in the 2022 GAO report, including increase in orbital debris, emissions into the upper atmosphere and disruption of astronomy (U.S. Government Accountability Office 2022).

3. Sustainability narratives

A useful distinction between different kinds of narrative was made by the sociologist Max Weber (1864-1920), who drew a distinction between objective truth, which deals in scientific facts, and narrative truth, which tells stories (Campion 2016b; Kalberg 1980). Spence (1982) wrote that "Interpretations are persuasive not because of their evidential value but because of their rhetorical appeal." The questions then, are "what value do we put on space exploration?", and "how does it express our values?" The problem for opponents of satellite constellations is that SpaceX adopts sustainability narratives. For example, it is indeed working to lower satellite brightness (Crider 2022). Other commentators report on satellite technology's potential to assist with research into climate change (Urban 2021). For wider context, oil companies, banks, airlines and other operations which are the target of environmental concern increasingly promote their 'green credentials', a phenomenon known as 'greenwashing' (Das 2022). The wider concern is with greenwashing amongst the new private actors in space (Christensen 2022). And, even if Starlink is successful in reducing light pollution from satellite constellations, that still leaves the space junk problem. This has already disrupted space flight (Jones 2023).

For wider narratives, then, we can look at wider issues. For example, D'Amato (2021) identifies five kinds of sustainability narrative. Of these, SpaceX/Starlink could well appeal to four: the circular economy, the bioeconomy, the sharing economy and sufficiency. The fifth, green economy, is the one which SpaceX would have the most difficulty with as it emphasises the total picture: 'Protecting and enhancing ecological functions to support human well-being (e.g., ecosystem services accounting, ecosystem conservation and restoration, nature-based solutions and green infrastructures). It is therefore not enough for SpaceX/Starlink to take mitigation factors over brightness without addressing launch pollution and space junk problems, and the fact that the stated aim is to reduce light pollution, rather than eliminate it may help, but does not solve the problem of potential damage to the wider ecosystem (United Nations Environment Programme 2020).

One important sustainability narrative, which deals in positive solutions has been developed by the United Nations - 'Harmony' (United Nations undated):

> the Earth and its ecosystems are our common home, and expressed their conviction that it is necessary to promote Harmony with Nature in order to achieve a just balance among the economic, social and environmental needs of present and future generations.

'Harmony' proposes a model of the Universe in which, because everything is interconnected, an action in one part has impacts in other parts (Campion 2020; Wales, Juniper & Skelly 2010). It has an ancestry in western thought in classical models, which Marcus Aurelius knew very well, which saw the world as a single functioning organism (Plato 1931). In modern environmental thought close parallels are found in chaos theory (Gleick 1998) and 'deep ecology' (Naess 1973). And, as expressed through economic affairs (Chrysopoulou 2020):

A well-being economy recognizes that people need to restore a harmonious relationship between society and nature, enjoy a fair distribution of resources, and live in healthy and resilient communities.

This is, then the test for SpaceX/Starlink and other actors in the field: whatever your intentions, what is the effect of your actions on all our communities? Does sky pollution harm our collective heritage in the sky - also mentioned in the La Palma Declaration?

The fundamental narrative of Harmony is positive: that because there is an inherent balance in the environment, we can act to make things better. This, at least, is the classical conception. The fundamental proposition of Harmony is that all things are interconnected and that we therefore have an inherent right to have a say in what happens in our environments, including the space environment. It is not up to powerful corporations to impose new technology on the rest of us just because they can, and without regard to the consequences. In this context, Greenpeace challenges companies planning to mine the Pacific Ocean with the slogan 'Our Pacific: not yours to destroy' (Greenpeace International 2021). We might equally say, in response to plans to populate space with constellations of 400,000 satellites, or mine the Moon or asteroids, 'Our Space: not yours to destroy'.

In this paper I have discussed the nature and value of narratives which do not rely on science. It may well be that national governments have the biggest say in whether satellite constellations continue, if they become fully aware of the space-junk risks to their own space programmes. But, in the meantime, it is up to the rest of us to use what narratives we can.

References

Aurelius, M. 1964 in Meditations, trans. M. Staniforth (London: Penguin Classics), 112

Besley, J. C., & Dudo, A. 2022, Strategic Science Communication: A Guide to Setting the Right Objectives for More Effective Public Engagement (Baltimore: Johns Hopkins University Press)

Campion, N. 2011, ASP Conf Series, 441, 415

Campion, N. 2016a, in Blair, A., Sark in the Dark: Wellbeing and Community of the Dark Sky Island of Sark (Lampeter: Sophia Centre Press), xvii

Campion, N. 2016b, in Campion, N. (ed.), Heavenly Discourses (Lampeter: Sophia Centre Press), xiv

Campion, N. 2020. (ed.), The Harmony Debates: Exploring a practical philosophy for a sustainable future (Lampeter: Sophia Centre Press)

Christensen, I. 2022, Space News, `https://spacenews.com/op-ed-guarding-against-greenwashing-in-space/`

Chrysopoulou, A. 2020, Stanford Social Innovation Review, `https://doi.org/10.48558/9SXJ-C595`

Crider, J. 2022, Teslarati,

Cruddas, S. 2020, Politico, `https://www.politico.eu/article/space-final-frontier-wild-west/`

D'Amato, D. 2021, Circular Economy and Sustainability, 1, 231

Dahlstrom, M. F. 2014, PNAS, 111(supplement 4), 13614

Das, L. 2022, Greenpeace, `https://www.greenpeace.org.uk/news/what-is-greenwashing/`

De Witte, M. 2022, Stanford News, `https://news.stanford.edu/2022/09/06/power-awe-cosmos/`

Downs, J. S. 2014, PNAS, 111(supplement 4), 13627

Gleick, J. 1998, Chaos: The Amazing Science of the Unpredictable (London: Vintage)

Greenpeace International 2021, `https://www.greenpeace.org/international/press-release/47077/deep-sea-mining-industry-confronted-sea-first-time-greenpeace/`

Jones, I. 2023, Palatinate, https://www.palatinate.org.uk/astronauts-stranded-by-space-debris-return-home/
Jones, M. D. & Anderson Crow, D. 2017, Palgrave Communications, 3, 53
Kalberg, S. 1980, The American Journal of Sociology, 88(5), 1145
Muindi, F. J., Ramachandran, L., & Tsai, J. W. 2020, Trends in Molecular Medicine, 26(3), 249
Naess, A. 1973, Inquiry, 16, 95
O'Callaghan, J. 2022, Scientific American Website, https://www.scientificamerican.com/article/satellite-constellations-could-harm-the-environment-new-watchdog-report-says/
Plato 1931, Timaeus, trans. Bury, R. G. (Cambridge Mass., London: Harvard University Press)
SpaceX 2022, https://www.spacex.com/updates/#sustainability
Spence, D. P. 1982, Narrative truth and historical truth: Meaning and interpretation in psychoanalysis (New York: W. W. Norton), 32
Starlight Initiative 2007, https://fundacionstarlight.org/docs/files/33_english-declaration-in-defense-of-the-quality-of-the-night-sky-and-the-right-to-starlight.pdf
Tyson, N. 1999, Natural History Magazine, https://neildegrassetyson.com/essays/1999-10-holy-wars/
Tyson, N. & Lang, A. 2018, Accessory to War: The Unspoken Alliance Between Astrophysics and the Military (New York: W. W. Norton)
United Nations Environment Programme (UNEP) 2020, https://www.unep.org/news-and-stories/story/global-light-pollution-affecting-ecosystems-what-can-we-do
United Nations (undated), Harmony with Nature, http://harmonywithnatureun.org/
United Nations 2015, United Nations General Assembly Transforming our world: the 2030 Agenda for Sustainable Development, UN Doc A/Res/70/1, 23, https://www.un.org/en/development/desa/population/migration/generalassembly/docs/globalcompact/A_RES_70_1_E.pdf
Urban, K. 2021, American University, https://www.american.edu/sis/centers/security-technology/how-satellites-track-climate-change.cfm
U.S. Government Accountability Office 2022, Large constellations of satellites: Mitigating environmental and other effects, https://www.gao.gov/products/gao-22-105166
Venkatesan, A., Lowenthal, J., Prem, P., & Vidaurri, M. 2020, Nat Astro, 4, 1043
Wales, HRH the Prince of, Juniper, T., & Skelly, I. 2010, Harmony: A New Way of Looking at our World (London: Harper Collins)
Whittaker, I. 2018, The Conversation, https://theconversation.com/falcon-heavy-spacex-stages-an-amazing-launch-but-what-about-the-environmental-impact-91423

Astronomy and Satellite Constellations: Pathways Forward
Proceedings IAU Symposium No. 385, 2026
P. Grimley, J. Barentine, D. Galadí-Enríquez & C. Walker
doi:10.1017/S1743921324001066

Summary of SatHub, and the current observational status of satellite constellations

Mike W. Peel[1,2], **Siegfried Eggl**[1,3], **Meredith Rawls**[1,4], **Michelle Dadighat**[1,5], **Piero Benvenuti**[1,6], **Federico Di Vruno**[1,7] **and Connie Walker**[1,5]

[1]IAU Centre for the Protection of the Dark and Quiet Sky from Satellite Constellation Interference, France.
email: sathub@cps.iau.org

[2]Imperial College London, Blackett Lab, Prince Consort Road, London SW7 2AZ, UK.
email: email@mikepeel.net

[3]Department of Aerospace Engineering/Department of Astronomy, University of Illinois at Urbana-Champaign, Illinois, USA.

[4]Department of Astronomy/DiRAC/Vera C. Rubin Observatory, University of Washington, Seattle, Washington, USA.

[5]NSFs NOIRLab, Tucson, Arizona, USA.

[6]Emeritus, Department of Astronomy, University of Padova, Italy.

[7]SKA Observatory, Jodrell Bank, Macclesfield SK11 9FT, United Kingdom.

Abstract. SatHub is one of the four hubs of the IAU Centre for the Protection of the Dark and Quiet Sky from Satellite Constellation Interference (CPS). It focuses on observations, data analysis, software, and training materials to improve our understanding of the impact of satellite constellations on astronomy and observers worldwide. As a preface to more in-depth IAUS385 sessions, we gave a summary of some recent work by SatHub members and the current status of satellite constellations, including optical and radio observations. We shared how the audience can join or get more involved, e.g., via the CPS Slack for asynchronous collaboration. We also touched on what a future with hundreds of thousands of constellation satellites might look like.

Keywords. SatHub, satellites

---

1. Introduction

SatHub is part of the International Astronomical Union (IAU) Centre for the Protection of the Dark and Quiet Sky from Satellite Constellation Interference (CPS). It is a public, community driven, coordinated hub focused on the collection and analysis of artificial satellite observations, as well as carrying out training and outreach related to satellite constellations. It was formed following the recommendations from SATCON2 (Rawls, Barron, Birdwell et al. 2021), and adopted the outline structure established in that report, see Fig. 1. SatHub is still under construction, and has built-in flexibility to change and evolve based on community feedback and needs.

SatHub is co-led by three people: Meredith Rawls (University of Washington, WA, USA; remote attendee of IAUS385); Mike Peel (Imperial College London, UK; in-person attendee); and Siegfried Eggl (University of Illinois at Urbana-Champaign, USA; in-person attendee). In addition, Michelle Dadighat (NOIRLab, in-person attendee) is a software engineer working on SatHub software development.

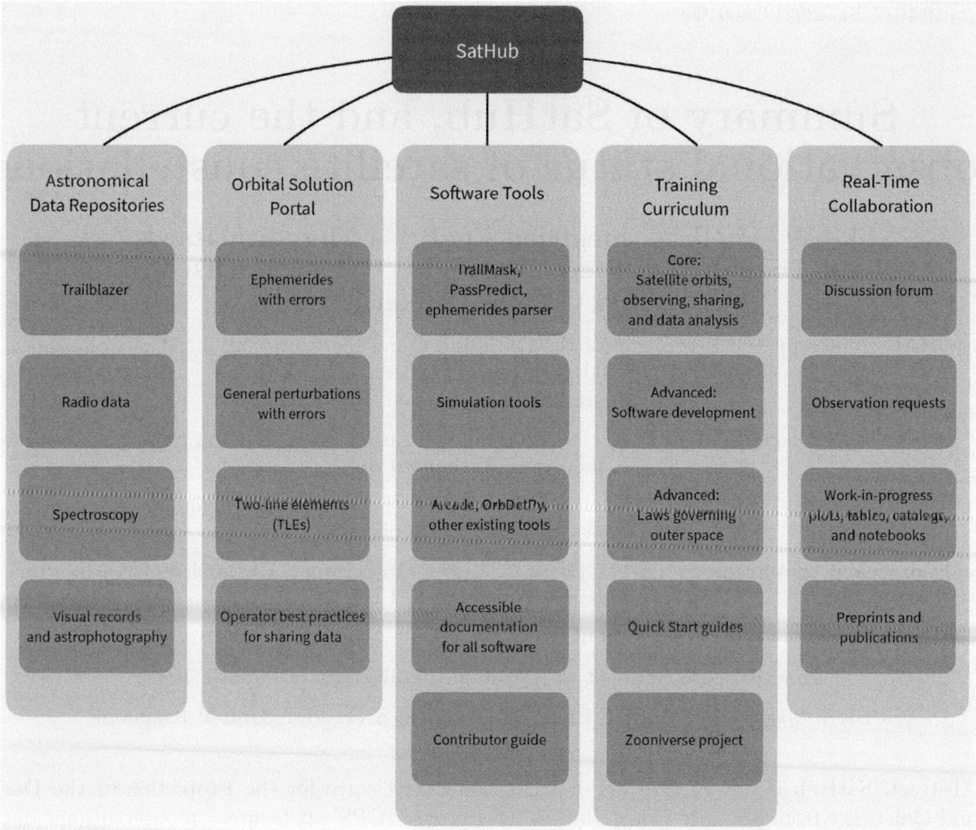

Figure 1. SatHub outline from the SATCON2 Observations Working Group (Rawls, Barron, Birdwell et al. 2021).

2. Software development

A central part of the work of SatHub involves software development to make it easier to identify, avoid, and remove satellites from astronomical observations across the electromagnetic spectrum. Software development is actively underway, and we have marked in the section below where things are already working well / up-and-running as "done"; in development as "developing"; and still in need of doing as "to be done".

2.1. *Data repository and exchange*

One of the primary purposes of SatHub is to serve as a Hub for Data Exchange from brightness measurements. This will be a central place where data can be deposited and exchanged with others. Such a system needs to be publicly available, easily accessible, user-friendly, and well documented.

The repository will enable the submission and retrieval of satellite brightness measurements (made visually by eye, observed with optical telescopes, etc.; developing); collections of optical/near-IR images with satellite streaks (developing); spectra contaminated with reflected solar spectrum (to be done); and radio data affected by satellite interference (to be done).

We already welcome contributions from everyone who is willing to share observing campaign plans and data. We aim to have easy-to-use interfaces for both professional and amateur astronomers to upload and download satellite-affected images (developing). Software developers/contributors are welcome to contact us to help with this work!

2.2. *Orbital solutions portal*

A critical need for astronomy is high accuracy predictions of where satellites will be seen from any observatory around the world, which we aim to enable through the Orbital Solution Portal. This will provide public, standardised access to orbital solutions of artificial satellites. NOIRlab has recently hired a software developer (Michelle Dadighat) to work with SatHub on this and other tools.

The portal will convert general perturbation-style two-line element ("TLE") solutions to on-sky positions, with automatic synchronisation with complementary services and databases (done). It is updated every 8 hours for Starlink satellites, with others updated every 24 hours (due to the availability of published TLEs). It will include error bars with all orbital solutions (developing), and have built-in robustness by using multiple backend systems to provide TLEs and position predictions (CelesTrak†, Privateer‡, etc.; developing). Ultimately it will include the full history of all TLEs, so positions can be predicted for any satellite at any time, which is critical when analysing archival and long-duration survey data (to be done; funding applied for).

At present we have a prototype SatChecker that does some of this§. We welcome contributions to this work. In particular, we need industry to cooperate and share the highest accuracy orbital information about their satellites in a timely, regular manner. We also need astronomers to check that orbits are accurate by observing and measuring satellites, and reporting inaccuracies back to us and to industry.

2.3. *Other software tools*

In addition to the two key tools above, we are also working on other tools, as described in the SATCON2 report (Rawls, Barron, Birdwell et al. 2021). We aim to make sure that all software tools have user-friendly documentation, support, and are actively maintained (developing). In addition, we plan to have a standard test suite supporting a wide range of instrument and satellite signature properties to support software development (to be done).

We encourage contributions to open-source software development, particularly through the tools developed within CPS. We also want to test the software with data from many different telescopes and cameras. We are looking for funding to work on software development.

In addition to software developed within CPS, we are also seeking to understand existing tools and their availability, so that we can reduce duplication and avoid recreating what already exists. SatHub is running a software survey¶, with one submission per tool requested. Tools inputted into this survey before and during IAUS385 were then discussed and worked on at the SatHub hackathon on the last day of the symposium.

3. Real-time collaboration

SatHub is working on establishing communication and collaboration across the astronomy community about satellite constellations in an open and collaborative way. We mostly do this through the IAU CPS Slack, which all members of CPS can access, with

† `https://celestrak.org/`
‡ `https://www.privateer.com/`
§ `https://satchecker.readthedocs.io`
¶ `https://forms.gle/FzkT1FekoVb2Fxqi6`

the main #sathub channel and a host of sub-channels on specific topics (for example, #sathub_optical_observing).

We also have a SatHub GitHub organisation†, which anyone can use to share open-source code, and we encourage documentation that follows "Write the Docs"‡ best practices (developing). We plan to have a website that will direct to new and existing resources (to be done).

We offer a paper review process, which provides free peer reviews within two weeks for satellite-related papers. The aim is to provide feedback from peers who are also working on satellites within CPS, to ensure accuracy and connections between publications produced by different groups. The reviews are carried out by CPS members, for CPS members, and while being a member is not strictly required to use this service, we encourage everyone interested in the topic to join CPS. Papers that have gone through this peer review process can include CPS as an affiliation and acknowledgement. A recent example of a paper that has gone through the process is Mallama, Hornig, Cole et al. (2023). We are currently working on documenting the process, to be published soon on the CPS website.

4. Potential impact at radio frequencies

As well as optical reflections, satellites actively transmit at radio frequencies. We don't yet have a complete picture of the impact that radio transmissions from satellite constellations will have across the full frequency range used by radio astronomy, and we need observations to assess actual impact. Various talks at IAUS385 covered this topic, and are described later in these proceedings.

Satellite constellations actively transmit at radio frequencies to provide internet connectivity services. This is currently at ∼10–20 GHz (12.7–14.7 GHz for Starlink, others use different frequencies), various satellite constellations are looking at using higher frequencies for higher bandwidth in the near future (∼40–200 GHz). Typically there are also octaves of radio emission at, e.g., 2× the base frequency (or even 3, 5, 10× etc.), which can also impact radio astronomy since the main signals are so bright. In addition, telescopes are not only sensitive to the direction they are observing, but also to other directions through sidelobes (reflections from different parts of the telescope other than the main mirror). If these couple to satellite transmissions, this can be a concern, particularly for CMB experiments who are mapping large angular scale emission and could see interference over large parts of the sky away from the satellite positions. This can be seen in data from the QUIJOTE MFI instrument (Rubiño-Martín, Rebolo, Tucci et al. 2010; Rubiño-Martín, Guidi, Génova-Santos et al. 2023, Génova-Santos in prep.) in 2012 observing at 10–14 GHz (Fig. 2), where geostationary satellites appear as bright as the Sun, and sidelobes were also initially visible—these were later avoided by adding additional baffles around the telescope aperture.

At lower frequencies (∼100-200 MHz), unintended emission has been observed by LOFAR (Di Vruno, Winkel, Bassa et al. 2023) and SKA low (Grigg, Tingay, Sokolovski et al. 2023). This emission comes from the electronics on board the satellites, and was not foreseen by the satellite operators before being detected by radio telescopes.

The types of detectors used in radio telescopes, particularly at high frequencies, can make it difficult to filter out satellite emission. Radiometers require FPGAs rather than broad-band detectors to be able to excise frequencies affected by interference - and these are much more expensive than simple detectors. At high frequencies (>30 GHz), superconducting detectors like Transition Edge Sensors (TESs) and Kinetic Inductance Devices (KIDs) are inherently broad-band and can only have optical filters to remove unwanted

† https://github.com/iausathub
‡ https://www.writethedocs.org/

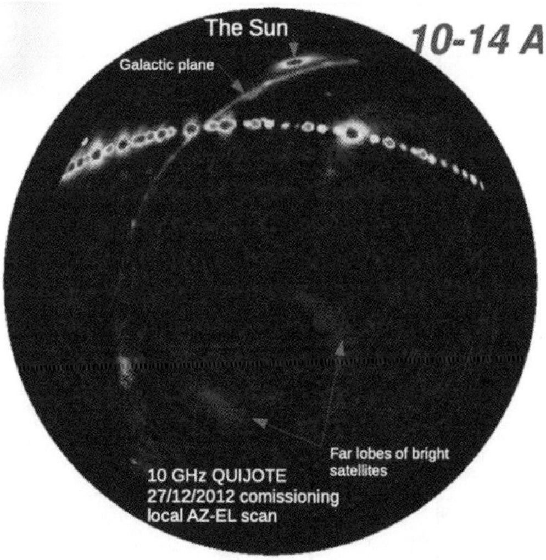

Figure 2. QUIJOTE local map of the radio sky above Tenerife, with relevant features highlighted. Geostationary satellites appear as a band across the sky, similar in brightness to the Sun (top of the image), and also appearing in telescope sidelobes in the bottom half of the image.

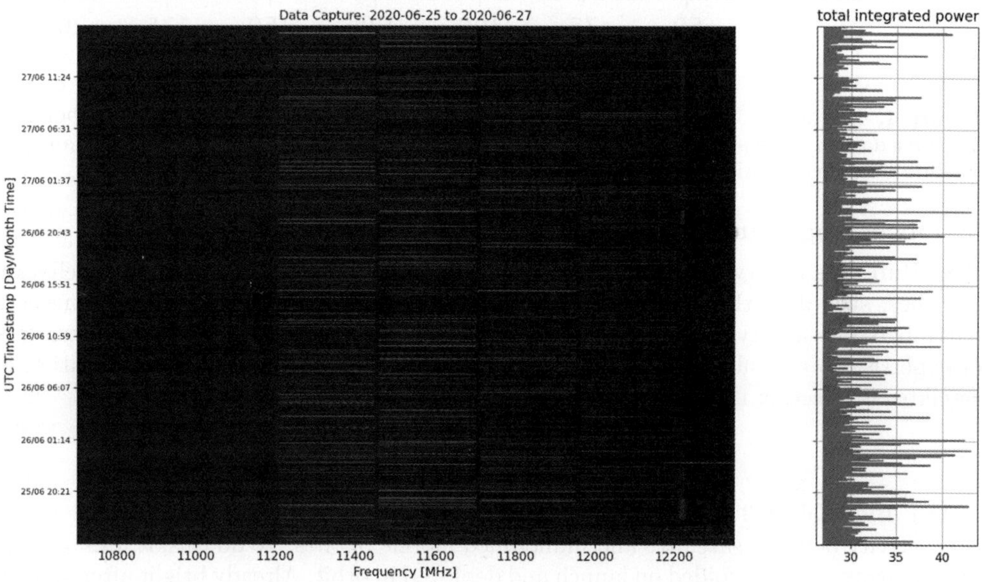

Figure 3. Observations of Starlink with a satellite dish in 2022, demonstrating broad bandwidth and high variability, courtesy of F. Di Vruno.

emissions - mainly high and low-pass filters that define the overall bandwidth. In such systems, interference can only be removed by flagging the time-ordered data.

In addition, satellite emission can be highly variable (see Fig. 3), both because of their high angular velocities, and because they can be changing transmission direction, e.g., using phased arrays. It is important to accurately know satellite positions, otherwise such emission could be mis-identified as transient astronomical signals.

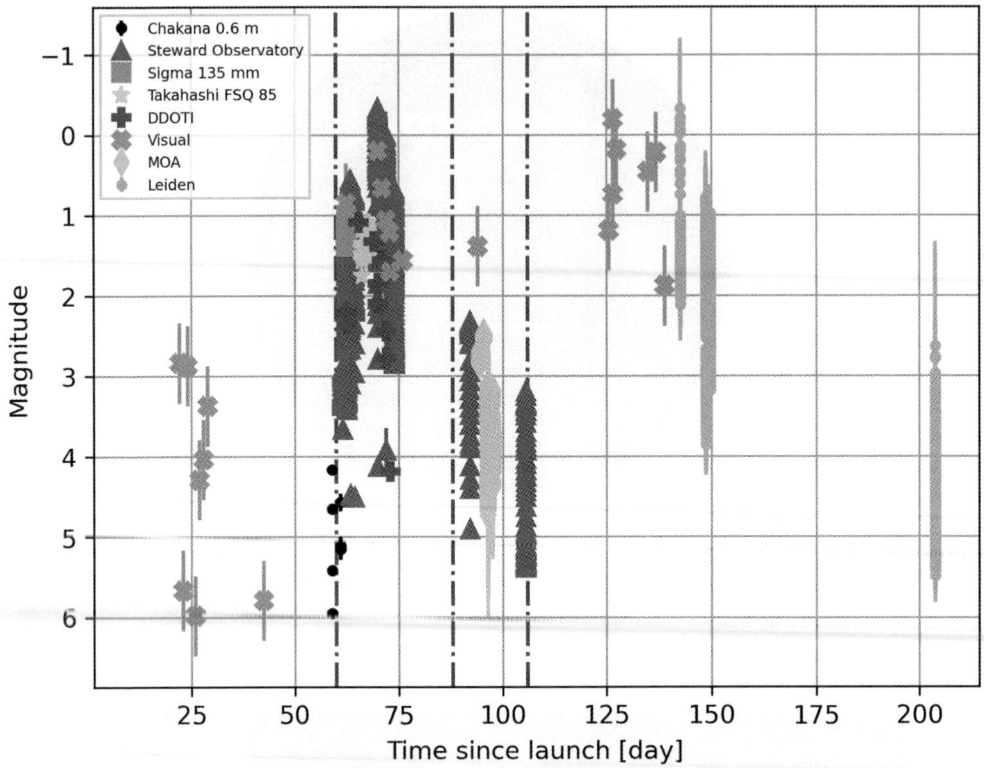

Figure 4. Brightness of BW3 over time, from Nandakumar, Eggle, Tregloan-Reed et al. (2023). The first dashed vertical line shows the point where BW3's phased array was deployed and the satellite became exceptionally bright.

5. Observing campaigns

SatHub runs observing campaigns for specific high-importance satellite constellation launches, such as a the launch of a new type or generation of satellite. A strength of SatHub is our ability to recruit observers from across the globe with a wide variety of expertise to assemble a more complete picture of how a satellite or constellation of satellites is affecting the sky.

5.1. *BlueWalker3*

AST BlueWalker 3 is a prototype satellite that was launched in 2022 to test direct connectivity between satellites and unmodified mobile phones. To do this, it uses a $64\,\mathrm{m}^2$ phased array that was folded on launch and deployed in orbit. Already bright after launch, the satellite became one of the brightest objects in the sky once unfolded (Fig. 4), and is also large enough to be imaged from the ground (Fig. 5 † ‡). The launch vehicle adapter was also bright, and initially untracked with public TLEs. In addition, we demonstrated that the position accuracy decays between releases of TLEs, highlighting the importance of regularly releasing updated TLEs.

The observing campaign was published in Nature (Nandakumar, Eggle, Tregloan-Reed et al. 2023), with an online announcement§. The announcement included a video of BW3,

† `https://www.astrobin.com/tffts1/`
‡ `https://noirlab.edu/public/images/ann23027b/`
§ `https://noirlab.edu/public/announcements/ann23027/`

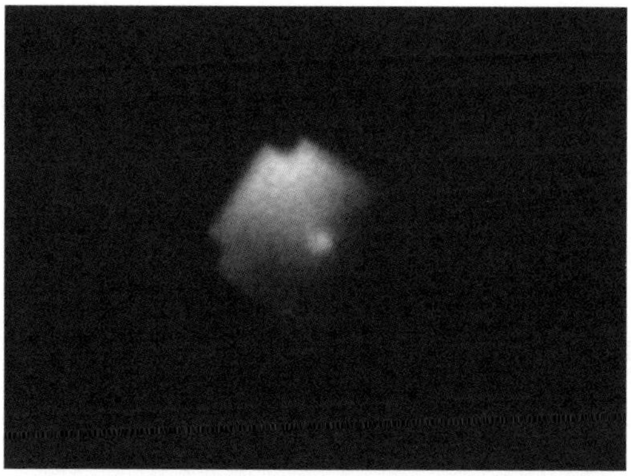

Figure 5. Image of BlueWalker 3 in orbit by M. Tzukran (CC-BY-4.0)

as well as Starlink satellites, as they transited the sky—showing how bright BW3 was compared with Starlink satellites and nearby stars.¶

5.2. *Ongoing*

Starlink started launching its Generation 2 mini satellites in 2023. These have optical mitigations using Bragg mirrors†, and CPS seeks to quantify how well these mitigations have worked. Early observations indicate mixed results so far, with some satellites being faint, but some remaining bright. This is concerning as the full Generation 2 satellites will be bigger than the minis (which were downsized to fit into the Falcon 9 launcher), and potentially will be correspondingly brighter. The results of the observing campaign are planned to be published in early 2024.

Observations are also continuing of BlueWalker 3, using SCUBA2/JCMT to observe the satellite (and also the International Space Station) at submm wavelengths, to see how thermally bright the satellites are. The observations were recently approved, and will start shortly; if successful these will lead to future observations of smaller satellites such as Starlink and OneWeb.

We are also preparing for other observing campaigns in the future, for example, of the Amazon Kuiper satellites. Participation in, and ideas for, observing campaigns are very welcome.

6. Get involved

SatHub needs the help of everyone interested in the impact of satellite constellations on astronomy. We are seeking sky observers, data analysts, software developers, industry experts, students, and others to help with this work. Additionally, as the satellite population changes, the evolving impacts that this will have on astronomy require ongoing observer-operator dialogs. Information in SatHub will be public, open, and accessible to support real-time collaboration. We aim to join innovation with existing solutions, prioritize ease of use, and enable coordination among multiple stakeholders.

To contribute to SatHub, first of all, join CPS! After that, join the SatHub channel at the CPS Slack, which is where most of the discussions take place. Please also share your observations, experience, relevant code and work plans with us, and contribute to

¶ `https://noirlab.edu/public/videos/ann23027a/`

† `https://api.starlink.com/public-files/BrightnessMitigationBestPracticesSatelliteOperators.pdf`

open-source software and documentation development! There are no requirements, prerequisites, or fees for membership—you just have to follow our collaboration agreement‡ and our code of conduct§. You can apply for membership at `https://cps.iau.org/`.

Acknowledgements

MP, SE, MD acknowledge travel funding from IAU CPS to attend IAUS385.

References

Di Vruno, F., Winkel, B., Bassa, C. G., et al. 2023, A&A, 676, A75

Grigg, D., Tingay, S. J., Sokolowski, M., et al. 2023, A&A, 678, L6

Mallama, A., Hornig, A., Cole, R. E., et al. 2023, arXiv e-prints, arXiv:2309.14152

Nandakumar, S., Eggl, S., Tregloan-Reed, J., et al. 2023, Nature, 623, 938

Rawls, M. L., Barron, D., Birdwell, I., et al. 2021, SATCON2 Observations Working Group Report. Technical Report 10.5281/zenodo.5608826, Zenodo

Rubiño-Martín, J. A., Guidi, F., Génova-Santos, R. T., et al. 2023, MNRAS, 519(3), 3383

Rubiño-Martín, J. A., Rebolo, R., Tucci, M., et al. 2010, Astroph Space Sci Proc, 14, 127

‡ `https://docs.google.com/forms/d/e/1FAIpQLSe6w5rZhxvF0ozGe1dMssLTolua1Apq1_QjJ8aLWprGUGQGog/viewform`

§ `https://tinyurl.com/IAUCPSCoC`

Astronomy and Satellite Constellations: Pathways Forward
Proceedings IAU Symposium No. 385, 2026
P. Grimley, J. Barentine, D. Galadí-Enríquez & C. Walker
doi:10.1017/S1743921324000590

SatHub Panel: satellite interference in observatories around the world

Siegfried Eggl[1,11], Zouhair Benkhaldoun[2], Genoveva Micheva[3], Samuel T. Spencer[4,5], David V. Stark[6,7], Benjamin Winkel[8], Meredith Rawls[9,11] and Mike W. Peel[10,11]

[1]Department of Aerospace Engineering, University of Illinois at Urbana-Champaign 61801 IL, Champaign, USA
email: eggl@illinois.edu

[2]Oukaimeden Observatory, High Energy Physics and Astrophysics Laboratory, Cadi Ayyad University, Marrakech, Morocco
email: zouhair@uca.ac.ma

[3]Leibniz-Institute for Astrophysics Potsdam (AIP)
An der Sternwarte 16, 14482 Potsdam, Germany
email: gmicheva@aip.de

[4]Friedrich-Alexander-Universität Erlangen-Nürnberg
Erlangen Centre for Astroparticle Physics
Nikolaus-Fiebiger-Str. 2, D 91058 Erlangen, Germany
email: samuel.spencer@fau.de

[5]Department of Physics
Clarendon Laboratory, Parks Road, Oxford, OX1 3PU, United Kingdom

[6]Space Telescope Science Institute
3700 San Martin Dr, Baltimore, MD 21218, USA
email: dstark@stsci.edu

[7]Department of Physics & Astronomy, Johns Hopkins University,
3400 N. Charles Street, Baltimore, MD 21218, USA

[8]Max-Planck Institute for Radioastronomy,
Bonn, Germany
email: bwinkel@mpifr-bonn.mpg.de

[9]Department of Astronomy/DiRAC/Vera C. Rubin Observatory, University of Washington, Seattle, Washington, USA.
email: mrawls@uw.edu

[10]Imperial College London, Blackett Lab, Prince Consort Road, London SW7 2AZ, UK
email: email@mikepeel.net

[11]IAU Centre for the Protection of the Dark and Quiet Sky from Satellite Constellation Interference, Paris, France
email: sathub@cps.iau.org

Abstract. Satellite constellation interference occurs across astronomical disciplines. We touch on the challenges for existing regulatory frameworks in radio astronomy, and present examples of interference from radio and γ-ray astronomy to optical and spectroscopic interference in ground-based and space-borne facilities. In particular, we discuss the impact of artificial satellites on the Hubble Space Telescope (HST), the High Energy Stereoscopic System (H.E.S.S.), an Imaging Atmospheric Cherenkov Telescope array, as well as possible mitigation strategies for the European Southern Observatory 4-metre Multi-Object Spectrograph Telescope (ESO 4MOST).

Furthermore, we shed light on how ground-based optical telescopes such as the Oukaimeden Observatory contribute to IAU Centre for the Protection of the Dark and Quiet Sky from Satellite Constellation Interference (IAU CPS) efforts that quantify satellite brightness.

Keywords. satellite constellation interference, radio astronomy, γ-ray astronomy, spectroscopy, optical astronomy, Hubble Space Telescope, IAU CPS

1. Introduction

Satellite constellation interference with ground-based optical and radio astronomy facilities is well documented (e.g., Deshpande and Lewis 2019; Barentine, Venkatesan, Heim et al. 2023; Karpov and Peloton 2023; Nandakumar, Eggle, Tregloan-Reed et al. 2023; Anderson-Baldwin 2023; Grigg, Tingay, Sokolovski et al. 2023). The issue of satellites passing through the field of view (FoV) of observatories and potentially corrupting astronomical data is not limited to those domains, however. In this contribution we discuss examples of satellite interference across astronomical disciplines. After a brief overview of regulatory challenges and implications for potential mitigation strategies in Radio Astronomy (RA) in section 2 and ground-based spectroscopy in section 3, we discuss the impact of satellite constellations on ground-based γ-ray astronomy and space-based facilities in sections 4 and 5. We conclude this article with an example of how ground-based observatories can contribute to IAU Centre for the Protection of the Dark and Quiet Sky from Satellite Constellation Interference (IAU CPS) efforts to quantify satellite interference with ground based astronomy in section 6.

2. A radio astronomer's perspective

Large satellite constellations impact modern radio astronomy (RA) in various ways. Classically, impact assessment studies have focused on intentional transmissions and their by-products (e.g. spectral side-lobes) in the bands allocated and adjacent to the satellite radio services (Di Vruno, Winkel, Bassa, et al. 2023; Grigg, Tingay, Sokolovski et al. 2023). Recently, however, other interference vectors have emerged, such as electromagnetic leakage from onboard electronics (Di Vruno, Winkel, Bassa, et al. 2023; Grigg, Tingay, Sokolovski et al. 2023). Some satellite constellation and terrestrial cell-phone network operators are also teaming up to smoothly integrate both systems which is known as direct-to-device (D2D) or direct-to-cell (D2C) technology. This provides extra challenges to RA as the carefully constructed coordination scheme between terrestrial-only cell phone networks and the affected radio services does not include space based infrastructure.

In order to enable a productive co-existence of RA and satellite constellations, coordination and regulatory actions are required. All regulatory activities at the Radiocommunication Sector of the International Telecommunication Union (ITU-R) and other organisations are based on compatibility studies. Spectrum managers are often asked by other stakeholders whether they can simply "try things out", i.e. assess interference based on real-life testing. However, that may not always be possible, since RA protection criteria are extremely strict: after about an hour of integration, any interference must still be below 10% of the noise level. That means observational proof that protection levels are violated can require an extraordinary amount of observation time. Moreover, some observatories may not be capable of tracking artificial satellites.

Modeling is sometimes the only viable route to determining potential satellite interference with RA facilities. As large constellations can consist of thousands of satellites, full network simulations are necessary. Such simulations have to consider the dynamic behavior of transmitters and receivers, antenna patterns (including side-lobes and time-variable pointing), aggregation effects and – at higher frequencies – atmospheric effects. The ITU-R has developed procedures and algorithms that can assist such calculations. Nevertheless, the implementation is non-trivial, especially with respect to computational costs. Without accurate models and simulations, determining appropriate measures for the protection of RA and coordination with other services becomes essentially impossible. Observations are still required on top of such efforts to ensure the consistency of models and provided technical parameters, e.g., transmitter power levels, as well as for calibration purposes.

The use of official algorithms and models provided by the ITU-R is also strongly recommended in order to avoid unnecessary ambiguities that have to be dealt with in high-level regulatory and technical working groups. It is, unfortunately, not uncommon that contributions from stakeholders containing compatibility calculations are not accepted if they are not based on ITU-R models.

Another important point is that observations should produce results that are properly calibrated in order to enable a sound comparison with ITU-R standards. Detection experiments alone are usually not suitable for regulatory purposes.

3. Satellite constellation interference and mitigation in spectroscopy

Active avoidance, i.e. pointing the telescope "somewhere else" to avoid FoV crossings of satellites, is often proposed as a satellite interference mitigation strategy. Unfortunately, this strategy is not always feasible or economical (e.g., Hu, Rawls, Yoachin & Ivezić 2022). This is also the case for the European Southern Observatory 4-metre Multi-Object Spectrograph Telescope (4MOST) given the survey program's wide, 2.5-degree diameter FoV and extended exposure durations. Hence, other strategies are needed to identify potential corruption in 4MOST spectroscopic data. On average, each satellite that traverses the 4MOST FoV will affect approximately 1.3 of the 2436 fibers that feed the light to the spectrographs. While many of these satellite trails will be too dim to impact the 4MOST spectra significantly, some will be bright enough to subtly alter the data without clear indication that something went wrong. Without additional mitigation, this issue would make it impossible to rely on individual 4MOST spectra; only large statistical samples could be safely analyzed. This would, for instance, negatively affect the exploration of anomalous spectra for groundbreaking new scientific discoveries. Simulations with 4MOST indicate that satellite contamination can also subtly affect stellar metrics like radial velocity and elemental abundances, depending on the satellite's brightness level. These effects are so subtle that they evade detection by existing data processing and quality control pipelines. To address this, 4MOST plans to employ a specialized imager for real-time, continuous FoV monitoring to identify crossing satellites. However, the technical specifications for this imager present challenges that have yet to be resolved and add substantial complexity to the project.

4. Satellite constellation interference in γ-ray astronomy

High Energy Stereoscopic System (H.E.S.S.) is an Imaging Atmospheric Cherenkov Telescope (IACT) array located in Namibia, and is designed to perform ground-based γ-ray astronomical observations. It works by detecting the optical to near-Ultra Violet Cherenkov light emitted when γ-ray photons interact with Earth's atmosphere and initiate a particle cascade, a so-called "Extensive Air Shower" (EAS). Unlike in optical or radio astronomy, the presence of very-high-energy γ-rays is inferred by classifying and determining the properties of these EAS events. The effect of satellite trails on these IACT observations has previously been considered minimal, due to their approximately

10 nanosecond signal integration time. However, IACTs are still exposed to the night sky and sensitive to optical frequency light, and, therefore, can be affected by satellites. Given the continuing satellite launches, and the construction of the next-generation Cherenkov Telescope Array (CTA) observatory, it is necessary to quantify such interference. CTA construction efforts have already begun on La Palma with the inauguration of Large Sized Telescope 1 (LST-1); a second site for CTA will be located in Chile (near Cerro Paranal). Lang, Spencer & Mitchell (2023) developed a method of detecting satellites in H.E.S.S. data using night sky monitoring measurements, which H.E.S.S. cameras take concurrently with γ-ray observations. Lang, Spencer & Mitchell (2023) show that, whilst measurable, the effect of satellite trails on EAS event classification and reconstruction with H.E.S.S. is minimal. However, a mildly increasing trend in the number of trails was detected over the past three years, and new analysis techniques that aim to reduce the energy threshold of IACT observations with next generation instruments could be more severely affected.

5. Satellite constellation interference with space-borne facilities

Space-borne telescopes in low Earth orbit (LEO) can also be affected by satellite streaks. To demonstrate this, Stark, Grogin, Ryon & Lucas (2022) studied over 30,000 individual exposures taken between 2002 and mid-2022 using the Wide-Field-Channel (WFC) on the Advanced Camera for Surveys (ACS) aboard the Hubble Space Telescope (HST). Stark, Grogin, Ryon & Lucas (2022) find that the rate of satellite trails in ACS/WFC imaging data has increased by approximately a factor of two in the last two decades, with approximately 10% of images currently affected, but there is no clear systematic evolution in the typical trail brightness. The Space Telescope Science Institute has spear-headed the development of a new software package that was used to identify and mask satellite trails in HST astronomical imaging data (Stark, Grogin, Ryon & Lucas 2022). The approach is based around the Median Radon Transform, a modified version of the standard Radon Transform. The key advantage of this method is its sensitivity; it can detect linear features with mean brightness significantly below the background noise level of an image, and it is resistant to false detections caused by bright astronomical sources (e.g., stars, galaxies) in the majority of cases. The software has been incorporated into the "acstools" Python package that is publicly available.

Thankfully, current ACS/WFC observation and data reduction practices (e.g., taking multiple dithered exposures per field and rejecting compromised pixels in individual exposures when creating a final image) remain effective at mitigating the impact of satellite trails on final data products. While HST science has not yet been compromised by satellite trails, nor is it expected to be affected even with the increase in the number of LEO satellites in the near future, the rate of satellite trails in ACS/WFC data will continue to be monitored.

6. What can astronomers do?

The administration of space traffic is getting increasingly complex, needing growing participation from both professional and amateur astronomers. The Cadi Ayad University Observatory, in partnership with the IAU CPS and SpaceAble, is actively involved in satellite and space debris observation. The IAU CPS is an international organization dedicated to working with all stakeholders to tackle the challenges artificial satellite constellations pose to astronomy. Participation in IAU CPS observing campaigns allows observatories around the world to contribute, analyze and publish satellite brightness data, for instance, on AST SpaceMobile's BlueWalker 3 (Nandakumar, Eggle, Tregloan-Reed et al. 2023) and several generations of Starlink satellites (Krantz, Pearce & Block 2023; Mallama, Hornig, Cole et al. 2023). The Oukaimeden observatory is a partner of the IAU CPS center and as such, the observatory's researchers are directly involved in the coordination and acquisition of satellite observations through the deployment of a variety of professional and amateur telescopes. The observatory also provides facilities for accommodating and collaborating with other satellite observation programs, as well as an academic and professional training program to support observational activities. Furthermore, utilizing orbital data processing techniques, the observatory has developed a variety of unique platforms that aid in observation planning. These user-friendly interfaces enable more effective observation scheduling, thereby increasing the observatory's contribution in the various scientific objectives behind satellite observations. The extensive activities carried out by the Oukaimeden observatory in coordination with the IAU CPS include observation planning, satellite tracking, and the extraction of orbital parameters through image processing. These efforts are an excellent example of how astronomers can contribute to assessing the impact of satellite constellations on ground based-astronomy.

Acknowledgements

STS is supported by the Deutsche Forschungsgemeinschaft (DFG, German Research Foundation) - Project Number 452934793. SE and MP thank the IAU CPS for assisting with travel support.

References

Anderson-Baldwin, J. 2023, LEO Satellite Constellations in the MOA Database. Master's thesis, University of Auckland, Auckland, New Zealand

Barentine, J. C., Venkatesan, A., Heim, J., et al. 2023, Nat Astro, 7(3), 252

Committee on Radio Astronomy Frequencies, et al. 2023, Committee on Radio Astronomy Frequencies (CRAF) – Periodic Review 2011–2021. arXiv e-prints, arXiv:2310.13407

Deshpande, A. A. & Lewis, B. M. 2019, J Astron Inst, 8(1), 1940009

Di Vruno, F. & Tornatore, V. Large satellite constellations and their potential impact on vgos operations. In International VLBI Service for Geodesy and Astrometry 2022 General Meeting Proceedings 2023, 19

Di Vruno, F., Winkel, B., Bassa, C. G., et al. 2023, A&A, 676, A75

Grigg, D., Tingay, S., Sokolowski, M., et al. 2023, A&A, 678, L6

Hu, J. A., Rawls, M. L., Yoachim, P., & Ivezić, Ž. 2022, Satellite constellation avoidance with the rubin observatory legacy survey of space and time. ApJL, 941(1), L15

Karpov, S. & Peloton, J. 2023, The rate of satellite glints in ZTF and LSST sky surveys. arXiv e-prints, arXiv:2310.17322

Krantz, H., Pearce, E. C., & Block, A. 2023, The Steward Observatory LEO Satellite Photometric Survey. PASP, 135(1051), 095003

Lang, T., Spencer, S. T., & Mitchell, A. M. 2023, Impact of satellite trails on hess astronomical observations. A&A, 677, A141

Mallama, A., Hornig, A., Cole, R. E., et al. 2023, arXiV e-prints, arXiv:2309.14152

Nandakumar, S., Eggl, S., Tregloan-Reed, J., et al. Nature, 623, 938

Stark, D. V., Grogin, N., Ryon, J., & Lucas, R. 2022, Instrument Science Report ACS 2022-8, 8

Astronomy and Satellite Constellations: Pathways Forward
Proceedings IAU Symposium No. 385, 2026
P. Grimley, J. Barentine, D. Galadí-Enríquez & C. Walker
doi:10.1017/S1743921324000929

Strategies for spectrum coexistence between radio telescopes and satellite networks

Christopher G. De Pree[1], Anthony Beasley[1], Sheldon Wasik[1], Urvashi Rau[2], Rob Selina[2], Will Armentrout[3], Harvey Liszt[1], Bang Nhan[1], Daniel Dueri[4] and Matt Iverson[4]

[1]National Radio Astronomy Observatory, 520 Edgemont Road, Charlottesville, VA 22901
email: cdepree@nrao.edu

[2]NRAO, Socorro, NM

[3]Green Bank Observatory, Green Bank, WV

[4]Space Exploration Technologies Corp., Hawthorne, WA

Abstract. Passive services (e.g. radio astronomy, Earth observing) are slowly responding to new radio frequency interference (RFI) challenges to their science missions, particularly those arising from constellations of communication satellites. We will report on efforts at the National Radio Astronomy Observatory and Green Bank Observatory to detect, characterize and mitigate RFI impacts at our sites, including (a) preliminary results from a prototype spectrum monitor located at Green Bank; (b) results from a 1.5-year study of the impacts of Starlink transmissions (satellites and user terminals) on the Very Large Array, indicating minimal interference; (c) our plans for real-time status coordination of telescope control systems with satellite operators to avoid RFI conflicts with observing programs. Future scientific endeavors such as the next-generation Very Large Array will operate in a more complex spectral environment than the current generation of radio telescopes, and the ability to accurately detect, characterize and effectively mitigate RFI is critical.

Keywords. Satellite constellations, interferometry, spectrum management

---

1. Introduction

For decades, typical RFI issues faced by radio astronomy observatories came from onsite, or from the the horizon. These spurious signals arose from nearby electronics, or later from cell phone towers placed in the near environment of observatories. For these reasons, observatories typically have been situated in remote locations. Other sites were protected by specially defined zones. The most well-known of such zones in the US is the National Radio Quiet Zone (NRQZ), established by FCC Docket 11745. Some locations, such as the Very Large Array (VLA) in New Mexico have established coordination agreements with known local transmitters (e.g. White Sands). For a time, these arrangements provided sufficient protections. Satellite-borne signals have had significant impacts on radio astronomy at least since the advent of the GLONASS system (e.g. Combrinck et al. 1994), which severely impacted observations the 1612 MHz Hydroxyl line.

The advent of satellite constellations like Starlink, operated by SpaceX, have added a significant complication to this operating environment. Starlink satellites operate in Low Earth Orbit (LEO), and transmit downward from their orbits, in order to provide broadband internet to receivers located on the ground. This new environment will require

Figure 1. The ASM-1 device undergoing RFI testing in the Green Bank Observatory anechoic chamber in May 2023.

additional safeguards, and in particular coordinated and cooperative effort for both active and passive users to meet their different goals.

In this presentation, we briefly cover recent efforts on the part of the National Radio Astronomy Observatory (NRAO) to address these new spectrum challenges by:

(*a*) developing and deploying a new generation of spectrum monitoring devices (Advanced Spectrum Monitors),
(*b*) carrying out coordinated testing with SpaceX to gauge the impact of SpaceX Starlink User Terminals operating at the Alamo Navajo Reservation near the VLA
(*c*) performing coordinated experiments to study the impact of satellites passing close to boresight, and ways to mitigate these effects (boresight avoidance)
(*d*) developing a scheme that will allow NRAO telescopes to communicate their current operating status (e.g. position and frequency band) to SpaceX, so that SpaceX can adjust their network to minimize interference (Operational Data Sharing, or ODS).

2. ASM development and testing

As part of the NSF's Spectrum Innovation Initiative - National Radio Dynamic Zone (SII-NRDZ) program, NRAO has been developing an Advanced Spectrum Monitor (ASM). NRAO's goal in developing this device is to design and deploy a relatively cheap, well-calibrated spectrum monitoring unit for radio observatories and radio dynamic zones (RDZs). The device will be portable, weather proof, and its spectral coverage will range from 1-120 GHz. In Spring 2023, NRAO deployed and tested the ASM-1 "proof of concept" device at the Central Development Laboratory (CDL) and Green Bank Observatory (GBO). These deployments and tests have provided input for the second phase of development now underway (see Fig. 1).

As an example of the effectiveness of the ASM-1 device, Fig. 2 shows a comparison of the RFI environment at the CDL (Charlottesville) and GBO as measured by the device in May 2023. While the RFI environment at GBO is markedly better than CDL, there is still RFI apparent - especially in WIFI bands - in the NRQZ.

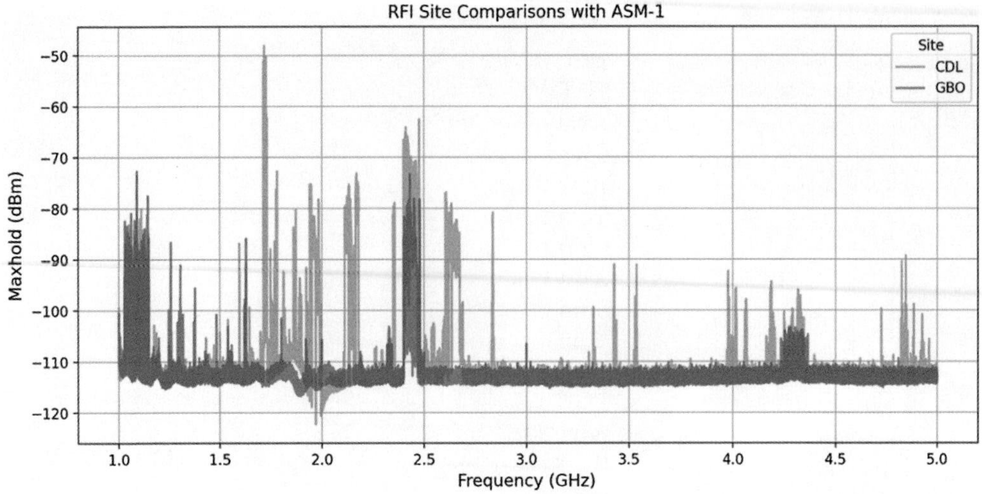

Figure 2. The RFI environment between 1 and 5 GHz at the CDL in Charlottesville (orange) and in the NRQZ (blue) as detected by the ASM-1 device in May 2023.

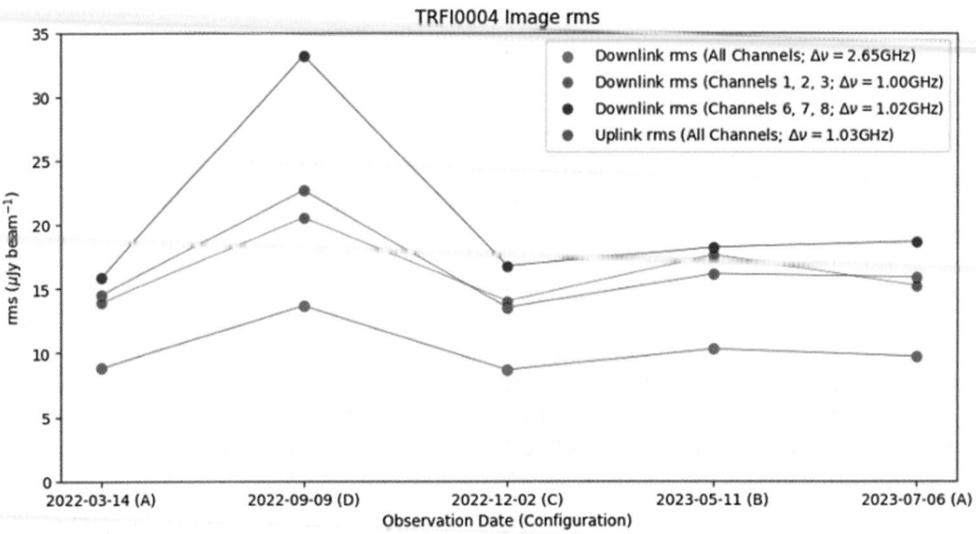

Figure 3. Long term monitoring of frequencies that coincide with the uplink and downlink frequencies of Starlink satellites and UTs over the past 1.5 years shows that (in particular when comparing similar constellation), the RMS noise in images has been relatively flat. There is some evidence that RMS noise is higher in downlink channels 6, 7, 8 versus 1, 2, 3. Theoretical noise in a ~30 minute observation is about 15μJy.

3. Curricular materials and publications

As part of the NRDZ program, NRAO has also developed curricular materials related to improving spectrum understanding for the general population, high school students and undergraduates. These materials are available for use at the SuperKnova site operated by AUI (`https://superknova.org`). The NRAO NRDZ team has also worked to solicit and capture stakeholder input related to the development of an NRDZ by holding Workshops and Lean Coffee Table events in the years 2020–2023. The major outcomes of these discussions are captured in Zheleva et al. (2023) and De Pree et al. (2023a)

4. Coordinated testing with SpaceX at VLA and GBT

Fall 2021 Coordinated tests began with SpaceX in September and October 2021 with the placement of a User Terminal (UT) at various locations near the VLA in New Mexico. With the device deployed and operational, the VLA was operated in bands capable of detecting both uplink (14.0–14.5 GHz) and downlink (10.7–12.7 GHz) signals. The results of these tests are described in detail in De Pree et al. (2023b). Phase I testing (September 2021) involved running the User Terminal in normal mode at a variety of locations near the Very Large Array and the Pie Town VLBA antenna. Phase II testing (October 2021) involved fixed channel downlink frequency illumination of the VLA by Starlink satellites.

Initial analysis of the Phase I testing indicated that UT uplink transmissions (between 14.0 and 14.5 GHz) had minimal impact on VLA and VLBA data integrity in locations where there is no line of sight from the UT to the telescope. Phase II testing demonstrated that analytical interference models have good correlation to real world observations, and as predicted by the model, satellite downlink transmissions were detected only on some of the shortest baselines in the B configuration. The impacted baselines change (as predicted) according to the location of the transmitting satellite at any given time. These interferometric effects are discussed in detail in Thompson (1982) and Perley (2002).

Spring-Summer 2022 In late 2021, NRAO began to consider how to help solve an ongoing issue with broadband internet access on the Alamo Navajo Reservation near the VLA site. By March 2022, in discussion with representatives from the Alamo tribe and SpaceX, NRAO developed a plan to help with the installation of ~60 Starlink UTs at family residences. This action was taken to help provide families on the reservation with decentralized broadband internet access for education. An equally important goal of this installation of Starlink UTs at a location that previously had no broadband internet access, was to evaluate the impact on normal VLA operations.

Since this installation was completed in Summer 2022, NRAO has been performing at first bi-weekly and then monthly monitoring of the frequencies utilized by SpaceX (and other satellite operators) for RFI in the uplink (14.0–14.5 GHz) and downlink (10.7–12.7 GHz) bands. A preliminary analysis of selected datasets from the first year of monitoring, and a description of the plan for ongoing monitoring and a full analysis of these data is presented in De Pree et al. (2023c). The purpose of these imaging tests is to examine if the data are affected by partially decorrelated RFI and to eventually check whether or not this RFI contributes to a noise floor.

Fig. 3 shows measured rms noise (μJy/beam) in images made from selected monitoring scans made from 2022 into 2023. Both downlink (10.7–12.7 GHz) and uplink (14.0–14.5 GHz) frequencies have been monitored. While these are limited tests (taking during VLA maintenance time, typically midday) that include about 1 hour of data, there is no indication that noise in images from data in the A-configuration show measurable change between March 2022 (before any UTs were installed) and July 2023 (after all UTs were installed).

Spring-Summer 2023 Coordinated experiments with SpaceX have continued into 2023, with the most recent experiments exploring the importance of satellites passing close to boresight of the GBT. In an April 2023 experiment, the GBT took data while pointed north at a variety of elevations. One of these scans produced a large signal (~4000 Jy), and it was determined that a Starlink satellite had passed within 0.05 degrees of boresight. This chance encounter led to a series of planned experiments that gauged the signal strength at a variety of planned distances from boresight. The results of these experiments are guiding the development of "boresight avoidance" mechanisms as described below. The April 26, 2023 encounter is shown in Fig. 4, the instantaneous and integrated GBT

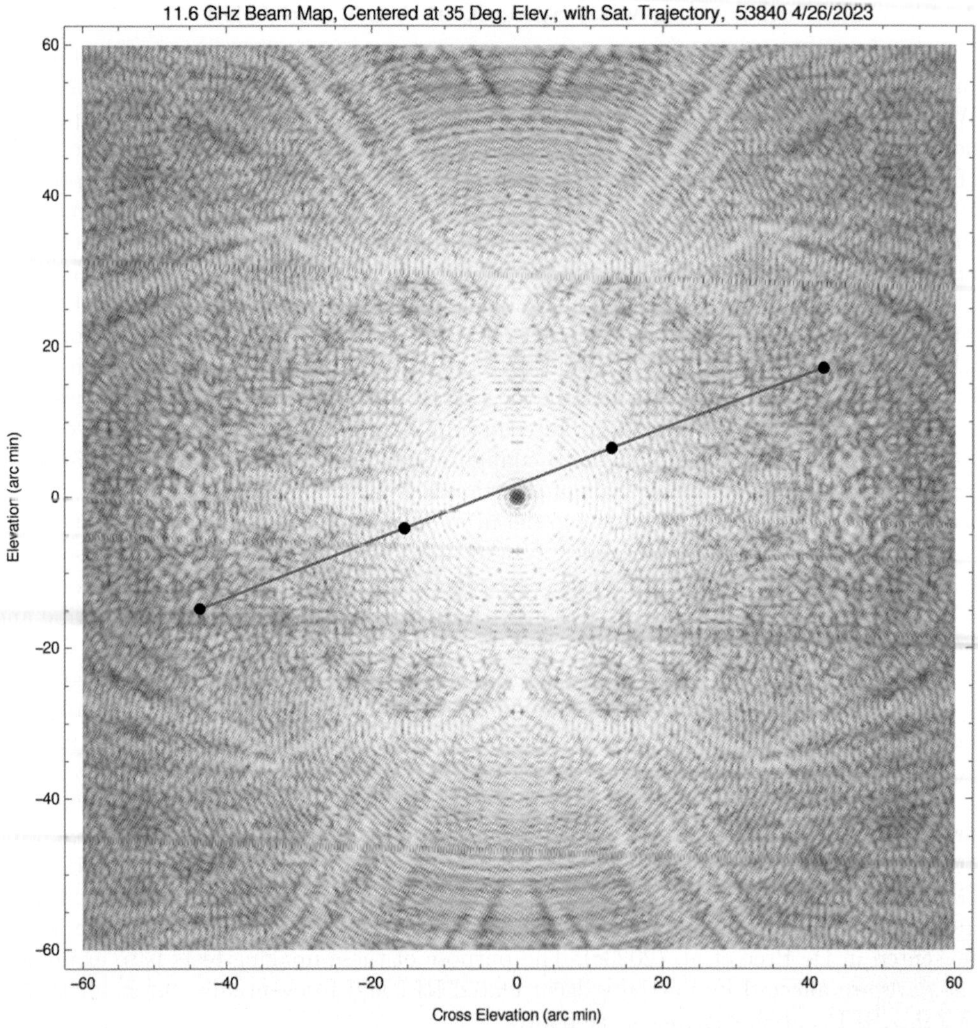

Figure 4. GBT beam map at 11.6 GHz, centered at the appropriate azimuth and elevation at the appropriate time on April 26, 2023. X-and y-axes are distance (arcmin) from boresight. The black dots indicate the position of a Starlink satellite at one second intervals. Note that this image covers a 2x2 degree square on the sky. The large signal shown in Fig. 6 is from the one second integration as the satellite passed closest to boresight.

gain along this trajectory is shown in Fig. 5. The 1-s GBT spectrum taken at the time of closest to boresight passage is shown in Fig. 6

5. Operational data sharing (ODS)

In order to find ways for active and passive users to more effectively share spectrum, NRAO has been developing a way to share current telescope status (e.g. frequency band, pointing position, observation time span) with a satellite operator (SpaceX) so that the constellation satellites that are operating (1) near boresight or (2) at the frequency being observed can adjust their operations to allow for minimal RFI to the radio telescope in question.

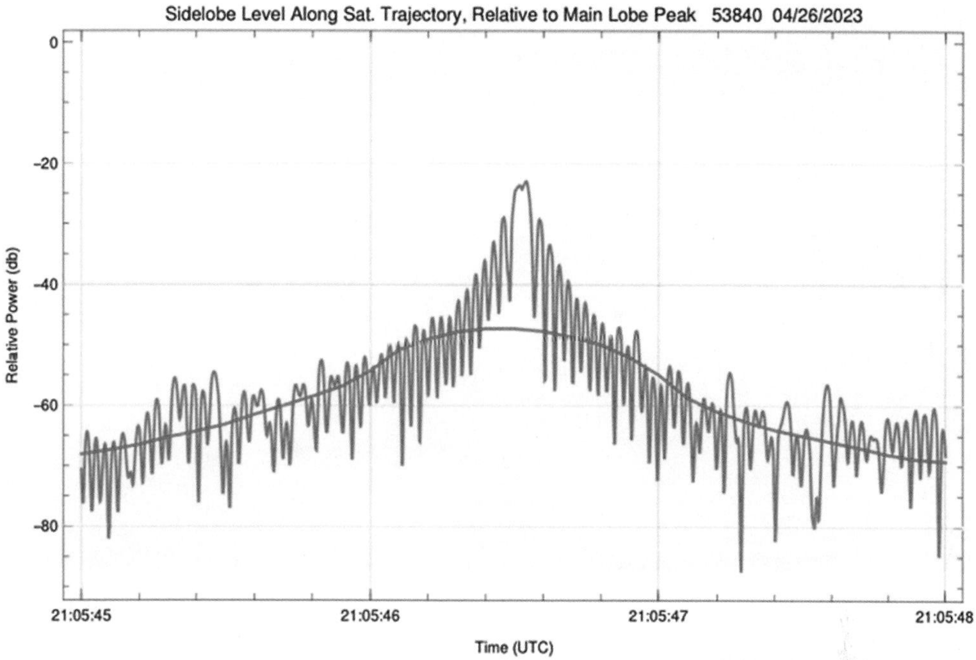

Figure 5. This figure shows the sidelobe relative power level along the trajectory of the close to boresight passage on April 26, 2023. Blue is the instantaneous sidelobe level, and the red line is a running 1-second average of the sidelobe level. Note that at the closest passage, the relative power is approximately 50 dB below the GBT gain (85 dB).

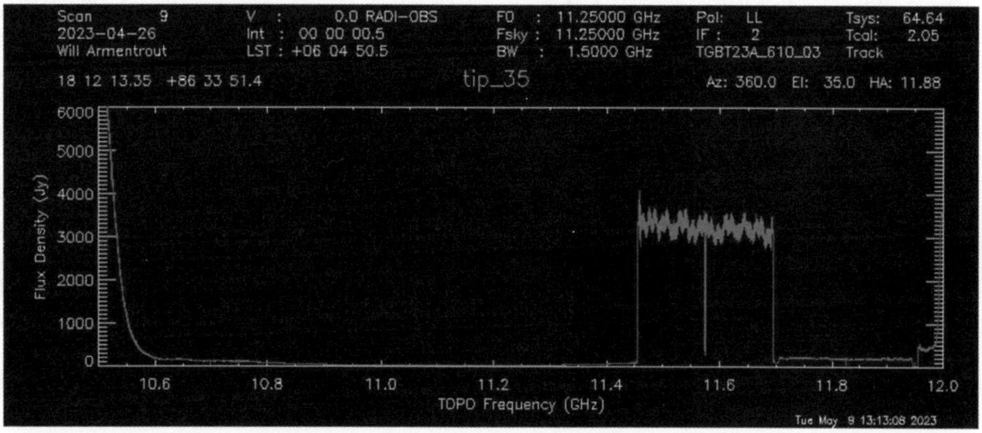

Figure 6. A 1-s scan taken with the telescope at 35 degrees elevation. This RFI is likely the result of a CH4 downlink interaction between the GBT-mainlobe and the SAT-sidelobe.

The system being developed is called Operational Data Sharing, or ODS, and NRAO is piloting its implementation at two telescopes: the VLA in New Mexico and the GBT in West Virginia. The ODS plan provides sufficient data to satellite operators so that they can have awareness of the current operations of radio telescopes in their vicinity. Fig. 7 shows a schematic representation of the ways in which ODS will work, by (1) Radio Zone Avoidance, (2) Frequency Multiplexing and (3) Boresight avoidance. All three of these mechanisms are available to the pilot satellite operator (e.g. SpaceX), but multiple

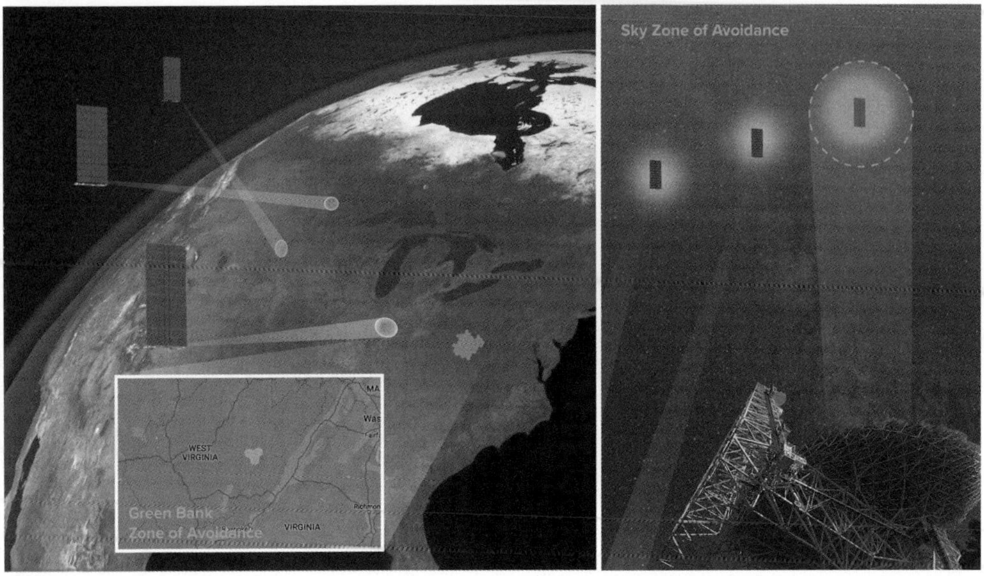

Figure 7. SpaceX Starlink satellites avoid direct illumination of the location of NRAO telescopes (left), and are able to change downlink parameters based on knowledge of current telescope operating conditions (right).

modes of coexistence should allow a variety of operators to eventually participate. Each of the modes is described briefly below.

Radio Zone Avoidance This mode is the most fundamental (when available), and radio zone avoidance simply means that downward facing active systems will not illuminate cells or regions containing radio telescopes with their main beams. This mode is possible for systems like Starlink that have smaller (~22 km diameter FWHM) beams, but may not be possible for all operators. Avoiding the illumination of the region containing telescopes is the most effective way to avoid main beam to main beam interactions that have the potential to produce very large signals in radio telescopes.

Frequency Avoidance or Multiplexing This mode entails the satellite network (1) knowing what frequency the radio telescope is operating in and (2) either operating in another frequency range or sharing the band in a pre-defined way. As an example of frequency avoidance, if the GBT is observing in X-band, then the satellite network might instruct nearby downlinking satellites to operate in Ku-band, or in a subset of X-band. Frequency avoidance is the simplest for both active and passive systems, but frequency multiplexing could be implemented once satellite networks have an understanding of the current operations of a radio telescope.

Boresight Avoidance Experiments over the past year have indicated that boresight avoidance might be the most effective way to protect sensitive radio telescopes from large RFI signals. Radio telescopes typically have beam patterns that give them very high sensitivity to a small area of the sky. For this reason, satellites that pass close to a radio telescope's boresight have the possibility of producing very large signals. Boresight avoidance involves an individual satellite knowing its position relative to the pointing of a nearby radio telescope, and adapting its downlink signals accordingly as it passes close to boresight (typically a few seconds in the case of an LEO satellite). Different satellite constellations will have different abilities to implement boresight avoidance techniques, but for constellations that are unable to change downlink beam patterns or disable downlinks, frequency avoidance (while close to boresight) might be another possibility. NRAO and SpaceX are exploring possibilities for boresight avoidance with the passive and active systems that we have access to.

6. Summary and conclusions

NRAO and SpaceX are piloting mechanisms that could allow active and passive users of the spectrum to more effectively coordinate spectrum use. Early monitoring and coordination results are promising, and NRAO is committed to continuing this work. NRAO hopes to expand these experiments to include other satellite networks as they come online. In conclusion, we emphasize the following:

(*a*) Major challenges and opportunities face radio astronomy in the coming decades. Resolutions to these challenges may result in significant changes to current operating models of radio astronomy and active services. It is important for all spectrum users to commit to exploring, adapting, and communicating.
(*b*) Spectrum management relies on spectrum measurement. NRAO performs regular spectrum monitoring with its telescopes, and the development of new standalone spectrum monitoring devices is continuing. The outcomes of ASM-1 testing (May 2023) are informing the design and deployment of ASM-2 in the coming year.
(*c*) The partnership with SpaceX in coordinated testing has led to important insights regarding how radio astronomy can adapt in the age of satellite mega-constellations. NRAO will continue to explore how to best define (global) standard operating procedures for radio telescope-satellite coordination.
(*d*) ALMA and NRAO telescopes that operate at higher frequencies are developing strategies to manage new challenges to spectrum coexistence that will arrive as higher frequency active systems are deployed. An analysis of RFI impacts on the ngVLA is given in Selina et al. 2023.
(*e*) An early version of the ODS system is working, and it will be tested in the coming months. This system is an important step towards building active/passive system RFI tools, and successful coexistence.

Acknowledgements

The authors acknowledge the work of many NRAO scientists and engineers and SpaceX engineers who have helped with the planning and execution of experiments over the past 2 years. Special thanks to James Robnett for his tireless work in deploying the UTs for families in the Alamo Reservation, and to Doug Knox and Mike Nicolls (SpaceX) who have devoted significant efforts to these coordination experiments.

References

Combrinck, W. L., West, M. E. & Gaylard, M. J. 1994, PASP, 106, 807

De Pree, C. G., Zheleva, M., & Anderson, C. 2023a, theconversation.com. https://theconversation.com/radio-interference-from-satellites-is-threatening-astronomy-a-proposed-zone-for-testing-new-technologies-could-head-off-the-problem-199353 (accessed November 2, 2023)

De Pree, C. G., Rau, U., Selina, R., Svoboda, B., & Beasley, A. 2023b, https://library.nrao.edu/public/memos/evla/EVLAM_222.pdf

De Pree, C. G., Rau, U., Selina, R., Svoboda, B., & Beasley, A. 2023c, EVLA Memo No. 223, https://library.nrao.edu/public/memos/evla/EVLAM_223.pdf (accessed November 2, 2023)

Perley, R. 2002, EVLA Memo No. 49, https://library.nrao.edu/public/memos/evla/EVLAM_49.pdf

Selina, R., Rau, U., & De Pree, C. G. 2023, EVLA Memo No. 109, https://library.nrao.edu/public/memos/ngvla/NGVLA_109.pdf
Thompson, A. R. 1982, IEEE Trans Antennas Propag, 30(3), 450
Zheleva, M., Anderson, C. R., Aksoy, M., Johnson, J. T., Affinnih, H., & DePree, C. G. 2023, IEEE Commun Mag, 61(6), 156

Astronomy and Satellite Constellations: Pathways Forward
Proceedings IAU Symposium No. 385, 2026
P. Grimley, J. Barentine, D. Galadí-Enríquez & C. Walker
doi:10.1017/S1743921324001054

Spectrum management at ALMA and synergies with other entities

Giorgio Siringo and Sean Dougherty

ALMA - Atacama Large Millimeter Array
email: giorgio.siringo@alma.cl

Abstract. An overview is presented of the activities of the ALMA Spectrum Management Office and the synergies with other entities that engage in similar scientific research activities or that develop actions aimed at the control and regulation of radio frequency emissions.

Keywords. ALMA, radio astronomy, radio-frequency interference, spectrum management

---

1. The ALMA site and the local environment

The Atacama Large Millimeter/submillimeter Array (ALMA) is a 66-element interferometer located on the high mountain plateau known as Llano de Chajnantor, at an altitude of 5000 m and remote from urban areas, the closest being Toconao (32 km) and San Pedro de Atacama (47 km, Figure 1). The ALMA operational spectrum spans from 35 GHz up to 950 GHz, divided into 10 non-contiguous frequency bands (Figure 4). The ALMA site has natural shielding from most typical sources of electromagnetic (EM) interference. Emissions at frequencies higher than ~100 GHz are attenuated strongly by atmospheric water vapour and do not pose a threat unless produced locally at short distance and within the ALMA site. The Chajnantor site is somewhat exposed to eventual emissions from the CH27 Paso Jama international road between Chile and Argentina, passing within ~10 km of the central area of the ALMA array and even less distant from some of the antenna foundations (pads) in Pampa La Bola.

1.1. *The ALMA concession and the Atacama Astronomical Park (AAP)*

In May 2003, Associated Universities, Inc. (AUI) and the European Southern Observatory (ESO) signed the acquisition from the Ministry of National Assets of Chile of land for the ALMA Operations Support Facility (OSF), 30 km from San Pedro de Atacama at an altitude of 2950 m. In November 2003, the Government of Chile, through the Ministry of National Assets, provided a 50-year land concession for the construction and operation of ALMA on the Chajnantor Altiplano at 5000 m. This area, known as the ALMA concession, includes two regions: an area of 161.6 km^2 for the radio telescope at 5000 m and an additional 16 km^2 for the road between the OSF and the Altiplano. The ALMA land concession at 5000 m is shown in dark-brown color in Figure 2.

1.2. *Protection and coordination zones*

The ALMA concession is subject to two identical Exempt Resolutions, to AUI and to ESO, issued by the Subsecretariat of Telecommunications (SubTel) of the Ministry of Transport and Telecommunications of the Republic of Chile.

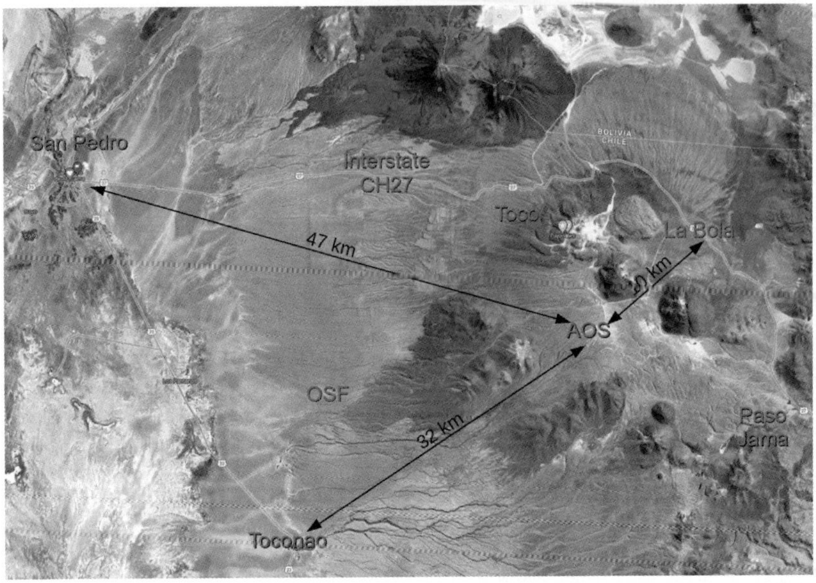

Figure 1. Google Maps image of the ALMA site seen from the sky, showing the distance of the closest urban areas and infrastructure.

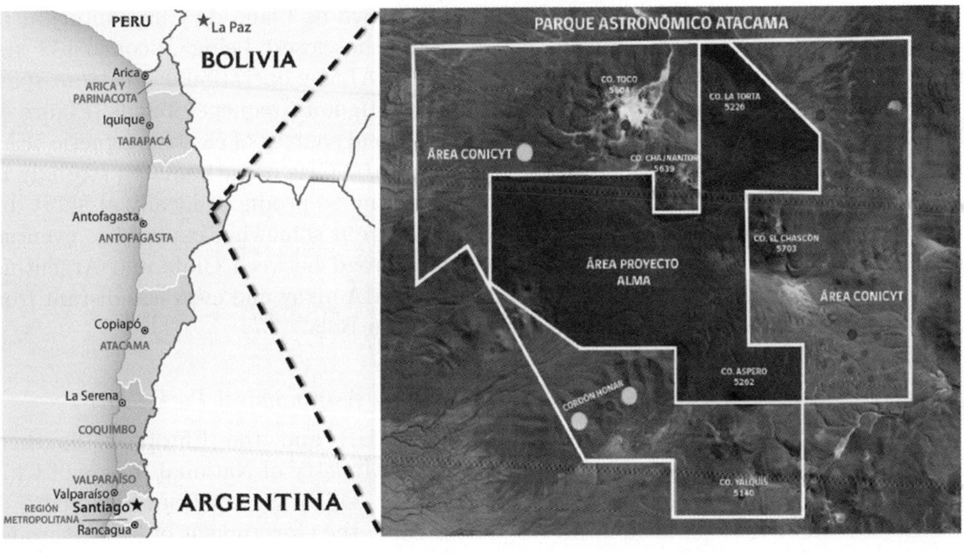

Figure 2. The ALMA concession (dark brown) on Llano de Chajnantor, within the Atacama Astronomical Park (from Bustos, R., Rubio, M., Otárola et al. 2014). The coloured dots show potential sites for other telescopes in the park.

• These resolutions give ALMA the authorisation as "receiving only antennas" to use the frequency bands defined in Table 1 under the assumption that "*protection cannot be guaranteed in the frequency bands, or part of them, that are not allocated to radio astronomy on a primary basis.*"

• They also define a "protection zone" centered at 23° 01' S, 67° 45' W (antenna pad A124 in the central part of the array, see Figure 3) "*with a radius of 30 km within national territory, inside which the installation of any other radio communications system will not be authorized to any third parties operating on the receiving frequency bands.*"

Table 1. Definition of the ten ALMA frequency bands in the original design. The values in parenthesis are offered on a best effort basis only and are not in the baseline specifications

Band	Wavelength of Reference [mm]	RF Range [GHz]	Sideband Scheme	IF Range [GHz]	LO Range [GHz]
1	7	35 – 50(52)	USB	4 – 12	31.0 – 40.0
2	4	67 – 116	2SB	4 – 18	79.0 – 104.0
3	3	84 – 116	2SB	4 – 8	92.0 – 108.0
4	2	125 – 163	2SB	4 – 8	133.0 – 155.0
5	1.6	(158)161 – 211	2SB	4 – 8	166.0 – 203.0
6	1.2	211 – 275	2SB	5 – 10	221.0 – 265.0
7	0.87	275 – 370	2SB	4 – 8	283.0 – 365.1
8	0.67	385 – 500	2SB	4 – 8	393.0 – 492.0
9	0.45	602 – 720	DSB	4 – 12	610.2 – 712.0
10	0.35	787 – 950	DSB	4 – 12	794.7 – 942.3

Band 1 50 – 52 GHz RF range offered on best effort
Band 2 in production phase, planning IF extension to 4 – 16 GHz
Band 5 158 – 161 GHz RF range offered on best effort
Band 6 upgrade planning IF extension to 4 – 18 GHz (maybe 20, TBD)
Band 9 upgrade planning IF extension to 4 – 18 GHz (maybe 20, TBD)

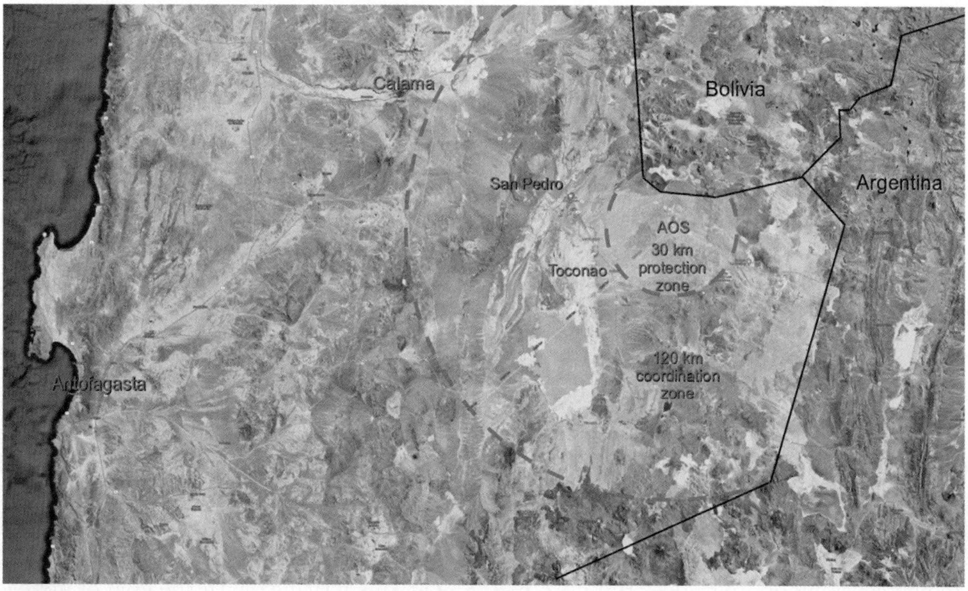

Figure 3. ALMA protection and coordination zones, as defined by the SubTel Exempt Resolutions, with a radius of 30 and 120 km respectively within the Chilean territory. The black line shows the border between Chile, Bolivia and Argentina.

• They also define a "coordination zone" centered at the same place "*with a radius of 120 km within national territory, where coordination being understood as the process whereby the opinion of the petitioners, ESO and AUI, will be sought regarding certain requests by third parties that this Sub-secretarial deems could interfere or affect the operation of the radio telescope. Likewise, in case such petitioners detect any emissions that affect the operation of the radio telescope, they will notify this Sub-secretariat for its coordination.*"

In addition to ALMA, there are several other astronomy projects in operation or in construction on the Altiplano, including the Atacama Pathfinder Experiment (APEX), the Atacama Submillimeter Telescope Experiment (ASTE), NANTEN-2, the Atacama Cosmology Telescope (ACT), the Cosmology Large Angular Scale Surveyor ((CLASS),

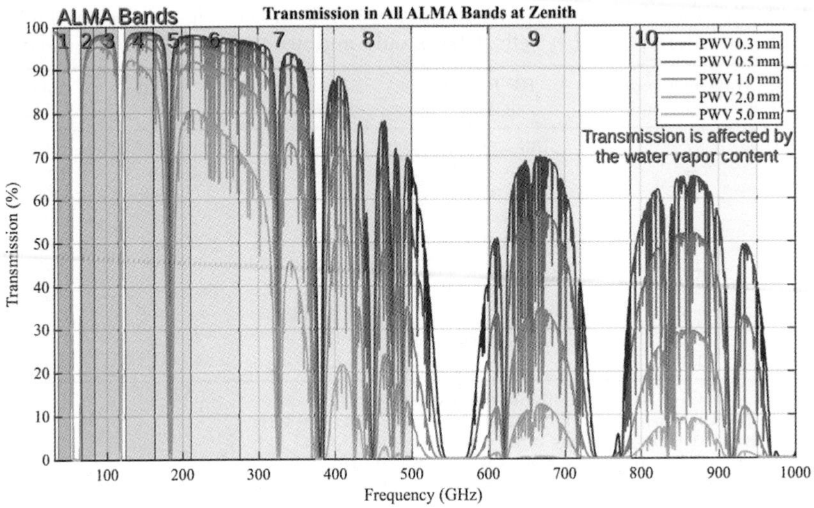

Figure 4. The atmospheric transmission windows at mm/submm wavelengths, here shown for different values of the precipitable water vapour (PWV), define the ten ALMA frequency bands.

the POLARization of the Background Radiation experiment (POLARBEAR), the Simons Observatory (SO), the Cerro Chajnantor Atacama Telescope (CCAT-Prime), and the University of Tokyo Atacama Observatory (TAO).

The AAP can host other types of scientific projects (e.g. meteorological or seismic stations, solar research, etc.) as long as they do not interfere with the astronomical projects. Coordination of activities between the various observatories and projects on the Altiplano takes place through the Chajnantor Working Group (CWG) of the AAP.

1.3. *The ALMA operational spectrum*

The ALMA receiver systems cover ten frequency bands that closely match the atmospheric transmission windows, those portions of the spectrum where the atmospheric transmission at millimetre/submillimetre wavelengths enables ground-based science observations at the corresponding frequencies (see Figure 4). In the first incarnation, the ten ALMA bands are defined as given in Table 1. Several receiver development groups are now working on upgrades (Bands 2, 6, 7, 8, 9 at this writing) with the goal of increasing the sensitivity of the receivers and doubling the intermediate frequency bandwidth (post-detection, hereafter IF), while keeping the same observing RF range. Consequently, when considering the ALMA spectrum it is important to consider the IF bandwidth will be soon extended from 8 GHz to 16 GHz per polarisation channel, spanning the possible frequency ranges of 2–18 GHz or 4–20 GHz depending on the receiver band.

2. Potential RFI from terrestrial emissions

2.1. *Emissions from nearby urban areas*

It is known that broadcast FM radio stations (87 to 108 MHz) from Calama and Antofagasta can be received on Llano de Chajnantor. These frequencies are far from the ALMA receivers spectrum and do not present any RFI to ALMA. Given the distance to these urban areas, any possible future millimetre/submillimetre emissions that overlap the ALMA spectrum with potential associated RFI will be attenuated by atmospheric water vapour and do not pose a risk of RFI to ALMA.

2.2. *Cell phones and wifi*

Some consumer market RF devices, such as the 5-GHz WLAN (IEEE 802.11 protocols) and the 5G cellular phones, could produce RFI at ALMA because they can operate at frequencies that fall within the IF range of the ALMA receivers' (4 – 12 GHz, see Table 1). Emissions from such devices have relatively low power and are attenuated by atmospheric water vapour. As a result they do not pose a risk to ALMA unless in use at the ALMA site in the proximity of the antennas, a practice forbidden in the Chajnantor Working Group regulations †

2.3. *Emissions from vehicles moving along nearby roads*

Citizen-Band radios (e.g. 27 MHz) used by truckers on the CH27 Paso Jama interstate road can easily reach the ALMA site. These frequencies are far from the ALMA receivers' spectrum and do not pose any risk of RFI to ALMA. Other devices operated by road users, such as cell phones or newer technologies such as 5G/6G, or cell phone repeaters along the highway, could produce emissions in the ALMA IF range at a relatively short distance from the ALMA antennas. At the typical frequencies and power levels, however, those emissions do not pose a significant risk of RFI to ALMA given the very good RF shielding of the IF signal chain inside the antennas. Other nearby roads (e.g. CH23, Paso Sico interstate) are much further away and do not pose any risk of RFI to ALMA. To date, there are no reports of RFI that can be attributed to emissions from nearby roads. The situation could change, however, if telecom operators were to increase frequencies and power levels.

2.4. *Automotive radar ranging systems*

There is growing use of precision ranging systems for automotive applications. This is motivated by increased safety capabilities and the development of autonomous self-driving vehicles. Relatively short wavelength and large bandwidth (e.g. 4 GHz) are needed to reach satisfactory measurement accuracy and real-time response. The need for larger bandwidth has pushed the emissions of these devices to higher frequencies, overlapping with the ALMA Band-2 receiver spectrum †. It is likely that these devices will use even higher frequencies in the near future. The range of operation is short (up to ~250 m) thus posing a risk of RFI only if used at the ALMA site in the proximity of the antennas.

2.5. *Emissions from inside the ALMA concession*

At both ALMA sites, the Operations Support Facility (OSF) near San Pedro de Atacama and the Array Operations Site (AOS) on the Chajnantor Altiplano, there are several devices that produce radio emissions, with the potential of producing RFI.

- **Communication radios**: ALMA has a radio system with several VHF handheld devices and stations.
- **Private cell phones, laptop computers**: staff may forget to switch off their phones or the wifi while working in the proximity of the antennas.
- **Vehicles** that have devices that produce RFI.
- **The ALMA Polarization calibration source** located on Cerro Honar ‡ emits CW or broadband white noise, linearly polarised, towards the ALMA central cluster in the frequency range 84–116 GHz or 211–373 GHz with maximum power of 1 μW

† See Sect. 9 of *CWG Policies*,ver.A, rev.2, 2008-02-27

† E.g. 81 GHz, see ITU-R M.2057-1, 2018

‡ At 23° 04' 10.4" S, 67° 45' 20.4" W

(-60 dBW). The operation of the source is coordinated strictly with the Chajnantor Working Group spectrum managers.

• **The ALMA Water Vapor Radiometer** (WVR): this device is inside each 12-meter ALMA antenna and is known to produce RFI in two ways:

a) emission from the Voltage Controlled Oscillator in the WVRs that produce CW signals of 15.275 GHz and corresponding harmonics;

b) emission through the horn antenna from the mixer.

Such RFI from the WVR has been observed affecting different ALMA receiver bands. For example, the 3rd harmonic at 45.825 GHz is affecting B1 observations near the methanol maser at 45.843 GHz, the 6th harmonic at 91.7 GHz that affects B3 observations. RFI from the WVR is particularly strong because it is produced directly inside the 12-m antennas, very close to the front-end receivers.

• **Holography towers**: at the OSF there are two 50-metre-high towers, each equipped with a 104.02 GHz, +13 dBm (20 mW) transmitter. These transmitters are powered off most of the time and only used when needed for antenna surface assessment at the OSF by holography techniques.

2.6. *Emissions from other observatories*

The other observatories on the Altiplano near the ALMA site have devices that can produce RFI that may affect ALMA operations:

• **APEX holography system**: the APEX telescope has a transmitter on the shoulder of Cerro Chajnantor, emitting a narrow (1 kHz) but powerful signal at 92.4 GHz. This is known to introduce RFI in the ALMA Band 3 and strict coordination between the observatories is mandatory.

• **Cerro Toco holography**: The Astro-Engineering group at the Universidad Catolica of Santiago is developing a holography system for surface assessment of the telescopes at Cerro Toco using a drone to lift a radio source in the near field of the telescopes.

• **Microwave links to San Pedro**: Both Cerro Toco and APEX use microwave links for remote operation from their base-camps in San Pedro. APEX has a transmitter on Cerro Chico that transmits at 7.3 GHz. There are no reports of RFI from these systems: they are strongly directional and pointed towards San Pedro, away from the ALMA antenna locations.

3. Potential RFI from the sky: satellite emissions

Satellite systems in orbit around Earth are sources of RF emissions and are consequently potential sources of RFI to ALMA. The low water column in the atmosphere above the Altiplano, the main reason the site is one of the best in the world for submillimetre ground-based astronomy, does not attenuate strongly emissions in the millimetre/submillimetre (see Figure 4) and potential RFI from the sky is an important consideration for ALMA. Some of these services operate at frequencies common to the ALMA receiver systems. Besides telecommunication satellites, there are a number of other satellites for scientific application, some of which host active RF transmitters, such as the radars used for Earth exploration services.

3.1. *Satellite constellations*

Telecommunication companies around the world are rapidly expanding their business opportunities through satellite constellations. Today, there are many satellite operators building or planning to build large satellite constellations and this is now a very dynamic field with the number of satellites and their orbital and radio configurations changing very quickly, making it difficult to keep track of all the advances. Some web sites are trying to provide updated information e.g. Jonathan Space Page maintained by astronomer Jonathan McDowell.

In June 2022, the International Astronomical Union (IAU) launched the Centre for the Protection of the Dark and Quiet Sky from Satellite Constellation Interference (CPS)†, co-hosted by NSF NOIRLab and the SKA Observatory. Based on the information currently available (January 2024) we can summarise that 18 mega-constellations (i.e with plans for more than 1000 satellites) are planning 537,267 satellites in total, of which 6085 have already been launched, and 44 large-constellations (i.e. with plans for less than 1000 satellites) are planning another 5,731 satellites in total, of which 2204 have already been launched. These numbers are evolving rapidly and therefore conclusions that can be drawn today are likely not going to be valid for long. Consequently, it is important to review the situation frequently.

Unlike optical observations, which are impacted by satellites via reflected sunlight, for ALMA the impact is limited to those satellites that make use of certain transmission frequencies that have the potential to interfere with radio observations. For optical telescopes, sunlight reflection from satellites only happens during a limited number of hours immediately after sunset or immediately before sunrise, in contrast to radio telescopes such as ALMA that can observe 24 hours per day and therefore are, in principle, continuously exposed to RFI risks from satellites. The risk of RFI obviously increases with an increasing number of operational satellites.

There are four main types of emission from the satellites:

(1) Space-to-Earth (downlink) to "User Terminals": the stream of data to the final user, who can be anywhere on the planet.
(2) Space-to-Earth (downlink) to "Gateway Stations": the stream of data to one of the ground antennas used for operation. Starlink will have many gateway stations in the United States and just a few in the rest of the world (see details below).
(3) Earth-to-Space (uplink) from the "User Terminals": uplinks are not a concern for radio astronomy.
(4) Earth-to-Space (uplink) from the "Gateway Stations": uplinks are not a concern for radio astronomy.

Earth-to-Space (uplink) emissions cannot affect ALMA observations as long as the emitting devices (user terminals or gateway stations) are not placed in the proximity of the antennas. We are interested in understanding Space-to-Earth emissions (downlinks) from telecommunication satellites in non-geostationary (NGSO) low Earth orbits (LEO).

3.1.1. Impact of selected LEO satellite constellations

We evaluated the impact of six satellite constellations: Starlink Gen1, Starlink Gen2, OneWeb, Kuiper, Kepler, SpaceMobile. These satellite constellations use different frequencies for the different streams (user or gateway stations, down- or up-links) and received authorisation from some national or international regulatory entities (e.g. the ITU or the FCC in the United States) to occupy several channels in a number of radio bands, not always limited to the national territory of the given regulating authority. To avoid confusion between different radio bands and frequency definitions, we refer here to the *Waveguide Bands* standard. None of these constellations makes use of the full range of frequencies they have been approved to use. Also, they do not emit 100% of their power continuously over the whole band. They use discrete, narrow-band channels

† `https://cps.iau.org/`

(e.g. 50, 100, 250 or 500 MHz) of variable intensity, digitally modulated, with all the constellations making use of adaptive coding and modulation schemes to maximise the data-transfer rates and to optimise the use of the available spectrum.

LEO satellites move along orbits ranging from a minimum of 360 km to more than 1000 km from Earth's surface, grouped in *planes* or *trains* or *shells* at different altitudes and inclinations. They use a mix of mechanically or electronically steerable (phased array) antennas for communication with user terminals and gateway stations on the ground, depending on the manufacturer's design choice. Depending on the orbital altitude, antenna design, and off-nadir steering angle of each system (typically 40 – 60 degrees), the footprint on the ground of the single emitted radio beam can be a few tens of km up to a few hundreds of km and the per-satellite coverage area (not simultaneously irradiated) can have a diameter of more than 1000 km. Some satellites use phased-array antennas for the communication with the user terminals. In contrast, communication with the gateway stations can be based on point-to-point connections, where satellite and gateway antennas point directly at each other. For example, this is the technology used by Kuiper satellites.

3.1.2. Present day impact on ALMA

To summarise:

• Satellite constellations will use the K,Q,U,V,E bands, but only within specific frequency ranges and in narrow-band channels.

• All downlink frequencies foreseen to date to user terminals are up to 20 GHz, well below the ALMA frequency bands.

• Other channels in higher bands are used only for Earth-to-space communication and pose little concern for ALMA.

• At this time there is no risk to ALMA from the satellite constellations because all emissions from the existing systems that could eventually reach an ALMA receiver use frequencies well below the minimum ALMA frequency of 35 GHz.

3.1.3. Potential impact of future developments

At this time:

• The frequency range 37.5 – 42.0 GHz is foreseen to be used by Gen1 satellites for downlink from satellites to gateway stations, not to user terminals. This range overlaps with the ALMA receiver Band-1.

• The frequency range 37.5 – 42.5 GHz is foreseen to be used by SpaceMobile satellites for downlink from satellites to gateway stations. This range overlaps with the ALMA receiver Band-1.

• The frequency range 71.0 – 76.0 GHz is foreseen to be used by Gen2 satellites for downlink from satellites to gateway stations, not to user terminals. This range overlaps with the ALMA receiver Band-2 that will not be in operation until after 2025. By then, Gen2 specifications and modes of operation could change, e.g. they may feature laser inter-satellite links allowing them to communicate with each other even if there is no gateway station within sight.

4. JAO Spectrum Management Office (JSMO) and synergies

Spectrum management at ALMA was first described in the document "Frequency Management, EMC, and RFI Monitoring for ADE Maintenance Operations" †. Spectrum management at ALMA was originally a duty of the Observatory System Engineer, and

† OPER-00.00.00.00-011-A-PLA, Clint Janes, 2009-10-31

has now passed ad interim to the Senior RF Engineer. The duties include, among others, coordination of activities concerning RF emissions and potential sources of RFI with all the facilities operating on the Chajnantor Plateau.

4.1. *Remotely Piloted Aircraft Systems (RPAS) policy at ALMA*

Facing growing interest from the media and frequent visits from video producers, ALMA has defined the general policy and procedures to regulate the operation of Remotely Piloted Aircraft Systems (RPAS, e.g. drones) within the ALMA concession. This should be coordinated with all the facilities operating in the AAP. The ALMA document OPER-10.08.03.00-0052 B-PRO (2017) is based on directives from the Chilean Civil Aeronautic Authority and compliance with CE or FCC EMC requirements. Apart from safety concerns related to the flying object, the most important aspect related to RFI is that any emission above 4 GHz (lower edge of the ALMA receivers IF) is forbidden (e.g. use of the 5+ GHz wifi channel).

4.2. *Chajnantor Working Group (CWG) spectrum coordination*

The JSMO plays a crucial role in the coordination of all matters related to radio emissions on the Chajnantor Plateau that could impact operations of any of the observatories on the plateau e.g. the use of powerful transmitters for surface-accuracy measurements by holography technique (e.g. APEX, ACT) or operation of the ALMA artificial polarisation source. Coordination is carried out by means of a dedicated "CWG Spectrum Managers" group.

4.3. *Interaction with Chilean entities*

The JSMO is taking part in a number of Chilean initiatives related to RFI prevention and mitigation:

- Member of the SoChiAs Radio Committee, an extension of the Light Pollution Working Group of the Chilean Society of Astronomy (SoChiAs), to cover radio astronomy and RFI issues.
- Member of the Chilean Low Earth Orbit Satellite group (CLEOSat) coordinating observations to measure optical and radio brightness of satellites
- Participation in some activities of the AAP
- Participation in the Chajnantor Working Group
- Coordination of activities with the other radio observatories in the ALMA surroundings.
- Interaction with SubTel:

  - The JSMO gives advice on the assessment of new applications from satellite operators on the Chilean territory (e.g. for the installation of new gateway stations).
  - Active participation in the process to renew the SubTel Exempt Resolutions, together with the ESO and AUI legal representatives.
  - Invite SubTel representatives to participate in the annual CWG meeting.
  - Evaluate the possibility to extend the protection and coordination zones beyond the Chilean national territory by opening a dialogue with Bolivian and Argentinean equivalent institutions.
  - Ongoing dialogue about radio protection of the ALMA Concession from ground and sky emissions. Proposed to consider a minimum distance of gateway stations from ALMA to prevent risk of RFI from future installations.

4.4. *Interaction with international institutions*

The JSMO also has a number of RFI-related collaborations and interactions with international entities:

- Collaboration with the Committee on Radio Astronomy Frequencies (CRAF) of the European Science Foundation, that coordinates activities to keep the frequency bands used by radio astronomy and space sciences free from interference.
- Collaboration with the Centre for the Protection of the Dark and Quiet Sky from Satellite Constellation Interference (CPS) of the IAU.
- Active participation in the SatHub of the CPS. Planning monitoring of eventual RFI from satellites constellation with the use of ALMA antennas and receivers.
- Participation in the RFI Working Group of the National Radio Dynamic Zone (NRAO), with special interest in the development of the RFI measurement hardware (up to 120 GHz) that could be tested at the ALMA site.
- Participation in the preparation of new Agenda Items for the next World Radiocommunication Conference (WRC-27) of the International Telecommunication Union (ITU) whose Radio Regulations (RR) do not currently consider for protection any radio astronomy bands above 275 GHz. An update of the RR is considered a priority for the future of ALMA, as the lack of regulations or recommendations could have a direct impact on operations if new commercial applications will start to occupy frequencies above 275 GHz, in the pursuit of larger bandwidth for digital data transfer.

Reference

Bustos, R., Rubio, M., Otárola, A. & Nagar, N. 2014, PASP, 126, 1126

Astronomy and Satellite Constellations: Pathways Forward
Proceedings IAU Symposium No. 385, 2026
P. Grimley, J. Barentine, D. Galadí-Enríquez & C. Walker
doi:10.1017/S1743921324001261

Impact of large satellite constellations on VLBI radio astronomical observations for materialization of celestial and terrestrial reference frames and EOP estimation

Vincenza Tornatore

Politecnico di Milano, DICA, Piazza Leonardo da Vinci 32, I-20133 Milan, Italy
email: vincenza.tornatore@polimi.it

Abstract. Geodetic VLBI is an application of the radio astronomy observation technique to study planet Earth. It is one of the geodetic space techniques which provides information about Earth in space and its rotational behaviour, hence it is fundamental for the global geodetic reference frame (GGRF). In particular with this technique ICRF can be determined at radio wavelengths, further it contributes with other space techniques to ITRF materialisation. Daily geodetic VLBI measurements are vital for determining and predicting the time-varying alignment of the Terrestrial Reference Frame with respect to the celestial reference frame (Earth Orientation Parameters). The routine work is coordinated by the IVS as a service for IAG and IAU and carried out by different institutions such as mapping authorities, space agencies, universities and research centres. In 2015 a UN Resolution emphasised the importance of the GGRF for sustainable development and asked member countries to extend these activities. At the same time the telecommunication networks are expanded on ground and in space increasing the number of transmitters. Large satellite constellations can have an impact on Geodetic VLBI products since they almost all broadcast downlink high power signals in the K band, and uplink signals (e.g. 14.0-14.5 GHz for Starlink). Owing to the increasing number of large constellations, an aggregate effect should also be considered, since they use different frequencies, usually in the K band. Furthermore their unwanted electromagnetic noise degrades the VLBI observations of faint cosmic noise. The signal strength of radar systems can even be destructive to the high-sensitive receivers at VLBI radio telescopes. The strategy is to answer the question of how to protect the global geodetic network infrastructure (locations of VLBI telescopes) in the corresponding study group at ITU-R and to seek for protection of the observed bands (32 channelised frequencies) in geodetic VLBI (which exceed the bandwidths of those bands already allocated to RAS).

Keywords. ICRF, ITRF, EOP, Satellite

Astronomy and Satellite Constellations: Pathways Forward
Proceedings IAU Symposium No. 385, 2026
P. Grimley, J. Barentine, D. Galadí-Enríquez & C. Walker
doi:10.1017/S1743921324001376

Modelling steerable beams of satellite constellations for radio astronomy impact studies

Federico Di Vruno[1,2] **and Gary Hovey**[3]

[1]SKA Observatory, Jodrell Bank, Macclesfield SK11 9FT, United Kingdom
email: federico.divruno@skao.int

[2]IAU Centre for the Protection of the Dark and Quiet Sky from Satellite Constellation Interference, France

[3]Onsala Space Observatory, Chalmers University of Technology 412 96 Gothenburg, Sweden
email: gary.hovey@chalmers.se

Abstract. Modelling is essential for studies that quantify the impact from satellite downlinks on radio astronomy sites. To estimate this impact it is necessary to know not only the position and velocity of satellites but also their behaviour in the radio spectrum domain. As many large satellite constellations are using steerable beam antennas, deterministically predicting the transmitted power towards a defined direction (in this case where a radio telescope points) becomes an almost impossible task and therefore another approach has to be used. This work presents a method to simulate and estimate the percentiles of the radiation pattern of satellites with steerable beam patterns based on simulations and a comparison with measurements of Starlink satellites using the Onsala Twin Telescopes in Sweden.

Keywords. megaconstellations

1. Introduction

Large constellations of satellites deployed in low Earth orbit (LEO) have been a concern because of their impact on optical and radio astronomy and the night sky, as discussed by Walker, Hall, Allen et al. (2020), Walker & Hallet (2021), Walker et al. (2021), Lawrence, Rawls, Jah, et al. (2022), and Walker & Benvenuti (2022). In the case of the impact on radio astronomy, few studies have been published, such as ECC (2021), Di Vruno & Tornatore (2023) and JASON (2021). These studies considered the impact of the largest constellations being deployed, such as Starlink and OneWeb, but they have either not considered the strong in-band emissions from satellites or used significant simplifications when referring to satellite's emissions.

None of the reports mentioned has considered the complexity of a satellite system such as the Starlink, which uses steerable beams with footprints of 24 km in diameter on the ground when considering the -3dB power flux density (PFD) level †. While an observer on the ground can estimate the position of most satellites in the sky within a certain accuracy, there is no simple way of knowing how many beams per channel a particular satellite is producing and where are those beams are pointing to at a specific time.

This paper presents the results of a model used to simulate the statistics of the power flux density (PFD) produced by a Starlink satellite and results of measurements conducted with the Onsala Twin Telescopes at the Onsala Space Observatory in Sweden.

† https://fcc.report/IBFS/SAT-MOD-20200417-00037/2274316

2. Motivation

Radio telescopes rely on a clean radio spectrum to observe the faintest radio sources of cosmic origin. Within the radio frequency spectrum (from 30 kHz up to 3000 GHz according to the International Telecommunication Union) there are several frequency bands that are allocated to the radio astronomy service (RAS). These "reserved" frequency bands are spread throughout the radio spectrum and are either centred on frequencies where there are important astronomical emissions, such as the hydrogen line at 1420.4057 MHz which has allocated frequencies between 1400 to 1427 MHz, or allocated roughly every frequency octave to allow an efficient sampling of a very broadband continuum emission. These radio astronomy bands were defined back in the 1970s and have been fundamental for the RAS since then, but with the advance of technology and scientific techniques, there is an ever pressing need to use wider frequency bands and increase the sensitivity of instruments. This situation of course collides with the reality that the frequency bands not allocated to radio astronomy are heavily populated with radio signals from other users of the radio spectrum such as cell phones, WIFI and Bluetooth, and satellite transmissions. When any of these man-made signals is received by extremely broadband radio astronomy receivers it's flagged as "interference" by astronomers. While it is understandable that from the point of view of an astronomer these are interfering signals, it is necessary to recognise that often these are legitimate transmissions within the allocated frequency bands for those services.

One effective way of widening the available radio spectrum for radio astronomy is to locate radio telescopes in extremely remote areas, where the terrestrial use of the radio spectrum is minimal. This is the case for telescopes such as the Square Kilometre Array Observatory (see Swart, Dewdney & Cremonini 2022 and Labate, Waterson, Alachkar, et al. 2022) which consists of two radio telescopes, one in South Africa and the other in Western Australia, both located in nationally protected Radio Quiet Zones (RQZ). An RQZ is a nationally defined area where the use of the radio spectrum, across a very large swath of frequencies, is specifically managed to benefit radio astronomy observations (see ITU-R 2021). One of the limitations of RQZs is that their regulations are not effective against satellite transmissions. The recent development of large satellite constellations like Starlink and OneWeb has raised concerns in the radio astronomy community as telescopes can be impacted not only by the "in-band" downlinks of the satellites (their rightfully allocated frequencies), but by low-level "out-of-band" emissions (meaning emissions outside of their allocated frequencies as a sub product of the in-band transmissions), and by unintended electromagnetic radiation (UEMR; Di Vruno, Winkel, Bassa, et al. 2023).

The impact of large satellite constellations (on non-GSO constellations as defined by the ITU) on radio telescopes is calculated by using the equivalent power flux density (EPFD) method. This is a computational method that considers the movement of satellites in the local sky, their radio frequency emissions, and the reception characteristics of the observer (in this case a radio telescope) and yields the instantaneous and integrated received power (see Figure 1). The EPFD method was developed by the ITU to address the situation that all satellites in a nonGSO constellation are constantly moving in the sky, so the impact on an observer must be calculated taking this into account. Radio telescope observations differ significantly from a typical ground terminal of a satellite constellation, especially in their sensitivity, variable pointing directions and integration times. The EPFD method to consider the impact on radio astronomy from nonGSO satellites is considered in ITU-R (2007a,b). A detailed description of the EPFD method for radio astronomy can also be found in section 3 of Di Vruno, Winkel, Bassa, et al. (2023). ITU-R (2007a,b)

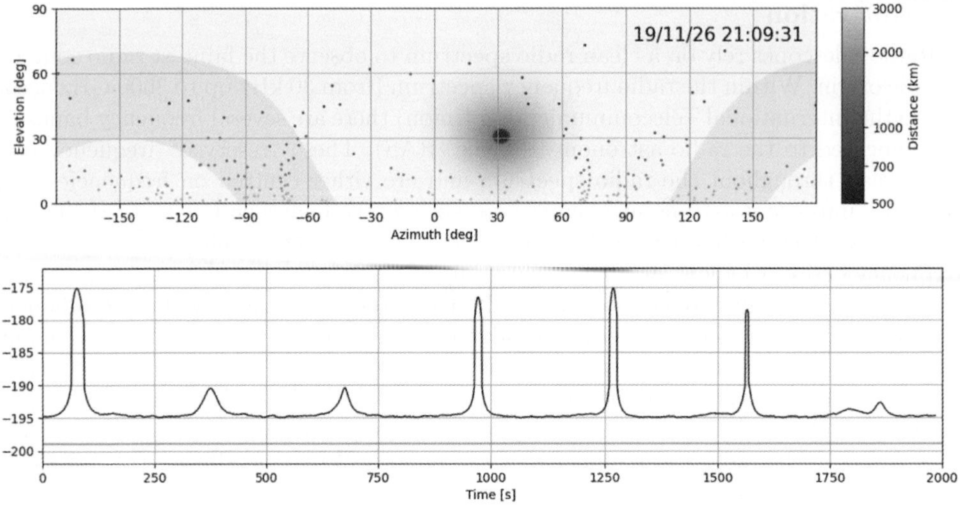

Figure 1. Instantaneous view of satellites in the sky (in azimuth and elevation) colour-coded in range to the observer, the shaded areas depict the gain pattern defined in recommendation ITU-R RA.1610 (top). Time domain total power received by the radio telescope, the high peaks are when a satellite transits the boresight (bottom). Source: Winkel, private communication.

The EPFD of a satellite constellation is a function of the following:

- The position of the observer in latitude, longitude, and height
- The complete radiation pattern of the radio telescope (or using a model)
- The position of all satellites in the constellation as a function of time
- The radiation characteristics of each satellite (centre frequency, bandwidth, radiation pattern characteristics, etc)

Given these parameters the EPFD distribution, in $\frac{W}{m^2.Hz}$, may be computed for each cell above the local horizon. With this distribution the cumulative density function (CDF) and other statistics maybe computed and compared with thresholds for harmful interference, such as defined in ITU-R (2007b).

2.1. *Spectral flux density and steerable beams*

While the first three parameters required to calculate the EPFD are readily available, the radiation characteristics of satellites with steerable beams, such as Starlink or Amazon Kuiper, have radiation patterns that are not static in time. Similarly their spectral usage is also dynamic as their downlink transmission are spread over the range of 10.7 to 12.7 GHz, divided into eight 250 MHz frequency channels. By using electronically phased transmitters these satellites can produce multiple beams in each frequency channel with relatively small footprints (24 km in diameter in the case of Starlink) as well as changing their pointing direction in short timescales to service different cells on the ground. As depicted in Figure 2 radio telescope and the satellite constellations interact in four different ways, making it impossible for the astronomer to accurately know the pointing direction of each antenna beam and therefore its impact on the observation. Note that this analysis is valid for in-band transmissions, where the beam-forming is designed to operate. The out-of-band response could be quite different and futher complicates the situation as the information is not available. In the case of UEMR, the emission is not even produced by the communication antenna so beam forming is out of the question.

The power received by the radio telescope from the satellite is a maximum on-axis when both beams are pointing at each other and is reduced as either is pointed away

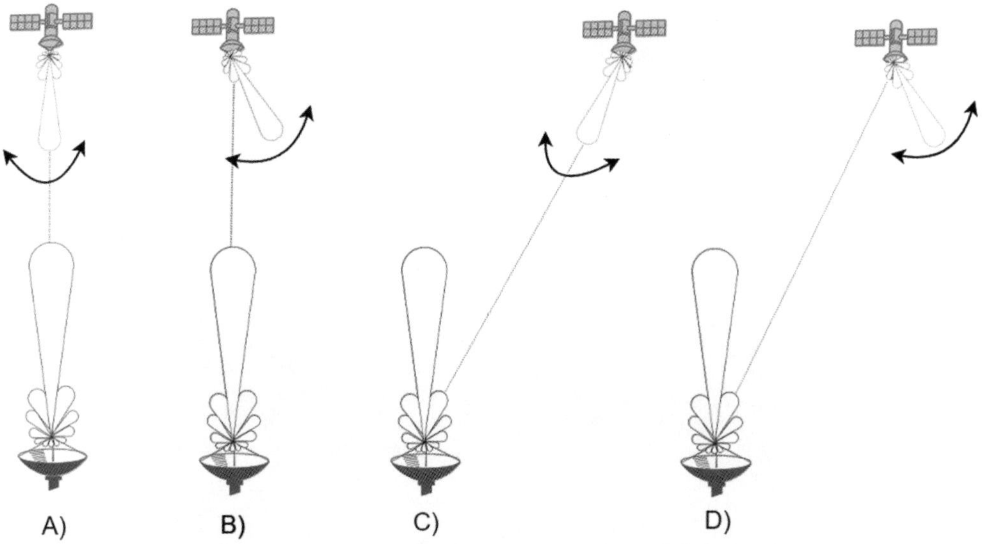

Figure 2. Radio telescope and satellite pointing possible scenarios when the satellite has steerable beams.

from the other as shown in Figure 2. The probability of a satellite being on-axis with a radio telescope (or in the telescope's boresight) is proportional to the number of satellites in LEO, their orbital characteristics, and the longitude of the observer. This probability continues to increase as more satellites are deployed.

It is important to understand that the purpose of a LEO satellite's downlink signals is to communicate reliably (with high signal to noise ratio, SNR) from space to earth stations some 500 km away. This requires high power levels which are constrained by the maximum power flux density (PFD) allowed by regulations. The maximum power flux density (PFD) that Starlink has declared in their 2020 FCC filling is $-146\ dB\frac{W}{m^2.4kHz}$, produced at the boresight of the steerable beam or in the "cell" that the beam is illuminating. This is equivalent to $-182\ dB\frac{W}{m^2.Hz}$, considering that emissions are uniform in a 4kHz band, and is millions of times stronger than a typically strong radio astronomy source at these frequencies. Consequently, LEO satellites can easily drive a radio-telescope signal chain into non-linear operation if the two beams are within a few beamwidths of each other, because even radio telescope antennas of a modest diameter have a gain of 60 dB or more at these frequencies.

3. Model

Ideally one would like to make a measurement of the radiation characteristics of each satellite in a constellation, so that the total power incident at a radio telescope antenna could be computed. This is very complex and almost impossible given the facts explained above (dynamic changing of pointing direction of beams) and the fact that nonGSO satellites are moving rapidly across the sky. Given this impossibility, a simulation model was developed. The model simulates the behaviour of each steerable beam by considering one static beam pattern and mathematically rotating it to point to any possible cell on the ground (see Figure 3). The simulation includes randomly sampling the pointing direction, with some constraints such as defining the maximum steering angle of a satellite and that a cell can be serviced only by one beam per channel at a time.

The radio telescope beam, or gain, is not considered in this simulation, as the primary parameter of interest is the PFD incident from each satellite. The gain of the radio

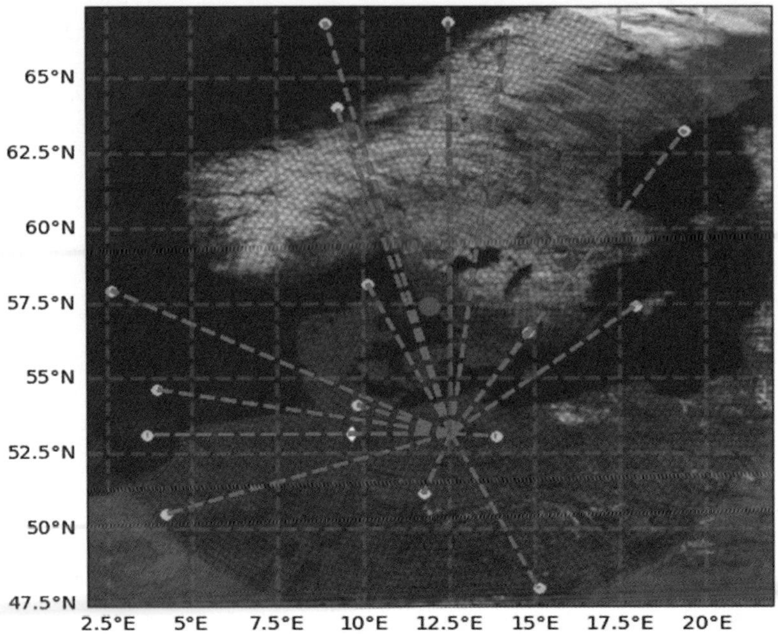

Figure 3. Example of 16 beams (red dotted lines) pointed towards different cells (blue circles) on the ground. The unconnected (green) dot represents the position of the radio telescope.

telescope antenna can be applied in subsequent stages to calculate the received power based on the satellite's position in its antenna beam pattern.

The orbital positions of the Starlink phase 1 constellation were calculated for 1000 seconds in 5-second timesteps (200 samples) using its orbital parameters and the open-source Python package cysgp4, which is available under GPL-v3 license (Winkel 2023). 1000 random pointing directions (of 16 beams per satellite) were calculated for each visible satellite, and the total power produced by them was added as a linear sum. This resulted in two million PFD samples per visible satellite. All samples of PFD as a function of satellite elevation are presented in Figure 4.

The simulation results reveal interesting patterns. One is that the PFD values reach a maximum around 30 degrees elevation and then falling at lower elevations. This is consistent with expectations, as satellites at elevations less than about 30 degrees cannot point their beams towards the observer cell or nearby cells owing to their maximum steering angle. Another interesting result is that the PFD decreases as the satellites rise in elevation, a reasonable result as a satellite high in the sky is more likely to point its beams to cells that are outwards from the radio telescope cell (see Figure 5). Satellites with low elevations have elongated beams which can spill over into the radio-telescope's cell. This is unavoidable to some degree, even if it is a region masked by the constellation.

4. Observations

Observations with the Onsala Twin Telescopes (OTT; Haas, Casey, Conway, et al. 2019), located at the Onsala Space Observatory (OSO) in Sweden, were carried out to validate the simulation model and gain further insights into the impact of satellite constellations on radio astronomy. The OTT consists of two 13.2m paraboloids with receivers that have a noise temperature of 15K at 2.2-14 GHz and 3-18 GHz. A spectrum analyser was connected to the output of each telescope capturing power in a sweep spectral mode with a resolution bandwidth of 1 MHz and a capture time of approximately

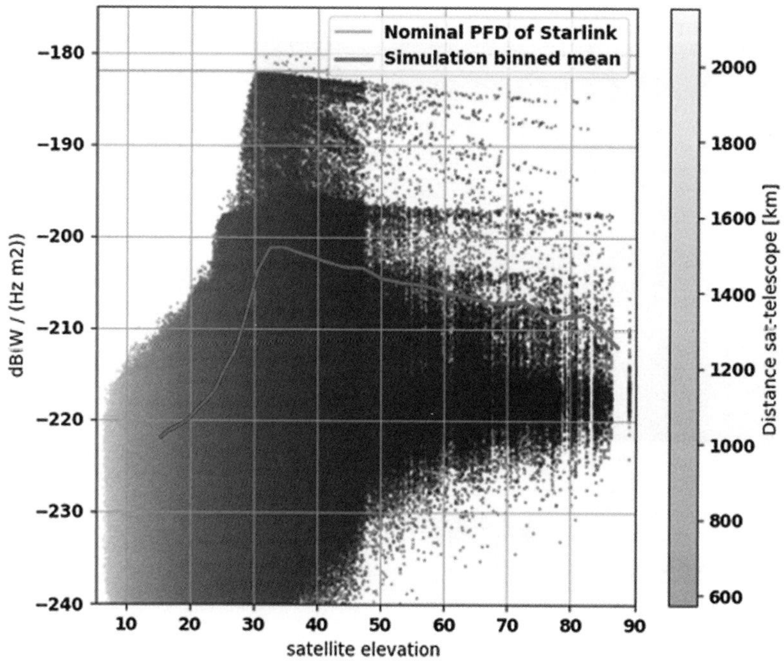

Figure 4. Calculated PFD of individual satellites for each random pointing as a function of satellite elevation.

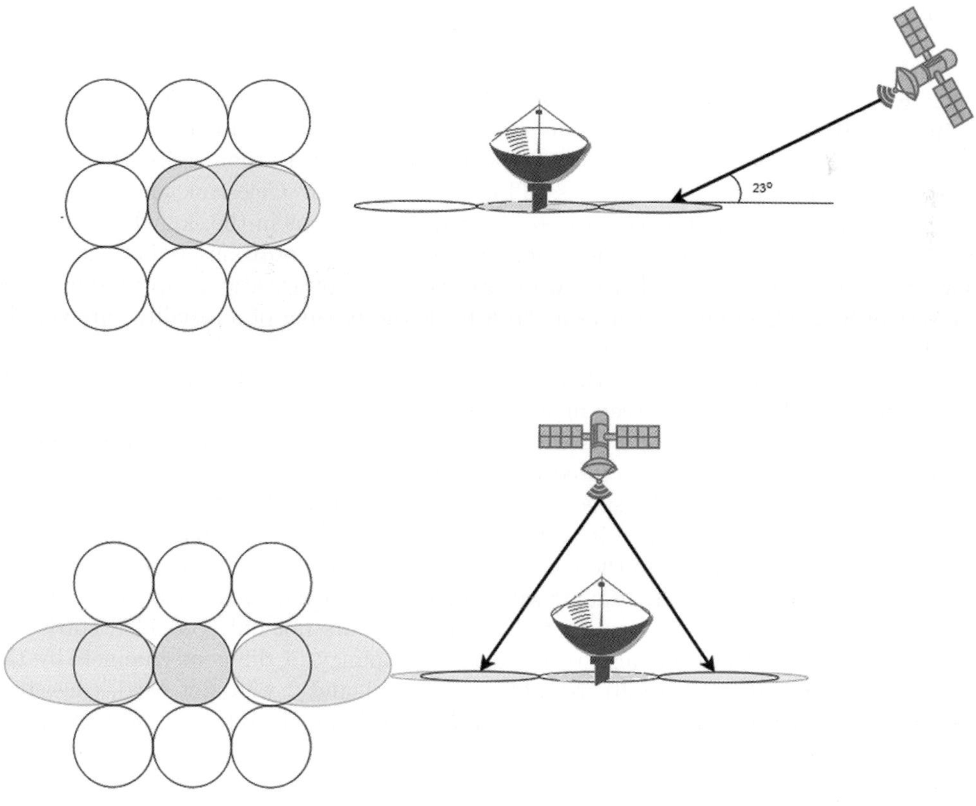

Figure 5. Top view of a rectangular arrangement of cells with the observer in the central cell (left); orange depicts the footprint of a circular beam being elongated as a result of the slanting direction of incidence. Side view of illumination of cells adjacent to the observer from a slant angle (top right) or overhead (bottom right).

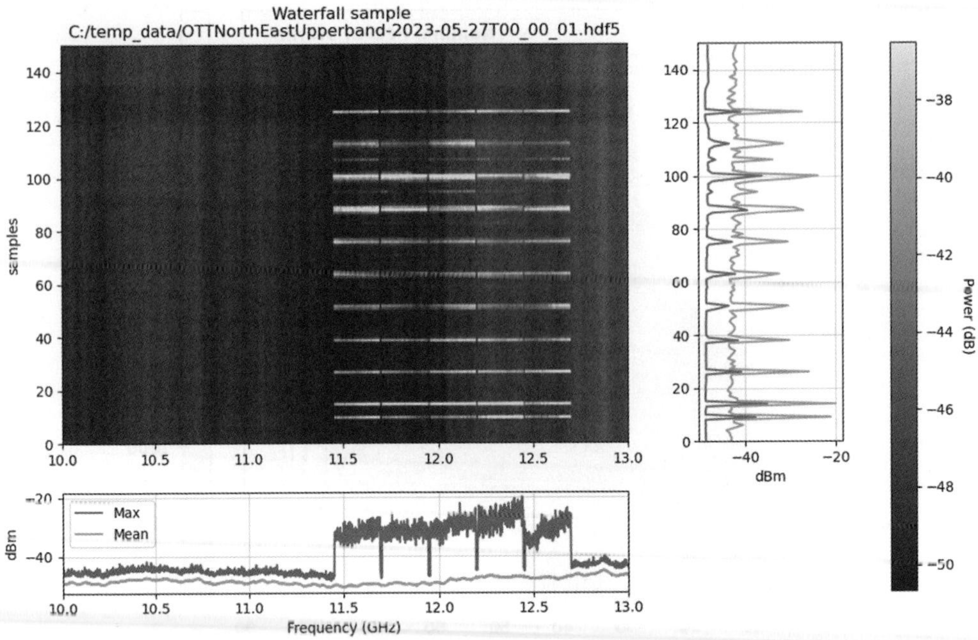

Figure 6. Waterfall plot of received power (top). The high frequency of occurrence of satellite downlinks in samples (of 1.5 s approximately) is due to the pointing schedule used, to maximise the satellite transits for an observation time. The bottom panel shows the integrated power in time as a function of frequency, clearly showing the five channels of 250 MHz used by the satellites. On the right panel is the integrated power in frequency as a function of samples (time).

1.5 s. Each frequency sweep was transferred from the spectrum analyser to a computer in units of dBm. An example of an observation run can be see in Figure 6.

The telescopes' pointing schedule was calculated using a Python script that utilised supplemental Two-Line Element sets (TLEs) calculated by Celestrak and propagated using cysgp4. Both radio telescopes were pointed in directions in the sky which intersect satellite traces. To account for inaccuracies in the satellites' position predictions, a window of +- 20 seconds was used as shown in Figure 7. The observations captured over 700 transits of Starlink satellites, where a "transit" is the passing of a satellite through the −3dB beam of the radio telescopes.

One important consideration is that the Starlink system does not illuminate the Onsala Space Observatory, as can be seen in their "Starlink availability map" available at †. To account for this the central cell of the simulation was marked as "passivated" in the software so that no beam was pointed towards it.

The power measured by the spectrum analyser is at a point in the signal chain and requires calibration to reference it to the power received by the antenna aperture and the incident PFD. To calibrate the measured power into PFD, the radio telescopes were pointed towards Cassiopeia A, a source with a known flux density at 11 GHz, for 2 minutes in every observation run. The calibrated observed results not only provided statistical information on the PFD but also revealed the occupancy of different channels by the observed satellites (see Figure 6). Notably, channels 1 and 2 were not used, consistent with the known practice of operators of avoiding these channels to protect adjacent radio astronomy bands (ECC 2021).

† `https://www.starlink.com/map`

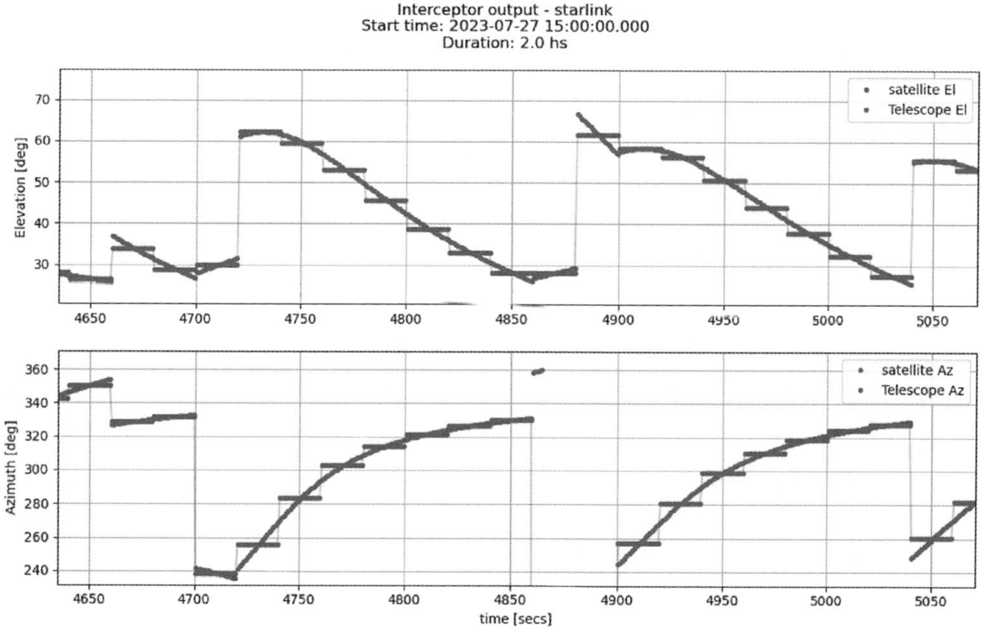

Figure 7. Pointing schedule as a function of time for the Onsala Twin Telescopes in Azimuth (top) and Elevation (bottom). The green dots represent the telescope position and the blue dots present the moving satellites positions.

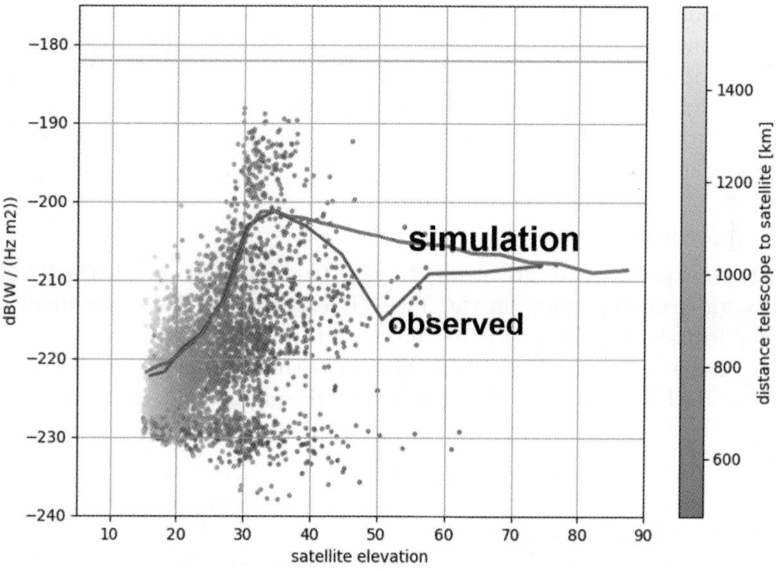

Figure 8. Distribution of the measured PFD for each one of the satellite transits detected. The green (lower) curve shows the averaged PFD for each elevation bin. The red (upper) curve is the averaged PFD for each elevation point from the simulation.

The observed mean results in Figure 8 closely match the mean result of the simulation for the lower elevations. At higher elevations this correlation does not hold we believe due to a lack of data points in the observations. Owing to the latitude of the Onsala telescopes, most satellites are located at low altitude as can be seen in Figure 9. To make

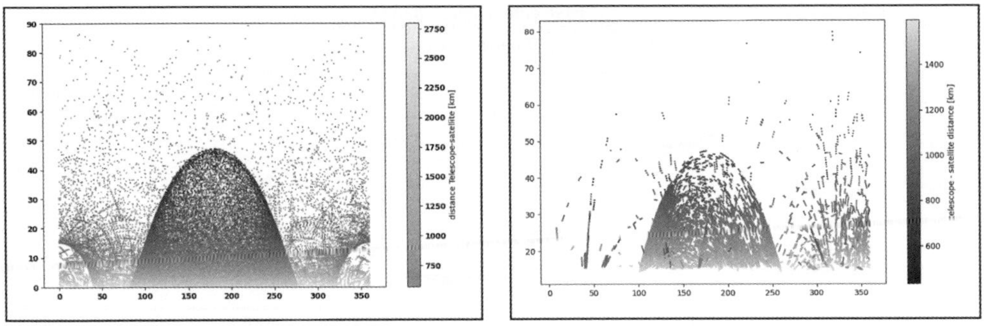

Figure 9. Positions in the sky of the satellites in the simulation (left) and observations with the OTT (right). The range from the observer to the satellite is colour-coded in km.

a substantial increase in observations at high altitudes we would require much more telescope time. The results show that the observations are in good agreement with the model predictions adjusting the parameters. These include the single beam pattern, number of beams per channel, maximum steering angle for one beam, cell size on the ground, network considerations, and avoidance cells on the ground. The successful alignment of the simulation with observations underscores the model's reliability.

5. Further applications

The developed model holds potential for further applications, such as for the simplification of EPFD calculations. When the size of the constellations is too large, adding another degree of freedom such as the pointing direction of beams could make the calculation very computation-expensive. This model can also be used to assess the overall occupancy of channels in radio astronomy receivers and their potential impact. By combining the statistical distribution of satellites in the sky with the model results, it becomes possible to estimate the total received power into the radio telescope, either as a full-sky average or as a function of the direction in the sky.

Acknowledgements

The authors express their gratitude to the Geodesy Group at the Onsala Space Observatory for providing time on the Twin Telescopes for this experiment, as well as their invaluable assistance in setting up and monitoring observations. Special thanks are extended to Starlink and the Monitoring Group of the Committee on Radio Astronomy Frequencies (CRAF) for constructive discussions during the development of the model.

References

Di Vruno, F. & Tornatore, V. 2023, Large Satellite Constellations and Their Potential Impact on VGOS Operations, arXiv:2307.04219

Di Vruno, F., Winkel, B., Bassa, C. G., et al. 2023, A&A, 676, A75

ECC 2021, Compatibility and sharing studies related to NGSO satellite systems operating in the FSS bands 10.7-12.75 GHz (space-to-Earth) and 14-14.5 GHz (Earth-to-space), ECC Report 271, Electronic Communications Committee, Copenhagen

Haas, R., Casey, S., Conway, J., et al. 2019, in Haas, R., Garcia-Espada, S., & López-Fernández, J. A. (eds.) Proceedings of the 24th European VLBI Group for Geodesy and Astrometry Working Meeting (Madrid: CNIG)

ITU-R 2007a, Interference Calculations between Non-geostationary Mobile-satellite Service or Radionavigation-satellite Service Systems and Radio Astronomy Telescope Sites, Recommendation M.1583-1 (Geneva: International Telecommunication Union)

ITU-R 2007b, Calculation of Unwanted Emission Levels Produced by a Non-geostationary Fixed-satellite Service System at Radio Astronomy Sites, Recommendation S.1586-1 (Geneva: International Telecommunication Union)

ITU-R 2021, Characteristics of radio quiet zones, Report RA.2259-1 (Geneva: International Telecommunication Union)

JASON 2021, The Impacts of Large Constellations of Satellites, The Mitre corporation Technical report JSR-20-2H

Labate, M. G., Waterson, M., Alachkar, B., et al. 2022, Journal of Astronomical Telescopes, Instruments, and Systems, 8, 011024

Lawrence, A., Rawls, M. L., Jah, M., et al. 2022, Nat Astro, 6, 428

Swart, G. P., Dewdney, P. E. & Cremonini, A. 2022, Journal of Astronomical Telescopes, Instruments, and Systems, 8, 011021

Walker, C., Hall, J., Allen, L., et al. 2020, BAAS, 52(2) (SATCON I)

Walker, C., & Hall, J. (Eds.) 2021, Report of the SATCON2 Workshop, 1216 July 2021, BAAS, https://baas.aas.org/report-of-the-satcon2-workshop-1216-july-2021

Walker, C., et al. 2021, Dark and Quiet Skies for Science and Society: Report and recommendations, Zenodo, https://doi.org/10.5281/ZENODO.5898785

Walker, C., & Benvenuti, P. (Eds.) 2022, Dark and Quiet Skies II Working Group Reports, Zenodo, https://doi.org/10.5281/ZENODO.5874725

Winkel, B. 2023, cysgp4: Wrapper for C++ SGP4 satellite library, Astrophysics Source Code Library, ascl:2303.001

Astronomy and Satellite Constellations: Pathways Forward
Proceedings IAU Symposium No. 385, 2026
P. Grimley, J. Barentine, D. Galadí-Enríquez & C. Walker
doi:10.1017/S1743921324000917

Systematic effects of LEOsats in Rubin Observatory LSST camera

Daniel Polin, Adam Snyder and Tony Tyson

Department of Physics & Astronomy, UC Davis, Davis, California, 95616
email: dapolin@ucdavis.edu

Abstract. Rubin observatory, and survey astronomy in general, will be greatly impacted by the proliferation of low Earth orbit satellite (LEOSat) constellations. Its large field of view guarantees that Rubin will observe satellite streaks in approximately every image passing through an average of 13 out of the camera's 189 CCD detectors. The impact of satellites on our science scales with the number of satellites launched, and with satellite brightness.

Keywords. Rubin Observatory, LSST, survey astronomy, satellite streaks

1. Survey Astronomy, Rubin, and Satellites

Vera C. Rubin Observatory will image the entire southern night sky to an unprecedented depth beginning in 2025. Over its 10-year run, Rubin will image the sky every three nights, accumulating two thousand exposures. Each pair of 15 second exposures will probe to a depth of 24.5 magnitude. Rubin also serves as a prime example of how survey astronomy will be negatively impacted by the proliferation of low Earth orbit Satellite (LEOSat) Constellations.

The main factors that make survey astronomy vulnerable to satellite interference are its requisite large field of view and high throughput. Surveying the sky demands viewing a broad swath of the sky in a small amount of time, making it difficult to avoid unwanted objects like satellites. Figure 1 shows the relative size of the Rubin Observatory field of view compared to the moon, the field of view of the Gran Telescopio Canarias (Álvarez et al. 1998) (the current largest diameter telescope in the world), and the largest field of view of JWST (Sabelhaus and Decker 2004). This highlights the difference in design of survey observatories to preference large scale field of view compared to targeted telescopes, making them more likely to image satellite streaks.

Rubin will take around 2000 images per night in 15s exposures. Any systematic delay in imaging caused by satellites and any data loss will compound greatly throughout the ten-year survey. Additionally, Rubin's depth per exposure means that it will be adversely impacted by brighter objects. The satellite constellations we are seeing are around 100 million times brighter than the objects Rubin is designed to observe (Abell et al. 2009).

Survey astronomy is traditionally not a field that can operate using space telescopes. The sheer scale of both hardware and data collected makes putting an operation like Rubin in space, and downloading the nightly 20TB generated virtually impossible to accomplish within the constraints of plausible funding. For comparison, the downlink from JWST is only around 270GB per day and that project already costs an order of magnitude more than Rubin. Space telescopes also have limited serviceability and have significantly shorter operational lifetimes than a ground based operation like Rubin.

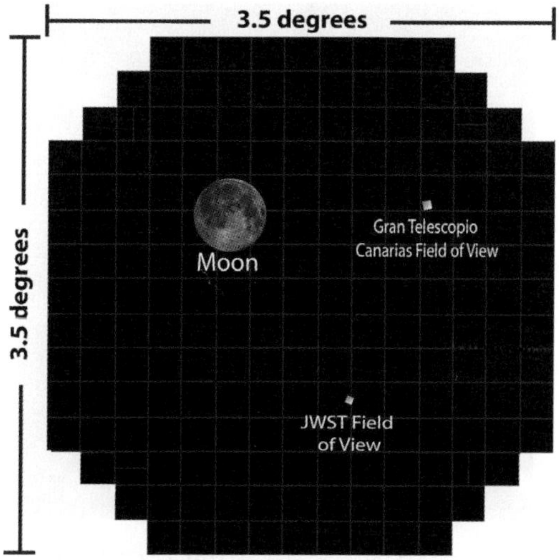

Figure 1. The field of view of the Rubin Observatory LSST Camera. The square divisions shown are the 189 CCD detectors that make up the camera. To scale are shown the moon, the largest field of view of the Gran Telescopio Canarias (3.5 arcmin) and the largest field of view of JWST (3 arcmin). Its large size means Rubin will be greatly impacted by satellite constellations.

2. Impact due to number of satellites

There are currently over one million satellites filed for launch with the ITU (Falle et al. 2023). Individual constellations propose numbers in the ten-thousands to hundred-thousands. Unsurprisingly, the impact of satellites on Rubin will scale with the number in orbit. If there are fewer than a couple thousand satellites, then avoidance is possible. That is, if we have perfect orbit location data for each satellite, we can choose to observe where the satellites are not. This comes with some loss of data while slewing or waiting when we could otherwise be observing.

As we increase the number to the tens of thousands range, avoidance becomes uneconomical. If we attempt to avoid satellites we will lose more data in waiting and finding satellite free spaces than we would by observing streaks. We will almost certainly have more than 40,000 satellites in orbit in the near future which will mean that on average we will observe one in about every image. A given streak will cover on average 12 CCD detectors out of the total 189 in the Rubin LSST Camera. Looking near the horizon this number scales by a factor of 10.

If 400,000 planned satellites are launched, we will have around 10 per image or 100 per image near the horizon. This would cover about 60% of our camera CCD detectors per image.

3. Impact due to brightness of satellites

The impact of satellites on survey astronomy is greatly dependent on satellite brightness (Barentine et al. 2023). In any image, the pixels where a satellite is present are unusable and must be masked out. For the Rubin LSST Camera specifically, our highly segmented camera design means that bright objects like satellites create coupled crosstalk images between amplifiers segments in each CCD. Figure 2 shows the crosstalk in a real Rubin CCD detector from a laboratory image of a simulated satellite streak. The brightness of the streaks in a crosstalk response segment does not scale linearly with brightness of the source streak. Each amplifier segment pair will require at least a three parameter fit to remove the crosstalk signal. For 189 CCDs with 16 detectors each, this means we

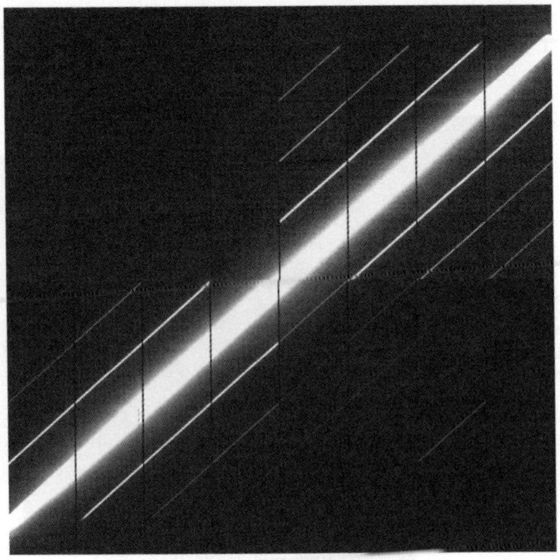

Figure 2. An image taken in a lab using a real Rubin CCD Detector with an image mask that simulates a satellite streak. Note the CCD is subdivided into 16 amplifier segments. Bright signals like those from the satellite produce a coupled crosstalk response between amplifiers that is read out as a false signal. These scale non-linearly with intensity and must be removed in the final dataset.

Table 1. Systematic error at different satellite brightness levels.

Stationary Satellite Magnitude	Peak Electron Count	Systematic Error
0-1	1,000,000	N/A
4	100,000	Large
6-7	10,000	20%
7-8	5,000	10%

will need to calculate three parameters for 22,680 amplifier pairs in the entire camera. On-sky calibration of CCD crosstalk coefficients can be done using selected LSST observations on the month timescale. For each CCD, in a 15-second exposure at an average 45 deg Galactic latitude in LSST operations, the useful calibration stars span 18-21 mag. With 800 exposure locations per night, over 30 nights, this yields a total of 5 million stars per CCD, assuming a density of 12 stars per CCD segment per exposure. However, many stars will be confused with the 15 times more numerous fainter objects. Calibrating and correcting crosstalk may be possible for LEOsats fainter than 0.5 magnitude. However, the systematic error in crosstalk removal scales with satellite brightness, and will remain a leading source of error due to satellites in our survey (Tyson et al. 2020).

Working backwards from the systematic error in crosstalk removal we can say how many electrons in the main peak would produce that error. For a given satellite orbit and profile in a specific band, we can then scale that to a stationary satellite magnitude. Here we will be assuming the same profile shown in figure 2. We will use an orbit height of 550km but should note that the same profile in a higher orbit would produce a brighter streak as the satellite takes longer to pass the field of view. We assume we are using an r-band filter, and a viewing angle of 40 degrees from zenith. These results are summarized in table 1.

If we want to achieve less than 10% systematic error in crosstalk removal, we will need a peak electron count of fewer than 5,000 electrons. This in turn means a stationary brightness of 7-8th magnitude. At this scale our faint science will still be impacted, especially science programs like near-earth orbit object detection which requires viewing faint objects in twilight close to the horizon where satellites are most present. Additionally faint science such as weak lensing to map out dark matter and dark energy will be affected.

Figure 3. An image of a bloomed CCD. When enough light enters the CCD the pixel wells of the entire detector become saturated and no useful data can be extracted. This will happen for a satellite in the 0-1 magnitude range such as the AST BlueWalker 3. If a satellite this bright is imaged, on average 12 out of the 189 CCDs will be unusable in that image.

If we scale to less than 20% systematic error in crosstalk correction, we need fewer than a 10,000 peak electron count. This corresponds to brightness of 6-7th magnitude. This is at the edge of brightness that is visible to the naked eye. At this scale many science programs will be affected. At the brighter end, our pixels reach saturation around a peak electron count of 100,000. At this scale, crosstalk is still correctable but with large systematic error. This corresponds to a 4th stationary magnitude satellite. Most of Rubin's science programs will be affected at this brightness. Finally, at around 1 million peak electron count, the CCD will bloom. Pixel wells will overflow electrons filling the entire detector. This makes removal impossible as the entire CCD will be unusable for that image. An example of a bloomed image is shown in figure 3. This will occur for a 0-1 stationary magnitude satellite, which is similar to the brightness measured for the AST BlueWalker 3 satellite currently in orbit (Nandakumar et al. 2023). On average, if a satellite this bright crosses our focal plane, it will make around 12 out of 189 CCDs unusable in that image. Note that AST SpaceMobile plan to launch larger satellites.

4. Conclusion

Rubin Observatory science programs generally require satellites to be fainter than visible to the naked eye can see. If they are not, some science may survive tens of thousands of LEOsats but with challenges in data processing and analysis. With over 100,000 bright LEOsats, many science programs are unachievable. Rubin Observatory cannot deal with more than about a thousand 0-1mag satellites like BlueWalker3 without major data loss corresponding to one year of observing.

We estimate that to calibrate the systematic ghosts caused by CCD amplifier segment crosstalk from satellites, we require around 5 million observations which could be accomplished with the data we collect on a month timescale. Calibrated correctable crosstalk may be possible for LEOsats fainter than 6.5 magnitude. The primary impact on LSST science is via systematic errors rather than lost sky area. These systematics originate from residuals from crosstalk ghost removal and from main streak masking.

The proliferation of LEOSat constellations will have a dramatic effect on survey astronomy in general and the Rubin observatory in particular. With the current satellites planned to launch there will be scientific discoveries that are simply not made due to LEOSats.

Acknowledgements

We acknowledge support from US NSF grant AST-2205095 and US DOE grant DE-SC0009999.

References

Abell, P. A., et al. 2009, *LSST Science Book*, version 2.0. *arXiv:0912.0201*, 1
Álvarez, P., Espinosa, J. R., & Sánchez, F. 1998, New Astron Rev, 42(6-8), 553
Barentine, J. C., et al. 2023, Nat Astro, 7(3), 252
Falle, A., Wright, E., Boley, A., & Byers, M. 2023, Science, 382(6667), 150
Nandakumar, S., et al. 2023, Nature, (6672), 1
Sabelhaus, P. A., & Decker, J. E. 2004, Proc SPIE, 5487, 550
Tyson, J. A., et al. 2020, ApJ, 160(5), 226

Astronomy and Satellite Constellations: Pathways Forward
Proceedings IAU Symposium No. 385, 2026
P. Grimley, J. Barentine, D. Galadí Enríquez & C. Walker
doi:10.1017/S1743921324000620

Estimating the impact to astronomy from the OneWeb satellite constellation using multicolour observations

Christian Adam[1], **Jeremy Tregloan-Reed**[2] **and Eduardo Unda-Sanzana**[1]

[1]Astronomy Center (CITEVA), Universidad de Antofagasta, Chile

[2]Instituto de Astronomía y Ciencias Planetarias (INCT), Universidad de Atacama, Chile
email: christian.adam84@gmail.com

Abstract. This study addresses the growing concern in the astronomical community regarding the brightness and interference caused by low Earth orbit (LEO) communication satellites. Utilising data from a global network of telescopes and a custom Python pipeline, we analysed 369 observations of 159 OneWeb satellites obtained in the *BVRI* bandpasses with the Danish 1.54-metre telescope at ESO La Silla, Chile, revealing significant variations in brightness across different wavelengths and a substantial proportion exceeding recommended brightness limits. Our preliminary findings, incorporating diffuse sphere phase models, offer further insights into the satellite's reflective properties and implications for future astronomical observations.

Keywords. Space vehicles, techniques: photometric, techniques: image processing, methods: data analysis, software: data analysis

1. Introduction

In recent years, the number of artificial satellites, especially in low Earth orbit (LEO), increased significantly, largely due to the deployment of large communication satellite constellations like Starlink and OneWeb. This promises an exciting future for space development, not only in communications, but also in areas such as navigation and remote sensing. However, the rapid increase in numbers - the future may include up to 500 000 satellites (McDowell 2024) - also poses new challenges, including but not limited to space traffic management, collision risks, environmental impacts (e.g. Boley & Byers 2021), and most notably, interference with astronomical observations (Walker et al. 2021; Walker, Hall, Allen, et al. 2020).

Typically, about a dozen bright satellites may be visible to the unaided eye, especially during twilight and in the early night hours. As a point of reference, a 1 m sphere with an albedo of 0.2 at orbital height of 500 km will have a visual magnitude of approximately 6.7 at the zenith during twilight (Hejduk 2011). Many new LEO satellites, however, are much brighter than this (see e.g, Tregloan-Reed, Otarola, Ortiz et al. 2020; Tregloan-Reed, Otarola, Unda-Sanzana et al. 2021; Mróz, Otarola, Prince et al. 2022; Nandakumar, Eggle, Tregloan-Reed et al. 2023) and can be brighter than 1st magnitude, such as the recently launched AST SpaceMobile BlueWalker 3 satellite (Nandakumar, Eggle, Tregloan-Reed et al. 2023).

A few such bright satellites might amount to a minor nuisance to stargazing or astrophotography. For astronomical research, however, one has to keep in mind that many more are still visible to telescopes and cameras that have higher sensitivity and can see fainter objects. Hence, the accidental passage of a LEO satellite may result not only in the potential obscuring of an object of interest, but also, in the event of a particularly bright satellite, may result in significant saturation of detector pixels, causing issues to pixels that are not traversed by the satellite such as cross-talk contamination, electronic ghosting, or row/column bleeding. The sheer quantity of these satellites exacerbates the matter, as their tracks are even more frequently captured in astronomical images (Ventakesan, Lowenthal, Prem et al. 2020). This may result in the discarding of additional images, the need for additional effort to retrieve them with software, or evading them completely, thereby necessitating a greater amount of observing time, processing time, and funding to complete scientific research projects.

In order to address these and other concerns, the astronomy community has convened workshops and technical reports including SATCON 1 and 2, (Walker, Hall, Allen, et al. 2020; Walker & Hall 2021) Dark and Quiet Skies 1 and 2 (Walker et al. 2021; Walker & Benvenuti 2022). In addition, in 2022, the International Astronomical Union also formed the Centre for Protection of the Dark and Quiet Sky from Satellite Constellation Interference (IAU CPS; IAU 2022), and through these meetings, the IAU CPS, astronomers, and satellite operators are working closely together to better understand the impacts on astronomical observations and create solutions to mitigate them. Key to understanding the impact on astronomy is determining how bright the satellites are in the night sky and how this brightness is influenced by factors like the Sun-satellite-observer geometry, and the satellite's distance, and reflective properties. With knowledge of satellite brightness and distribution, astronomers can determine if satellites will be visible to their instruments and how they will appear in their data, and assess the level of impact on their project, e.g. how many detector pixels are expected to be affected by the satellite. Furthermore, by obtaining a baseline measurement of satellite brightness, the community can assess the efficacy of their present mitigation efforts and establish objectives for the future.

Numerous articles, such as those by Horiuchi, Hanayama & Ohishi (2020), Mallama (2020, 2021), Tyson, Ivezić, Bradshaw et al. (2020), Tregloan-Reed, Otarola, Ortiz et al. (2020); Tregloan-Reed, Otarola, Unda-Sanzana et al. (2021), Boley, Wright, Lawler et al. (2022), Halferty, Reddy, Campbell et al. (2022), Mróz, Otarola, Prince et al. (2022), and Krantz, Pearce & Block (2023), have documented the brightness levels of communication satellites in constellations. These studies have been crucial in providing insights into the brightness of these satellites. However, there still remains a significant gap in independent monitoring and comprehensive assessment of their brightness and reflectivity across the spectrum, and how they correlate with the geometry and the satellite design.

To aid these efforts, we have established a global network of telescopes, in cooperation with the IAU CPS and with the endorsement of the Chilean Astronomical Society (SOCHIAS), composed of small to medium-sized telescopes located in Chile, Mexico, Spain, Vietnam, and the Republic of Korea, called the Chilean Low Earth Satellite† (CLEOsat) group. The CLEOsat group's primary mission is to assess the impact of LEO mega-constellation communication satellites on the night sky, and their effects on optical and radio astronomy.

In this study, we present the preliminary results of our analysis of LEO satellite observations taken with the Danish 1.54-metre telescope at ESO La Silla, located in Chile, and we introduce a custom end-to-end Python pipeline developed to provide a consistent way to process and analyse these and other satellite observations obtained through our

† https://docs.google.com/document/d/1qzSJzhdv0jM6qmNZe83JSuFDXGqHUAj_FVeV7pzVUqk/edit?usp=sharing

growing network. A more detailed description of this work will be given by Adam et al. (in preparation).

2. Data analysis

2.1. *OneWeb and their observation with the Danish 1.54-metre Telescope*

The Eutelsat OneWeb satellites are mainly characterised by their box-wing design featuring a trapezoidal main bus, approximately 1 m in size. The bus is covered in reflective multi-layer insulation and complemented by two angled antennae reflectors. Additionally, two solar arrays project from the opposite sides of the bus. These satellites are initially deployed into a 500 km orbit, subsequently adjusting to a 1200 km circular operational orbit using their onboard thrusters (Krebs 2023). As of early 2024, the OneWeb constellation consists of 640 satellites, 620 of which are operational (McDowell 2024), with plans to expand to 716 (Federal Communications Commission 2020) for completion.

The OneWeb observations analysed in this study were obtained in three campaigns from 2021 May–June, and 2021 September–October, as well as from 2022 June–August, respectively, with the 1.54 m Danish Telescope (DK154) operated by the Microlensing Network for the Detection of Small Terrestrial Exoplanets (MiNDSTEp) consortium (Dominik, Jørgensen, Rattenbury et al. 2010) and located at ESO La Silla, Chile. The instrument used was the Danish Faint Object Spectrograph and Camera (DFOSC) which is equipped with a 2048×2048 pixel thinned Ford-Loral CCD, with a field of view (FoV) of 13.7×13.7 arcminutes and a nominal pixel scale of 0.39 arcseconds pixel^{-1} (for a 1×1 detector binning). To perform the multicolour photometric observations, the Johnson-Cousins *BVRI* filters (Bessell 1990) available to the DFOSC system were utilised.

The satellites observed in this study were selected on the basis of their visibility at the observatory location. The visibilities were calculated using two-line element (TLE) sets of the OneWeb satellite obtained from Celestrak†. These TLEs, which encode a list of orbital elements of an Earth-orbiting object at a given point in time, were then utilised to compute the ephemerides of the satellites using the coordinates of the Danish telescope and employing a custom telemetry code. Finally, the obtained set of satellite visibilities was constrained and only those were selected satellites that have an altitude > 30 degrees above the horizon, while the Sun-zenith angle is within a range of 110 to 120 degrees.

Generally, each image was exposed for 10 seconds. This exposure time was chosen to balance the requirement for sufficiently bright sources for the photometric measurements and to account for the possible differences between the predicted and actual locations and arrival times of the targeted satellites due to the unknown uncertainties on the TLEs. Additionally, each target satellite position was observed twice or more in order to obtain a reference image without the satellite trail, either directly before or after the satellite's passage. These reference images were then used for removing sources along the satellite trail to minimise their effect on the photometric measurements.

2.2. *Image processing, satellite detection and analysis with leosatpy*

For the processing and analysis of low-earth orbit satellites, we have developed a custom end-to-end Python pipeline, called `LEOSatpy`. `LEOSatpy` uses Astropy (Astropy Collaboration, Price-Whelan, Lim et al. 2022) libraries and additional custom procedures to reduce, calibrate, detect and analyse the observations of LEO satellites from various small and medium-sized telescopes that are part of the global CLEOSat telescope network.

† `https://www.celestrak.com/NORAD/elements/supplemental`

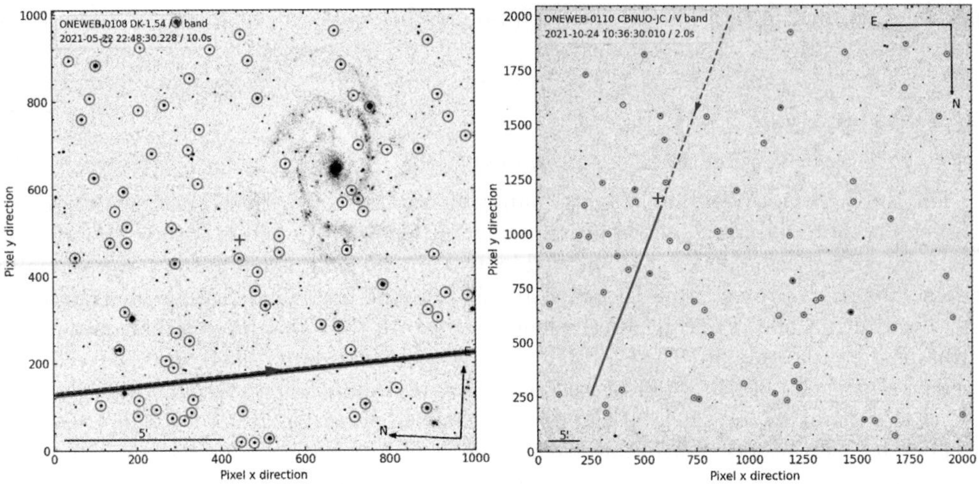

Figure 1. Result plot examples produced by our pipeline. The satellite trail is marked with a green box. The comparison stars used for the aperture photometry are indicated by the red circle. The predicted path from the TLE is shown as a blue dashed line, with the arrow indicating the direction in which the targeted satellite moved across the detector.

2.2.1. Reduction and plate solving

For basic processing of the OneWeb satellite observations, `LEOSatpy` uses `ccdproc` (Craig, Crawford, Seifert et al. 2023). The raw FITS images are registered and the associated, most timely calibration files are identified. These images are then combined to create master calibration images, specifically a master-bias and a master-flat for our OneWeb observations. The final reduced image is produced by removing the instrumental signature, i.e. all non-celestial counts, from the images by subtracting the bias frame and dividing by the flat-field image to correct for non-uniform sensitivity across the chip. The reduced FITS images are then calibrated with `LEOSatpy` to correct for any ambiguities in the instrument's pointing or tracking. The procedure was adopted from the `Astrometry` (Wenzl 2022) package, and uses phase correlation (see e.g., Reddy & Chatterji 1996; González 2011) to estimate the scale, rotation, and translation needed to match the position X and Y of sources on the detector with their known celestial positions RA and DEC on the sky, from the GAIA DR3 catalogue (Gaia Collaboration, Prusti, de Bruijne et al. 2016; Gaia Collaboration, Vallenari, Brown et al. 2023).

2.2.2. Satellite trail detection and analysis

The main purpose of our pipeline is the detection and characterisation of satellite trails. To identify the satellite trails in the calibrated FITS-images, `LEOSatpy` uses `photutils` (Bradley, Sipőcz, Robitaille et al. 2023) and extracts a segmentation map from the background subtracted image. The properties, e.g., position, eccentricity, and orientation, of each segment are determined, using the `regionprops` (Burger & Burge 2013) method from the `scikit-image` (van der Walt, Schönberger, Nunez-Iglesias et al. 2014) package. The set of image segments are then filtered, and the remaining segments grouped by their orientation and further analysed by applying a Hough Transform (HT) to them.

The HT is a common feature extraction technique in image processing that uses a voting procedure to associate a set of lines in the x-y image space to a pair of values in the θ-ρ parameter space (see Duda & Hart 1972), also referred to as Hough space. Once all pixels in the image space have voted for all possible lines in the Hough space, the resulting voting pattern is analysed to determine a complete set of parameters for these lines, i.e., their centroid position, orientation, length, and width, following Xu, Shin, Klette et al. (2015). This detection method enables us to detect multiple trails in an image, such as trails crossing each other or parallel satellite trails, as for example reported by Nandakumar, Eggle, Tregloan-Reed et al. (2023).

The highest votes in the Hough space are identified, and a window centred at a selected initial peak (θ_p, ρ_p) is extracted. For each voting angle θ_i, corresponding to a column i in the extracted region of the Hough space, the statistical mean μ_i and variance σ_i^2 over the range of ρ_j-values are then calculated and their functional relationship with respect to the voting angles is evaluated. Direction, length, and width of an assumed line segment are simultaneously computed based on the coefficients of a second-order polynomial fitted to all pairs (σ_i^2, θ_i). The midpoint (x_c, y_c) of an assumed line segment, on the other hand, is determined based on the coefficients of a linear fit applied to all pairs ($\mu_i/\cos\theta_i$, $\tan\theta_i$).

The magnitudes of the detected satellite trail(s) in the different bands were estimated using aperture photometry by comparing the instrumental magnitude of the trail to the well known magnitudes of a set of comparison stars in the image (see Brennan & Fraser 2022). For this, `LEOSatpy` compiles a set of comparison stars in the FoV and the given filter from the GSC 2.4.2 Guide Star Catalogue (Lasker, Lattanzi, McLean et al. 2021) and matches them with the sources detected in the image. Sources that are too close to the trail or to each other are then removed to avoid an overlap in the apertures used. In case multiple observations with the same pointing are available, the reference image without the satellite trail was used to mask sources along the trail.

To estimate the optimum aperture size for the photometry, we measure the signal-to-noise (S/N) ratio for different aperture sizes and use the radius at which the S/N is maximised as the new aperture radius for the image. Additionally, `LEOSatpy` estimates the magnitude correction required to compensate for the flux lost due to a finite aperture size by calculating the ratio of fluxes in the optimum aperture and a larger standard aperture. We use the estimated optimum aperture from the standard aperture photometry to measure the flux contained in the rectangular aperture around the detected satellite trail and together with the aperture correction calculate the observed magnitude m_x of the satellite in each filter.

However, the satellite typically only spends a fraction of the total exposure time on the detector. The observed flux is therefore corrected using the estimated time on the detector, which can be estimated from the ratio between the observed to the expected trail length according to the ephemerides from the TLE. The estimated magnitudes M_x furthermore depend on the distance R between the individual satellites and the observer. Thus, to be able to compare the different measurements, the magnitudes are then normalised to the nominal orbital height of the OneWeb satellites, 1200 km, by adding $-5\log(R/1200\,\text{km})$ to the apparent magnitudes. Using the ephemerides from the TLE data and in addition to the magnitudes, `LEOSatpy` also calculates and stores various other parameters such as the predicted positions during the exposure and the sun-satellite-observer geometry, e.g. the observer phase angle or the solar incidence and phase angles. Figure 1 shows two examples of the result plots produced by `LEOSatpy` for each dataset.

3. Results and discussion

In the presented study, we have so far identified and analysed 369 satellite trail observations from 159 individual OneWeb satellites obtained in the *BVRI* bandpasses. The mean estimated magnitudes scaled to an operational orbital height of 1200 km were found

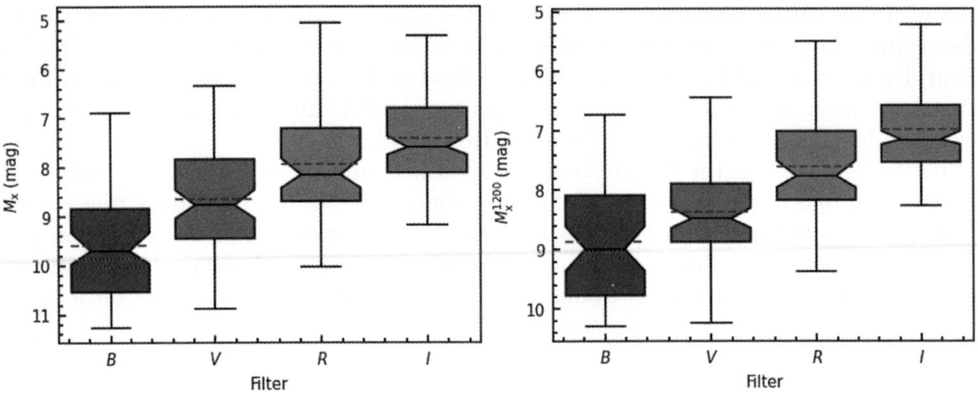

Figure 2. *BVRI* brightness distribution of the observed OneWeb satellites. (Left) The apparent brightness, corrected for exposure time. (Right) The corrected apparent brightness, scaled to an orbital height of 1200 km.

to be $B = 8.87 \pm 0.99$, $V = 8.37 \pm 0.83$, $R = 7.60 \pm 0.84$, and $I = 6.97 \pm 0.79$, demonstrating a noticeable increase in reflective brightness towards redder passbands (see Figure 2). This is in agreement with previous observations of the Starlink satellites, Starlink-1113 and Starlink-1130 (Darksat), which showed that both satellites got brighter towards the near-infrared (see Tregloan-Reed, Otarola, Ortiz et al. 2020; Tregloan-Reed, Otarola, Unda-Sanzana et al. 2021). Following the suggested estimation of a recommended V band limiting magnitude† set by the SATCON 1 (Walker, Hall, Allen, et al. 2020) and IAU/UNOOSA D&QS reports (Walker & Hall 2021), we found 27.4% (23.6% after range correction) of OneWeb satellites to be brighter than the recommended V band limiting magnitude ($V_{\rm Lim}$) of 7.85 mag. Furthermore, the colour indices derived from the observed apparent magnitudes differ significantly from the those obtained from the apparent AB magnitudes of the Sun by Willmer (2018) in all observed filters. This indicates that the apparent brightness is mostly determined by the wavelength dependent reflection and absorption properties of the surface materials and not by specular reflection of sunlight. The results also show a significant variation in brightness, possibly due to atmospheric variations across the multiple observation nights. Examining data from the same night could resolve this issue, but was not applied, given the small sample size ($N < 20$).

The general trend of an increase in reflective brightness is also visible in Figure 3. The figure shows how the normalized magnitude changes as a function of the solar phase angle, i.e. the angle between the Sun and the observer as seen from the satellite. While we found no apparent correlation of brightness and whether the satellites are in an orbital rising stage or have reached their designated orbit, an increase in the effective albedo ($\rho_{\rm eff}$; see e.g. Krantz, Pearce & Block 2023, and colours in Figure 3) with increasing phase angle was noted, which contributes to the overall reflective brightness of these satellites. As noted by Krantz, Pearce & Block (2023), the effective albedo is not a real measurement of the albedo, rather than a first order approximation in which the albedo, i.e. material properties, is scaled by the reflective cross-section, which itself depends on the orientation and design of the satellite. In other words, a higher than baseline effective albedo indicates where the satellites are brighter than they could be, assuming that a diffuse reflection is the minimum possible brightness.

Finally, we utilized various "diffuse-specular" sphere models (Pradhan, Hickson, Surdej et al. 2019; Boley, Wright, Lawler et al. 2022) to replicate general trends seen

† $V_{\rm Lim} = 7\,\text{mag} + 2.5 log(H_{\rm orb}/1200\ km)$

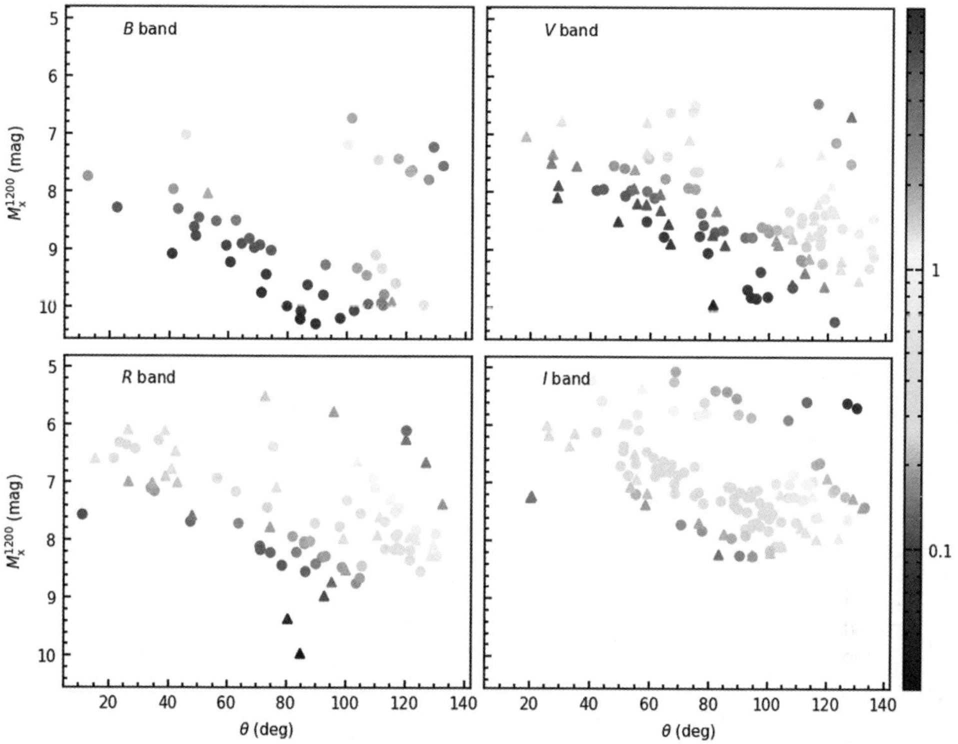

Figure 3. Range-corrected magnitude vs. solar phase angle. The effective albedo is represented as colours. Triangles indicate satellites that in an orbital rising stage, whereas circles indicate OneWeb satellites around the nominal orbital altitude of about 1200 km.

in the data, and to assess the reflective properties of the OneWeb satellites, depending on the Sun-satellite-observer geometry. For this, we employed a Markov-Chain Monte Carlo (MCMC, Foreman-Mackey, Farr, Sinha et al. 2019) routine to explore the parameter space for the mixing coefficient β, slope index p, and effective area ζ†. The best models, including the 1-σ intervals, are shown in Figure 4. A summary of the best-fit parameters is given in Table 1. Both models show a similar increase in effective area ζ towards redder wavelengths, consistent with the observed increase in brightness. The found parameter for the mixing coefficient β and the slope index p show deviation from a purely diffuse sphere model. Both models describe the general trend seen for when the phase angle is below 90°, but they fail to account for the increase in reflective brightness at phase angles above 90°. This indicates that the models used in this study, fail to fully account for satellite properties, such as its geometry or design. Noteworthy is that with the attitude of the satellite being an unknown property, there is the potential that a phase offset is needed to account for the divergence between the data and models. This would also take into account any deviation from diffuse to specular reflections due to the solar panels that occasionally and independent of the Sun-satellite geometry produce strong reflections. However, the present data do not allow for a direct determination of the type of material used, but when augmented with reflective properties of satellite materials such as Kapton, they may enable researchers to study the effective cross-section as a function of phase angles of the observations and create a 3 D model of the satellites' attitude.

† Effective area $\zeta = A \cdot \rho$, where A is the cross-sectional area and ρ is the albedo.

Table 1. Diffuse-specular sphere model results.

	Pradhan, Hickson, Surdej et al. (2019)		Boley, Wright, Lawler et al. (2022)	
Filter	ζ (m^2)	β	ζ (m^2)	p
B	0.09±0.01	0.26±0.16	0.10±0.02	1.05±0.05
V	0.15±0.01	0.71±0.09	0.17±0.02	1.02±0.03
R	0.22±0.02	0.46±0.13	0.26±0.03	1.02±0.03
I	0.32±0.02	0.62±0.10	0.36±0.03	1.02±0.03

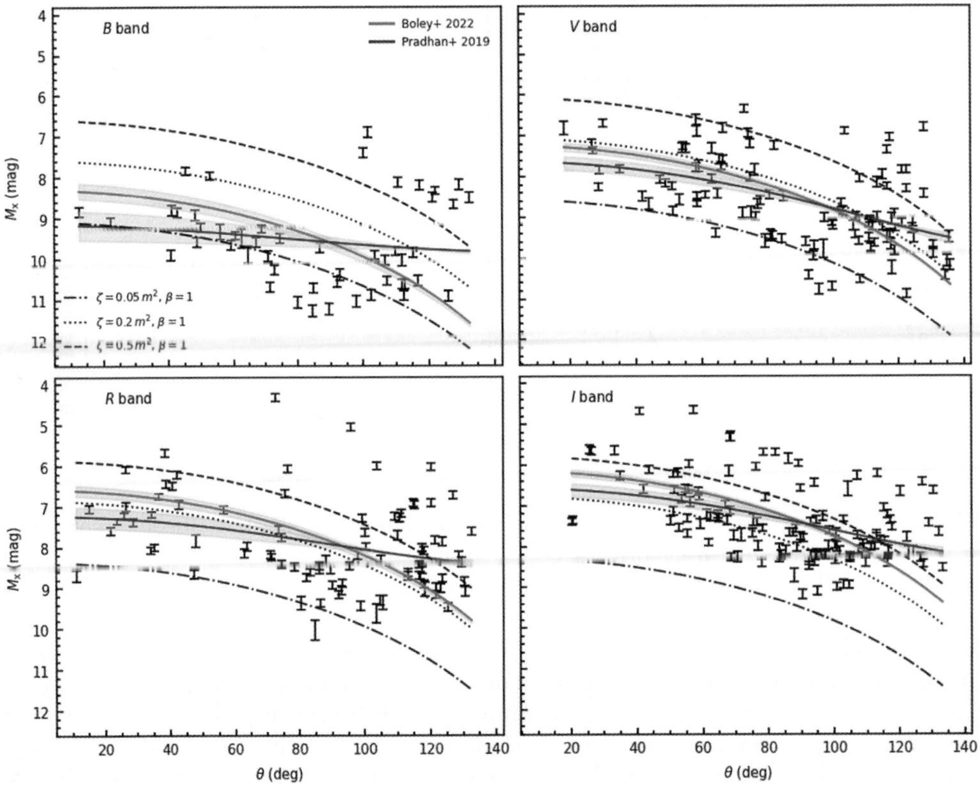

Figure 4. Preliminary diffuse-specular sphere model results. The expected changes in reflective brightness for a purely diffuse model ($\beta = 1$) for different effective areas ζ are shown as black dashed, dotted, and dash-dotted lines.

4. Conclusion

In conclusion, our findings underscore the need for improved modelling and forecasting of satellite reflective brightness. With updated data and a more nuanced understanding of factors like satellite geometry and attitude, we can develop more accurate models to predict and mitigate the impact of LEO satellites on astronomical observations. This research is not only crucial for the astronomical community, but also imperative for balancing the rapid advancement of satellite communication technologies with the needs and objectives of astronomical research.

Acknowledgements

CA acknowledges the Regional European Southern Observatory (ESO) Fund, and the 2021 China-Chile Joint Research Fund awarded by the Chinese Academy of Sciences

South America Center for Astronomy (CASSACA) and the National Astronomical Observatory of China (NAOC). CA furthermore acknowledges the support of ANID / Fondo 2022 QUIMAL / QUIMAL220005. JTR acknowledges financial support from ANID/FONDECYT (an Initiation in Research grant; Project Number 11220287), and this material is based upon work supported by the U.S. Air Force Office of Scientific Research (Award FA9550-22-1-0292).

References

Astropy Collaboration, Price-Whelan, A. M., Lim, P. L., et al. 2022, ApJ, 935, 167
Bessell, M. S. 1990, PASP, 102, 1181
Boley, A. C. & Byers, M. 2021, Sci Rep, 11, 10642
Boley, A. C., Wright, E., Lawler, S., et al. 2022, AJ, 163, 199
Brennan, S. J., & Fraser, M. 2022, A&A, 667, A62
Bradley, L., Sipőcz, B., Robitaille, T., et al. 2023, Zenodo
Burger, W., & Burge, M. J. 2013, Principles of digital image processing: advanced methods, (London: Springer)
Craig, M., Crawford, S., Seifert, M., et al. 2023, Zenodo
Duda, R.O., & Hart, P.E. 1972, Commun ACM, 15, 11–15
Dominik, M., Jørgensen, U. G., Rattenbury, N. J., et al. 2010, Ast Nach, 331, 671
Federal Communications Commission 2020. OneWeb Non-Geostationary Satellite System (LEO) Phase 2: Modification to authorized system. Retrieved from `https://fcc.report/IBFS/SAT-MPL-20200526-00062/2379706.pdf`
Foreman-Mackey, D., Farr, W., Sinha, M., et al. 2019, JOSS, 4, 1864
Gaia Collaboration, Prusti, T., de Bruijne, J. H. J., et al. 2016, A&A, 595, A1
Gaia Collaboration, Vallenari, A., Brown, A. G. A., et al. 2023, A&A, 674, A1
González, R. 2011, in Delmas, P., & Wuensche, B. (eds.) Image and Vision Computing New Zealand Conference 2011 Proceedings.
Halferty, G., Reddy, V., Campbell, T., et al. 2022, MNRAS, 516, 1502
Hejduk, M. 2011, Advanced Maui Optical and Space Surveillance Technologies Conference, `https://amostech.com/TechnicalPapers/2011/NROC/HEJDUK.pdf`
Horiuchi, T., Hanayama, H., & Ohishi, M. 2020, ApJ, 905, 3
IAU 2022, Launch of New IAU Centre Safeguarding Astronomy from Satellite Constellation Interference, IAU Announcements `https://www.iau.org/news/announcements/detail/ann22024/`
Krantz, H., Pearce, E. C., & Block, A. 2023, PASP, 135, 095003
Krebs, G. D. "OneWeb 1, ..., 900". Gunter's Space Page. `https://space.skyrocket.de/doc_sdat/oneweb.html`
Lasker, B., Lattanzi, M. G., McLean, B. J., et al. 2021, VizieR Online Data Catalog, I/353
Mallama, A. 2020, arXiv:2012.05100
Mallama, A. 2021, arXiv:2111.09735
McDowell J. 2004, Jonathan's Space Pages. `https://planet4589.org/space/con/ow/stats.html`
Mróz, P., Otarola, A., Prince, T. A., et al. 2022, ApJL, 924, L30
Nandakumar, S., Eggle, S., Tregloan-Reed, J., et al. 2023, Nature, 623, 938
Pradhan, B., Hickson, P., & Surdej, J. 2019, Acta Astronaut, 164, 77
Reddy, B. S. & Chatterji, B. N. 1996, IEEE Transactions on Image Processing, 5, 1266
Tregloan-Reed, J., Otarola, A., Ortiz, E., et al. 2020, A&A 637, L1
Tregloan-Reed, J., Otarola, A., Unda-Sanzana, E., et al. 2021, A&A 647, A54
Tyson, J. A., Ivezić, Ž., Bradshaw, A., et al. 2020, AJ, 160, 226
van der Walt, S., Schönberger, J. L., Nunez-Iglesias, J., et al. 2014, PeerJ 2:e453
Venkatesan, A., Lowenthal, J., Prem, P., & Vidaurri, M. 2020, Nat Astro, 4, 1043
Walker, C., Hall., J., Allen, L., et al. 2020, BAAS, 52(2) (SATCON I)
Walker, C., & Hall, J. (Eds.) 2021, Report of the SATCON2 Workshop, 12–16 July 2021, BAAS, `https://baas.aas.org/report-of-the-satcon2-workshop-1216-july-2021`

Walker, C., et al. 2021, Dark and Quiet Skies for Science and Society: Report and recommendations, Zenodo, https://doi.org/10.5281/ZENODO.5898785
Walker, C., & Benvenuti, P. (Eds.) 2022, Dark and Quiet Skies II Working Group Reports, Zenodo, https://doi.org/10.5281/ZENODO.5874725
Wenzl, L. 2022, Zenodo
Willmer, C. N. A. 2018, ApJS, 236, 47
Xu, Z., Shin, B.-S., & Klette, R. 2015, Pattern Recognition, 48, 4012

Astronomy and Satellite Constellations: Pathways Forward
Proceedings IAU Symposium No. 385, 2026
P. Grimley, J. Barentine, D. Galadí-Enríquez & C. Walker
doi:10.1017/S1743921324000607

Contamination of spectroscopic observations by satellite constellations

Olivier R. Hainaut and Sabine Moehler

European Southern Observatory, Karl-Schwarzschild-Straße, 2, 85470 Garching-bei-München, Germany
email: ohainaut@eso.org

Abstract. This study focuses on the impact of satellite trails on spectroscopic observations in the visible and near-infrared, which represent a large fraction of the observing time on large professional telescopes. We simulate the number of observations affected using an hypothetical large set of low Earth orbit satellite constellations. We evaluate the effect of this contamination on various spectrographs. We also measure the contamination on actual data from the ESO archive for a set of representative objects, after adding a scaled solar analogue spectrum to represent the satellite.

Even with a satellite population much larger than today's, the fraction of contaminated spectra will remain low, essentially thanks to the typically small field of view of these instruments. Furthermore, high-resolution spectrographs, and medium-resolution ones on smaller telescopes, are essentially blind to satellites. However, for a range of instruments the contamination will be at a level comparable to that of the science signal, making its identification difficult, and potentially affecting the astrophysical measurements derived from the data.

Keywords. Spectroscopy, Abundances, Light pollution

1. Introduction

The number of artificial satellites orbiting Earth has increased dramatically in recent years, from around 1500 in 2015 to over 5000 today (McDowell 2023a), essentially because of the launch of large satellite constellations that aim to provide global internet coverage. Several constellation operators have plans to launch over 500 000 satellites to low-altitude orbits in the near future (McDowell 2023b).

The environmental impact of this large number of satellites, as well as the impact on astronomical observations is being studied, and described by Walker, Hall, Allen, et al. (2020), Walker & Hall (2021), Walker et al. (2021) and Walker & Benvenuti (2022). Hainaut & Williams (2020) and Bassa, Hainaut & Galadí-Enríquez (2022), quantified the effect of the contamination on a series of representative instruments and telescope combinations.

Here, we focus on spectroscopic observations: in most cases, the data from a spectrograph lack all (fibre-fed or short-slit spectrograph), or most (long-slit spectrograph) spatial information. Therefore, the data will not contain the typical tell-tale trail that identify a satellite on an image.

The details of the methods and results are presented by Hainaut & Moehler (2024).

2. Effective magnitude

To be observable in the visible or near-infrared, a satellite must be illuminated by the Sun, and the Sun must be below the horizon. Hainaut & Williams (2020) and Bassa,

Hainaut & Galadí-Enríquez (2022) provide the details. This implies that satellites will mostly affect twilight observations, and that their numbers dwindle in the middle of the night.

The apparent visual magnitude of a satellite is a complex function of the illumination and viewing geometries and of the shape and surface of the satellite. To proceed, we evaluate the magnitude from a simplistic model that scales a magnitude at 1000 km at zenith $M_{1000\text{km}} = 6$ and account for the distance to the observer. While this model is clearly not realistic, it provides magnitudes that are conservatively brighter than the measured ones.

Because of the very fast apparent motion of the satellites (of the order of a degree per second), the satellite will remain for only a very short time in the aperture of the spectrograph. The effective magnitude of the satellite as seen by the spectrograph must therefore account for the ratio of that effective exposure time and the total duration of the exposure.

For a satellite to be detected by an instrument, this effective magnitude must be brighter than the limiting magnitude for the exposure. Because of this, essentially all high-resolution spectrographs (with r of the order of 100 000) are blind to all satellites. Medium-resolution spectrographs on large telescopes (> 8 m) and most low-resolution instruments will register the spectrum of a typical constellation satellite or anything brighter.

3. Effect on astrophysical measurements

There will therefore be cases where a spectrum is contaminated by a satellite whose contribution is at about the same level as that of the science target. As this contamination is not immediately obvious, the spectrum is likely to be processed and analysed without noticing the issue. To evaluate the effect of these blind contamination, we selected a series of spectra of representative objects from the ESO Science Archive, and contaminated them with the spectrum of a solar analogue scaled to the effective magnitude corresponding to various exposure times. The original and contaminated spectra were then measured using standard tools to extract equivalent widths, abundances, etc, and compare the contaminated value to the original one.

The rather obvious conclusion of this exercise is that the astrophysical parameters measured from longer exposures with better signal-to-noise ratio (S/N) are less affected than those from shorter exposures or exposures with poor S/N. A tell-tale sign of contamination is a S/N higher than expected (caused by the extra light from the satellite).

4. Mitigation and conclusions

While the contamination of a low-resolution spectrum by a very bright satellite will be obvious, and that of high-resolution data by a normal constellation satellite is negligible, a (fortunately small) fraction of intermediate cases will present blind contamination. That contamination, if not identified, will affect the astrophysical measurements at the level of a few percent.

While machine learning could potentially be trained on contaminated spectra (possibly simulated ones), and unleashed to help identifying the problematic data, the best protection will remain to split the total exposure time in various sub-exposures. Comparing them will allow the astronomer to identify discrepancies and reject the affected data. Keeping the absolute and apparent magnitude of the satellites fainter than $V = 7$, which is critical for visual observations and to avoid saturation of the Vera C. Rubin Observatory camera, will also help minimise the contamination of spectra.

References

Bassa, C. G., Hainaut, O. R., & Galadí-Enríquez, D. 2022, A&A, 657, A75

Hainaut, O. R. & Williams, A. P. 2020, A&A, 636, A121

Hainaut, O. R. & Moehler, S. 2024, A&A, in press, astro-ph doi: `10.48550/arXiv.2401.09976`

McDowell, J. 2023, Satellite statistics, `https://planet4589.org/space/stats/stats1.html`

McDowell, J. 2023, Enormous statellite constellations, `https://planet4589.org/space/con/conlist.html`

Walker, C., Hall., J., Allen, L., et al. 2020, BAAS, 52(2) (SATCON I)

Walker, C., & Hall, J. (Eds.) 2021, Report of the SATCON2 Workshop, 12–16 July 2021, BAAS, `https://baas.aas.org/report-of-the-satcon2-workshop-1216-july-2021`

Walker, C., et al. 2021, Dark and Quiet Skies for Science and Society: Report and recommendations, Zenodo, `https://doi.org/10.5281/ZENODO.5898785`

Walker, C., & Benvenuti, P. (Eds.) 2022, Dark and Quiet Skies II Working Group Reports, Zenodo, `https://doi.org/10.5281/ZENODO.5874725`

Astronomy and Satellite Constellations: Pathways Forward
Proceedings IAU Symposium No. 385, 2026
P. Grimley, J. Barentine, D. Galadí-Enríquez & C. Walker
doi:10.1017/S1743921324000875

Spectroscopy of human-made space objects: from low Earth orbit debris to satellite constellations and exotic outliers

Danica Žilková[1], Jiří Šilha[1], Julián Rodríguez[2], Thomas Schildknecht[2], Pavel Vojtek[1], Julia de León[3] and Katarína Sabolová[1]

[1]Comenius University in Bratislava,
email: danica.zilkova@fmph.uniba.sk

[2]University of Bern,

[3]Instituto de Astrofísica de Canarias

Abstract. As the commercial space industry advances, the number of artificial objects orbiting the Earth rises exponentially. To categorize the reflectivity of bright Low Earth Orbit (LEO) objects, a number of spectroscopic observations of such objects was performed in collaboration with the Astronomical Institute of the University in Bern, Switzerland. Supported by laboratory measurements of various aerospace materials, spectra of space objects were analyzed to search for correlations with material samples. On top of that, near-Earth object 2020 SO originally discovered as asteroid was later identified as Surveyor 2 rocket debris. Spectroscopic observation of this object was conducted with the OSIRIS camera-spectrograph at the 10.4m Gran Telescopio Canarias (GTC) located in La Palma (Spain). Spectrum of this object is compared with spectra of upper stage rockets observed by the ZimMain telescope to investigate correlations within their material properties and search for signs of space weathering.

Keywords. Reflectance spectroscopy, LEO regime satellites, space debris

---

1. Introduction

Spectroscopy is a widely used technique in astronomy to investigate the chemical composition and surface properties of space objects. With immense rise in space traffic, especially in LEO, endangering future space missions and night sky pollution, spectroscopy could offer a unique way to obtain information about the surface reflectance properties of artificial space objects. Earth satellites reflect sunlight in diffuse or bright specular reflections which can be detected by a ground observer. By processing recorded spectra, information about the radiating surface elements can be extracted. Highly reflective materials will have a dominating imprint within the spectrum, and if the spectrum is observed during a specular reflection phase, particular materials can be identified (Žilková, Šilha, Matlovič & Toth 2023). Regarding the targeted optical spectroscopic observations, when diffuse spectra are detected incoming radiation is the sunlight reflected from multiple surface elements, depending strongly on objects attitude, geometry and surface materials. The illuminated features will then imprint their reflective properties in the recorded spectrum. Since satellites, upper stage rockets and other artificial space objects are complex structures, it is challenging to identify each reflecting component within the spectrum. A number of studies already examined the reflectance spectra of in-orbit satellites and debris, specifically in Geosynchronous Earth

Orbit (GEO), such as Schildknecht, Vannanti, Krag & Erd (2009), Seitzer, Lederer, Cowardin et al. (2012), Bedard & Wade (2014) or Abercromby, Abell & Barker (2009).

Various methods can be applied to analyse the spectra. Initially we have started by recording spectra of different aerospace samples in a controlled laboratory environment and aim to identify similarities between laboratory and observed spectra to define a scope of parameters which will describe the spectral space. Presented here will be the first set of observations with 80-cm telescope ZimMain operated by the Astronomical Institute of University in Bern (AIUB) and LISA Shelyak low-resolution spectrograph and first samples of laboratory measurements of aerospace materials obtained by Faculty of Mathematics, Physics and Informatics (FMPI) of Comenius University. Initial correlation between the spectral gradient and relative flux will be investigated. Finally, in a similar manner, we will analyze the spectrum of 2020 SO measured by the OSIRIS spectrograph on 10.4-m GTC telescope located at the El Roque de los Muchachos Observatory (La Palma, Spain).

2. Observations with ZimMain 80-cm telescope at AIUB, Switzerland

Three observational campaigns were conducted in spring and autumn 2022 at the Astronomical Institute of University of Bern (AIUB) in Switzerland, (Rodríguez-Villamizar, Vannanti, Schildknecht et al. 2023). AIUB operates an 80-cm telescope ZimMain at Zimmerwald observatory, (Cordelli, Lauber, Prohaska et al. 2019), which is equipped with a slit-based spectrograph LISA manufactured by Shelyak Instruments, who also provided the imaging and processing software Demetra for data recording and reduction (Shelyak Instruments 2020). The objectives of the observational campaigns were to establish a workflow for routine spectroscopic observations using the ZimMain telescope and to obtain spectra of various space debris types to assess the system's capabilities. Categories of space objects were selected, to test if we can observe spectra of different types of objects:

- **GEO satellites:** stable, box-wing structure, less bright since further away
- **Calibration spheres:** spherical single-material satellites - Lageos 1 & 2, Etalon (too faint to detect)
- **Bright objects with specular flares:** to be observed during, before and after glint - Iridium 921 with glint, ERS-2 with glint
- **Box-wing satellites:** to observe changes with phase angle - Glonass 86, Glonass 134
- **Upper stage rockets:** with different age, ideal to observe material degradation - CZ-3B (NORAD 43110, 43247, 43624, 45864)
- **Satellite constellations:** Starlink (NORAD 45362), Oneweb (too faint to detect), Iridium 33, Iridium 921

Within the identified groups, target objects were selected according to their magnitude, phase angle and overall visibility during the specific night. Sample spectra of these types of objects are presented in Figure 4. However, a detailed analysis will be performed on this spectral set in the future.

2.1. *Spectral reduction*

Spectra were recorded and processed by the Shelyak software Demetra and the reduction was verified by cross reduction in IRAF yielding the same results. Spectra were corrected by bias, flat and dark images in standard spectral reduction and calibrated into wavelengths using helium-neon calibration lamps. Spectra were also flux calibrated to the response spectrum of star Vega, recorded at the night of the observation, using the Demetra algorithm.

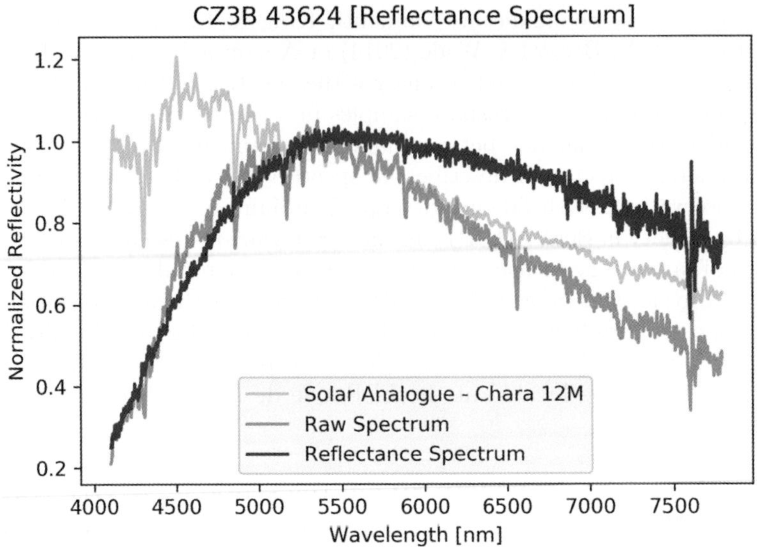

Figure 1. Recorded raw spectrum of CZ-3B (43624), solar analogue and final reflectance spectrum of CZ-3B (43624) recorded on 10 June 2022, by 80-cm telescope ZimMain at the Zimmerwald observatory.

To obtain the reflectance spectra, the measured raw spectra must be corrected for sunlight reflected from the object's surface. Real-time solar spectrum observation is challenging, so solar analogue stars with similar spectral shapes are observed for correction. The measured spectrum is divided by the star's spectrum to derive the satellite's surface properties. Ideally, the solar analogue star is observed in the same sky location or, at least, the same airmass as the target, to correct for the atmospheric extinction effects. As demonstrated in Figure 1, the raw spectrum is divided by the spectrum of the solar analogue and it is clear, that the Balmer lines and atmospheric extinction lines are eliminated from the final reflectance spectrum.

In this manner, all spectra demonstrated in Figure 4, were reduced and corrected to extract the reflectance spectrum.

3. Laboratory measurements at Department of Experimental Physics, FMPI Slovakia

To extend our investigation into spectroscopic observations and the reflection of light by different materials, we conducted a series of laboratory experiments. The experiments involved testing samples of diverse materials, such as aluminum, silicon, or copper foil. Many satellites are coated with multi-layer insulation foil, typically composed of polyimide (Kapton). An original 15 × 15 cm DuPont Kapton sample layered on 0.5 nm aluminum foil was acquired from the Goodfellow distributor.

Additionally, Slovak tech company Needronix supplied us with a one-to-one model of a CubeSat satellite structure along with various material components (Laszlo, Musilova & Koleda 2023), including Dural with different compositions and a gallium arsenide solar cell. During the experiments every sample had been illuminated with a standard laboratory halogen lamp to retrieve the reflectance spectrum, the raw spectrum of a sample was divided by the spectrum of the halogen lamp. Every sample was rotated to observe changes in the spectrum before, during and after the glint. Furthermore, spectra of different combinations of materials were measured as well. Altogether, a quite large database of spectra was recorded and will require a timely analysis. But the intention is to publish

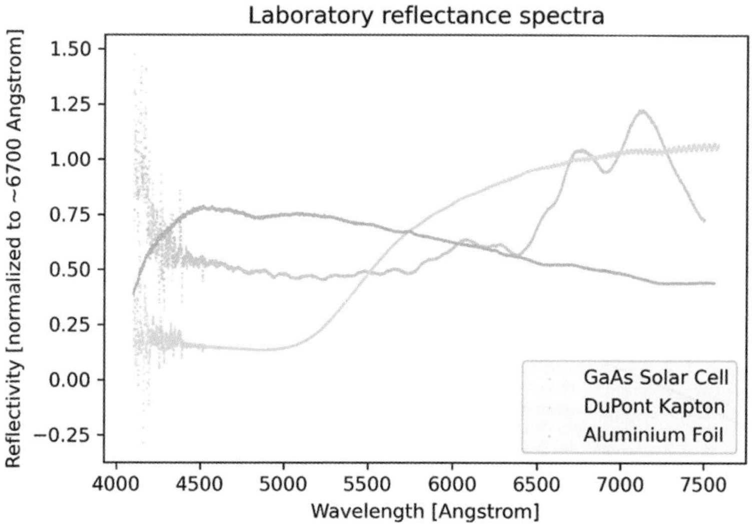

Figure 2. Samples of DuPont Kapton, Aluminium foil and GaAs solar cell measured with Shelyak LISA spectrograph at the Department of Experimental Physics, FMPI.

a unified public database of laboratory measured samples. Hereby, we are presenting the reflectance spectra of DuPont Kapton, Aluminium foil and gallium arsenide solar cell, Figure 2 which will be used to correlate our ZimMain recorded spectra of space debris and other LEO objects with laboratory measurements. It is important to note that the laboratory spectra were acquired with the same optical instrument, Shelyak spectrograph LISA, and the same camera, to ensure the same system sensitivity, optics, resolution and the same wavelength range.

Laboratory measurements are the key to discovering how different aerospace materials behave during diffuse and specular reflections and to decode the information hidden within the spectra recorded during the observational campaigns.

4. 2020 SO spectrum measured with OSIRIS at GTC, Instituto de Astrofísica de Canarias, La Palma

On 17 September the PanSTARRS 1 telescope in Hawaii observed an unusual object with low orbital velocity and Earth bound orbit. Initially identified as an asteroid, hence the naming convention, the object was too faint to be identified with traditional near-IR characterization. Various observations of the object, from astrometry (Petrova, Devyatkin, Gorshanov et al. 2021), colour observations, light curves (Campbell, Furfaro, Reddy et al. 2022) to visible spectral observations are in support of the theory that the unusual object is Surveyor 2, Atlas-Centaur upper stage rocket booster.

Object 2020 SO was observed on November 11th, 2020 using the OSIRIS camera-spectrograph at the 10.4-m GTC telescope (La Palma, Spain), together with three solar analogue stars. Spectra were bias and flat-field corrected, extracted, collapsed to 1D array and wavelength calibrated. The spectrum of the of the object was divided by the spectrum of each solar analogue and the resulting reflectance spectra of the object ware averaged. In Figure 3, the reflectance spectrum of 2020 SO is compared with spectra of Chinese upper stages CZ-3B with NORAD numbers 43110, 43247, 45864. An indisputable correlation can be observed between the spectrum of 2020 SO and CZ-3B 43110. However, more detailed analysis is required, taking into account attitudes, material properties and time spent in orbit of these objects to verify the spectral match.

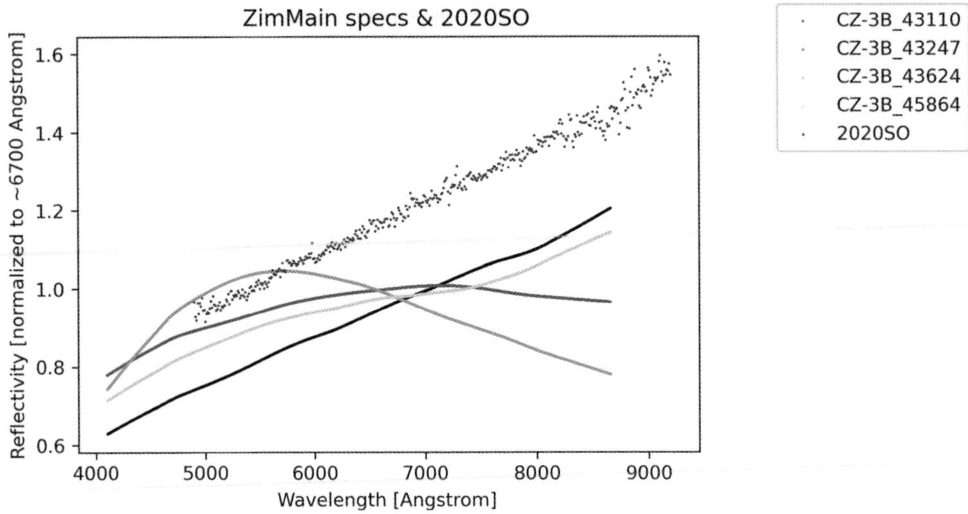

Figure 3. Spectrum of 2020 SO (blue dotted line) compared with four spectra of CZ-3B. While 43247, 43624, 45864 have a bit concave shape and smaller slope increase towards red wavelengths, there is almost identical match between the slope of 2020 SO and 43110 spectra.

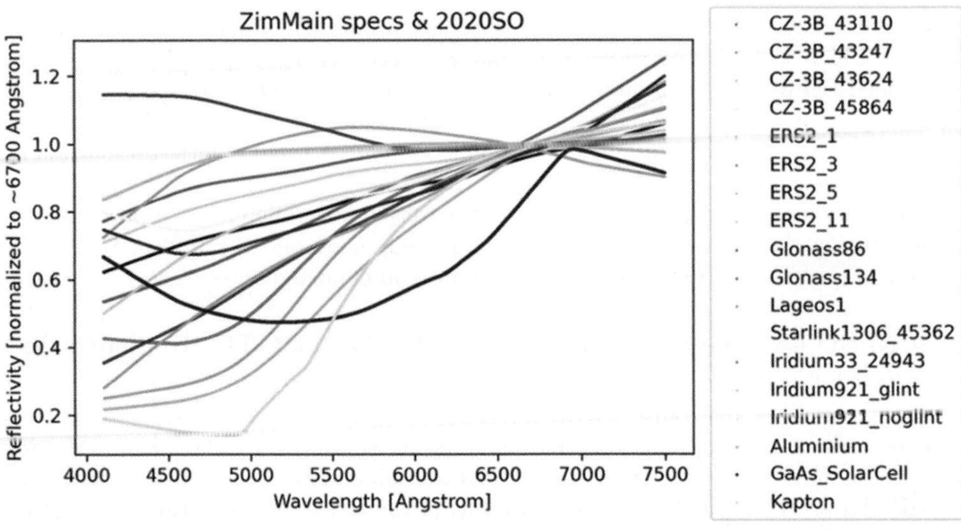

Figure 4. Fifteen spectra of different artificial space objects measured by 80-cm telescope ZimMain at Zimmerwald Observatory and three spectra of sample space materials measured in a laboratory at FMPI.

5. Relative flux and spectral slope relationship

A number of spectra of different space debris, LEO objects and satellites in LEO constellations, had been observed to study their reflectance properties across the wavelengths of visible light. However, these spectra are most often smooth curves, as can be seen in Figure 4, with little features to help connect the spectral characteristics to specific materials. Hence a different approach must be considered when trying to analyze these types of spectra.

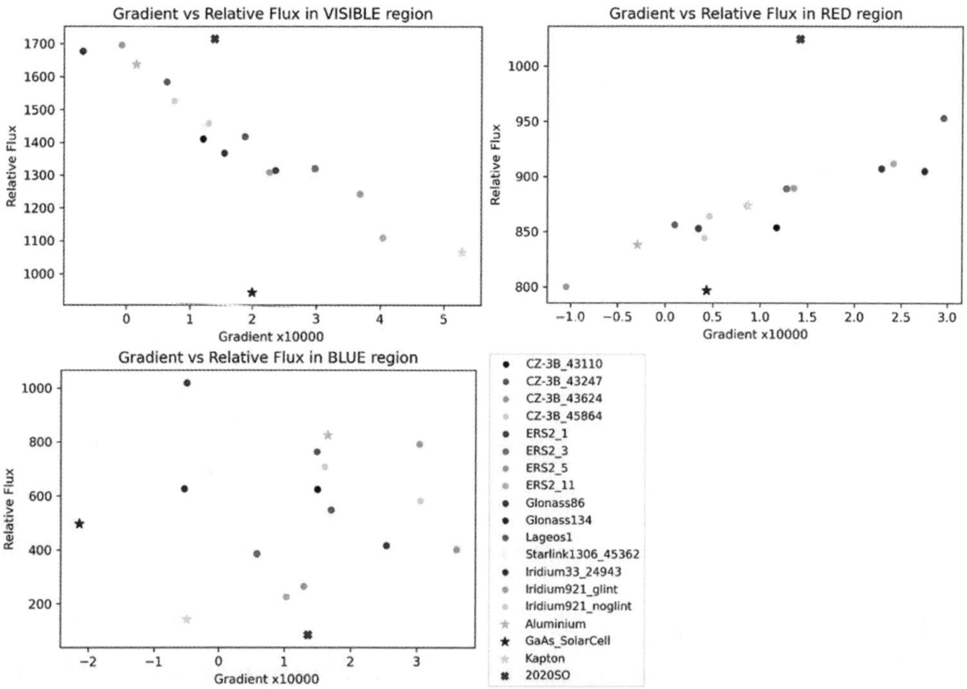

Figure 5. Relationship between the gradient and the relative flux of spectra of different artificial space objects and laboratory measured spectra of aerospace materials, plotted in regions of blue (400 - 500 nm), visible (500 - 660 nm) and red (660 - 760 nm) wavelengths.

5.1. *Splitting the wavelength regions to investigate distinct features*

In Figure 4, clear differences can be observed in the slopes of the spectra which correlate to the amount of increase towards the longer, redder wavelengths. Some spectra are increasing rather steadily while others remain more or less constantly flat throughout the visible wavelengths. However, some spectral functions are not flat, for example the spectrum of the solar panel is convex. Therefore, in order to examine the amount of increase or decrease of the spectrum without missing essential information, each spectrum will be split into three regions derived approximately from pass-bands of colour photometry - blue, visible and red region.

- **Blue region:** 400 - 500 nm
- **Visible region:** 500 - 660 nm
- **Red region:** 660 - 760 nm

To quantitatively evaluate the spectral slopes in each region, spectra had been fitted by a polynomial of a first degree within the given region, and their gradients had been computed.

Next parameter to consider is the overall reflectivity throughout the wavelengths. Some spectra have rather high reflectivity on y-axis, above 1.2, while for others it decreases below 0.2. Since the spectra observed by the ZimMain telescope are reduced by the Demetra software and are flux calibrated to Vega, we can evaluate and compare their relative intensities on y-axis. Hence, the relative flux was computed as an integral of intensities across the wavelengths. Finally, the relationship between the spectral gradient and the relative flux for each spectra at selected wavelength regions had been plotted in Figure 5.

Analyzing each region, we can note certain conglomerations of objects. For example, CZ-3B 43110, 43247, 43624, 45864 are within a very close vicinity to each other and to the laboratory spectrum of aluminium in visible region but differ slightly in blue and red regions either in flux or in slope.

Similarly can be noted for the spectra of ERS-2. They are marked ERS-2_1, ERS-2_3, ERS-2_5 and ERS-2_11 according to the changing satellites rotation throughout the series. First spectrum, ERS-2_1 appears to be very different from the rest of the spectra, occupying similar region as aluminium-type objects with constantly flat spectra. While other three ERS-2_3, ERS-2_5 and ERS-2_11 are allocated on the opposite side of the plot near the spectrum of copper-coloured Kapton. Suggesting that as the satellite ERS-2 rotated, the properties of the reflecting material changed significantly.

Another interesting conclusion can be derived from the position of Lageos satellite within plots of Figure 5. Lageos is a spherical satellite with surface structure constructed of aluminium with reflecting prisms made of fused silica glass and germanium (ESA eoPortal 2023). In blue wavelength region it behaves very similarly to aluminium. However, as the wavelengths increase so does the gradient, differing significantly from aluminium spectrum, especially in the red region. More observations of this satellite are required to prove whether this trend is changing with the phase angle or reflecting geometry as the reflectance spectrum is affected by the reflectors, or there could be some effects of space weathering influencing the spectrum, since Lageos had spent already over 30 years in space environment.

The spectrum of 2020 SO had been incorporated into this analysis as well, although it had been recorded by a different instrument and is not flux calibrated. Hence, the value of relative flux on y-axis is strictly informative and does not comply with the rest of the data. Considering only the gradient and its evolution in blue, visible and red regions, similar trend can be observed as for the Lageos satellite, although not as significant. If the 2020 SO is indeed the Surveyor 2 upper stage, its surface material is made of stainless steel which had presumably degraded in the space environment over the course of 57 years that the object had spent in space. This could be related to the shape of the spectrum and the decrease in reflectivity in blue wavelength regions.

Lot more assumptions can be drawn from this type of analysis, however, more observational and laboratory data must be added to better define the gradient-flux space. The location of object's spectrum within this space can be defined by a set of parameters which will ultimately help with spectral characterization and material identification.

6. Conclusions and future plans

To conclude, the application of spectroscopy to targeted observations of space debris, LEO objects and LEO constellations had been investigated. Various LEO objects had been observed at the Zimmerwald Observatory in Switzerland to test the efficiency of this type of observation on fast, bright and fast-rotating targets.

Observational data was supported by laboratory measurements of aerospace samples with the same optical instrument to aid with spectral characterization and identification. The full data set of laboratory spectra of aerospace materials will be published in the future and made accessible to the general public.

Furthermore, spectrum of 2020 SO was observed at Observatorio del Roque de los Muchachos at La Palma, Canary Islands and had been correlated with spectra of CZ-3B upper stage rockets, demonstrating some clear similarities.

Finally, relative flux and spectral slope relationship have been analysed over regions of blue, visible and red wavelengths for laboratory measurements, spectra of space objects observed in Zimmerwald and 2020 SO spectrum measured at La Palma. Some preliminary assumptions can be estimated already, however more data is required to define the gradient-flux space.

This analysis will be enhanced in the future by incorporating spectra from all laboratory measurements. Spectra measured at different incident angles and in different material combinations will be included to derive a spectral map which will fully describe the gradient-flux space. Consequently, spectrum of any artificial space object can then be evaluated according to where it fits within this spectral map. And if the object is observed multiple times, during targeted observational campaigns, as its attitude and reflecting geometry changes, the spectral behaviour can be mapped to the behaviour simulated in a laboratory.

References

Abercromby, K., Abell, P., & Barker, E. 2009, Reflectance spectra comparison of orbital debris, intact spacecraft, and intact rocket bodies in the GEO regime, European Space Agency, (Special Publication) ESA SP, Volume 672

Bedard, D., & Wade, G. 2014, Interpretation of Spectrometric Measurements of Active Geostationary Satellites, Advanced Maui Optical and Space Surveillance Technologies Conference, EID E31

Campbell, T., Furfaro, R., Reddy, V., et al. 2022, J Astronaut Sci, 69, 95

Cordelli, E., Lauber, P., Prohaska, M., et al. 2019, Recent Developments at the Swiss Optical Ground Station and Geodynamics Observatory Zimmerwald, ESA Space Safety Programme Office, Proceedings of 1st NEO and Debris Detection Conference

ESA eoPortal 2023, LAGEOS (Laser Geodynamics Satellite-I) / LAGEOS-II, `https://www.eoportal.org/satellite-missions/lageos#spacecraft`

Laszlo, R., Musilová, M., & Koleda, M. 2023, Needronix, `https://needronix.eu/`

Petrova, N. S., Devyatkin, A. V., Gorshanov, D. L., et al. 2021, JPCC, 2103

Rodríguez-Villamizar, J., Vannanti, A., Schildknecht, T., et al. 2023, Update on the Attitude State and Motion of Space Debris from Ground-Based Observations at the SwissOGS, 2nd NEO and Debris Detection Conference

Schildknecht, T., Vannanti, A., Krag, H., Erd, C. 2009, Reflectance Spectra of Space Debris in GEO, Advanced Maui Optical and Space Surveillance Technologies Conference, EID E24

Seitzer, P., Lederer, S. M., Cowardin, H., et al. 2012, Visible light spectroscopy of geo debris, in Proc. of AMOS Technical Conf., Maui, Hawaii, USA

Shelyak Instruments 2020, Demetra software, `https://www.shelyak.com/software/demetra/?lang=en`

Žilková, D., Šilha, J., Matlovič, P., & Tóth, J. 2023, Adv Space Res, 71 (8)

Astronomy and Satellite Constellations: Pathways Forward
Proceedings IAU Symposium No. 385, 2026
P. Grimley, J. Barentine, D. Galadí-Enríquez & C. Walker
doi:10.1017/S1743921324000504

Space surveillance as a by-product of space-based astronomical observations with CHEOPS

Nicolas Billot[1], Stephan Hellmich[2] and Willy Benz[3]

[1]Observatoire astronomique de l'Université de Genève, chemin Pegasi 51, 1290 Versoix, Switzerland
email: nicolas.billot@unige.ch

[2]EPFL Laboratory of Astrophysics (LASTRO), Chemin Pegasi 51, 1290 Versoix, Switzerland

[3]Physikalisches Institut, University of Bern, Sidlerstrasse 5, CH-3012 Bern, Switzerland

Abstract. The CHaracterising ExOPlanet Satellite (CHEOPS) is a partnership between the European Space Agency and Switzerland with important contributions by 10 additional ESA member States. It is the first S-class mission in the ESA Science Programme. CHEOPS has been flying on a Sun-synchronous low Earth orbit since December 2019, collecting millions of short-exposure images in the visible domain to study exoplanet properties.

A small yet increasing fraction of CHEOPS images show linear trails caused by resident space objects crossing the instrument field of view. CHEOPS' orbit is indeed particularly favourable to serendipitously detect objects in its vicinity as the spacecraft rarely enters the Earth's shadow, sits at an altitude of 700 km, and observes with moderate phase angles relative to the Sun. This observing configuration is quite powerful, and it is complementary to optical observations from the ground.

To characterize the population of satellites and orbital debris observed by CHEOPS, all and every science images acquired over the past 3 years have been scanned with a Hough transform algorithm to identify the characteristic linear features that these objects cause on the images. Thousands of trails have been detected. This statistically significant sample shows interesting trends and features such as an increased occurrence rate over the past years as well as the fingerprint of the Starlink constellation. The cross-matching of individual trails with catalogued objects is underway as we aim to measure their distance at the time of observation and deduce the apparent magnitude of the detected objects.

As space agencies and private companies are developing new space-based surveillance and tracking activities to catalogue and characterize the distribution of small debris, the CHEOPS experience is timely and relevant. With the first CHEOPS mission extension currently running until the end of 2026, and a possible second extension until the end of 2029, the longer time coverage will make our dataset even more valuable to the community, especially for characterizing objects with recurrent crossings.

Keywords. Satellite, space debris, mega-constellation

1. The CHaracterising ExOPlanet Satellite (CHEOPS) mission

The European Space Agency (ESA) and the CHEOPS Consortium have partnered up to build and operate the first S-class mission of the ESA Science program. The CHEOPS Consortium is composed of 11 European countries, led by the University of

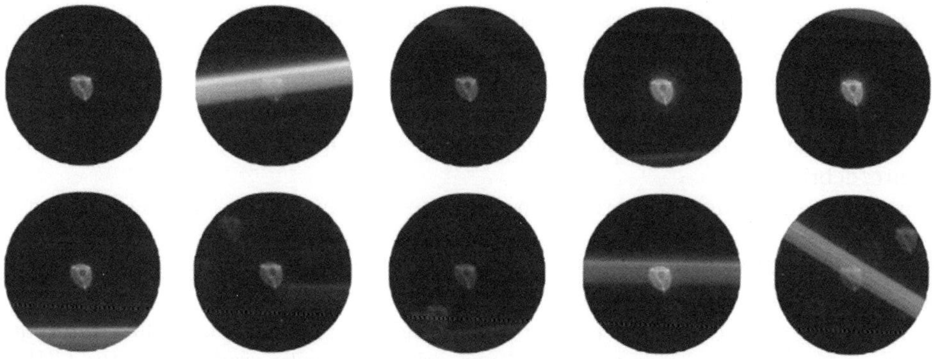

Figure 1. Gallery of satellite/debris trails observed in CHEOPS science images showing bright, faint, grazing, partial and multiple trails.

Bern, Switzerland. CHEOPS was launched from the European Space Port in Kourou, French Guyana, by a Soyouz rocket on 18th December 2019. CHEOPS was successfully injected into its nominal Sun-synchronous orbit at 700 km altitude with a Local Time of Ascending Node of 6 a.m..

CHEOPS was designed to observe and characterize exoplanets with an exquisite precision of 10 parts-per-million (ppm) on targets brighter than GAIA magnitude 9. CHEOPS is equipped with a telescope of effective aperture 30 cm, operating in the visible range between 400 and 800 nm. CHEOPS' observing mode is to stare at one star at a time, for periods typically ranging from hours to days, to acquire a series of short-exposure images from which the target flux is measured precisely. Resulting light curves are exploited to detect variations that are primarily caused by planet(s) orbiting the host star. Planetary parameters such as radius, refined ephemeris, atmospheric properties can then be extracted from such light curves.

A complete description of the CHEOPS mission, including its platform, payload, ground-segment organization as well as its science program is given in Benz, Broeg, Fortier et al. (2021).

2. Resident space objects crossing CHEOPS' field of view

In early January 2020, only a few hours after having opened the telescope tube cover, still during the commissioning phase, we observed for the first time a bright object that had crossed the field-of-view of the instrument at high velocity leaving behind a trail across the entire image. Since then, we have observed thousands of those, several per day, affecting the science images of CHEOPS. Figure 1 shows a gallery of such trails that we have collected.

The feature at the center of each image in Figure 1 is the target star that was being observed at the time of the object crossing. As the CHEOPS instrument has been intentionally defocused by design to enhance its sensitivity, the Point Spread Function (PSF) spreads over 16 arcsec in radius and it has a complex structure, with a roundish halo and three bright spots due to the 3-legged structure that holds the secondary mirror. Details of the CHEOPS PSF and the overall instrument performances can be found in Fortier, Simon, Broeg et al. (2024).

3. Automatic detection algorithm

In the present analysis, we are considering all and every CHEOPS science observation carried out over a period of over 3 years covering March 2020 until May 2023. This represents over 1.25 million images collected over this period as part of the science program.

These images are 200 arcsec across. They have been acquired with exposure times ranging from 1 millisecond for the brightest target stars to 60 seconds for the faintest ones, with a typical value around 30 seconds.

To cope with the large number of images, we have set up an automatic pipeline that scans each individual image and searches for linear features characteristic of satellite/debris trails.

Some pre-processing steps are required to ease and make the trail detection process more robust:

- Apply a median filter to the image to remove hot pixels
- Normalise the image to increase its contrast
- Apply a Canny edge detection to further increase the contrast. This step requires the fine tuning of thresholds to detect appropriate intensity gradients in the CHEOPS images and obtain the expected results.

We finally execute a Hough transform on the pre-processed images to detect and characterize trails (extract position angle and location on the image). A dedicated completeness analysis of the trail detection process is underway. We have nevertheless already identified the root cause of most false positive detections:

- Excessive straylight from the Earth dayside may generate structures in the image background, e.g., linear ripples. Those images are excluded from the analysis.
- Edge-on cosmic ray hits lead to linear features in images. Those features can easily be filtered out as they are about 1-pixel wide, significantly thinner than a trail caused by a crossing debris.
- Crowded stellar fields, in combination with the extended structure of the CHEOPS PSF, may lead to false detections. These erroneous trail detections occur solely at 45 and 135 degrees, they are thus straightforward to filter out.
- Rare cases of saturation lead to vertical features that can be mis-identified for debris trails. Filtering out trails with position angles equal to zero solves this issue.
- CHEOPS' camera being shutterless, images containing very bright stars suffer from vertical smearing features. We can remedy this issue as above, filtering out images with 0 degree position angles.

The last caveat we have identified in our detection process concerns multiple trails. Those are currently considered as single trails with erroneous position/angle. Note that those represent only a small fraction of all trails seen in CHEOPS' field-of-view, estimated to a few percent at most, thus not biasing the statistical analysis presented below. Out of the 1.25 million images scanned with our pipeline, and after filtering out false positives, we have identified a total of 3200 images that contain a trail due to a satellite or space debris crossing the field-of-view. This represents a mere 0.26% of the whole dataset, and this confirms that the CHEOPS Science program is not currently at risk due to these unwanted trails. This relatively low value, compared to the few percent quoted in e.g. Kruk, García-Martín, Popescu et al. (2023) for the Hubble Space Telescope, can be explained by the very small field-of-view of CHEOPS as well as the relatively short exposure times compared to those typically used with Hubble.

4. Dataset analysis

With thousands of trail detections, we have a large enough sample to explore possible trends on these crossing events. In the following sub-sections, we use the metadata available from CHEOPS images, like the line-of-sight (LOS) relative to the Earth horizon, to enrich our dataset and highlight statistically significant features.

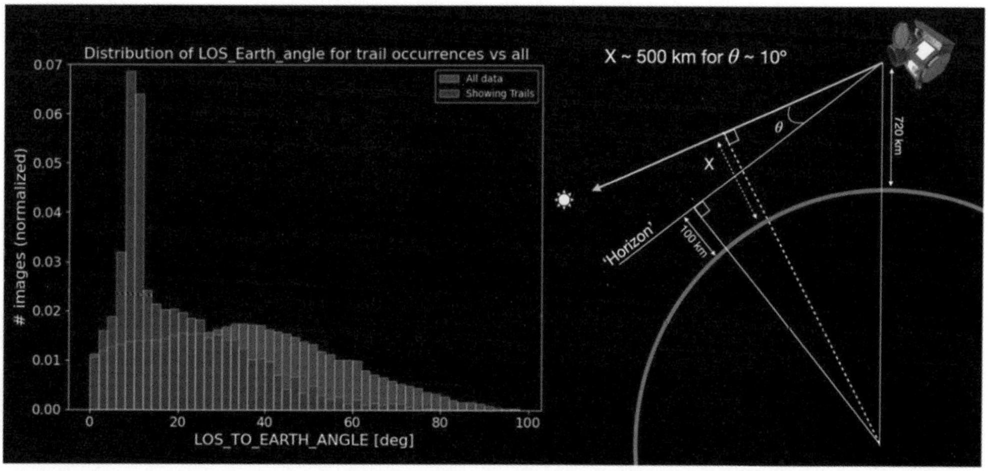

Figure 2. Distribution of angular distances between line-of-sight (LOS) and Earth horizon.

4.1. *Shell of debris in low Earth orbit*

Along its revolution around the Earth, CHEOPS' attitude (i.e., telescope line-of-sight) remains constant for the duration of an observation, typically for a few hours/days. CHEOPS therefore probes various angles above the Earth horizon during the course of an observation. Figure 2 shows the distribution of the angular distance between the line-of-sight and the Earth horizon.

The blue histogram shows the distribution for the entire dataset of 1.25 million images. The green histogram shows the distribution exclusively for images affected by trails. Both distributions are strikingly different, which means that trails occur preferentially at some specific angles above the horizon. In particular, the green histogram of Figure 2 shows a peak at around 10 degrees above the horizon which denotes an excess detection rate at a specific 'angular altitude' above the horizon. Depending on where along the line-of-sight the object is crossing, as depicted in Figure 2, an elevation of 10 degrees corresponds to an object orbiting at 500 km or above, typically between 500 and 700 km. This peak in the green histogram traces a high-density shell of satellites and debris, the so-called Low-Earth Orbit, that has been nicely probed by the CHEOPS line-of-sight over the past 3 years.

4.2. *Higher detection rates for small phase angles*

CHEOPS' orbit is Sun-synchronous such that the Sun always shines on the backside of the spacecraft to minimize straylight contamination (and power up solar arrays). Figure 3 shows the distribution of the angular distance between the line-of-sight and the Sun direction. Allowed values range from 120 to 180 degrees, meaning that we only observe within 60 degrees of the anti-Sun direction.

The blue histogram shows the distribution for the entire dataset of 1.25 million images. The green histogram shows the distribution exclusively for images affected by trails. Both distributions are different, with an excess detection rate at low phase angles, i.e., when the crossing object is illuminated from a direction close to the line-of-sight. We interpret this feature as meaning that crossing objects look globally brighter when the Sun shines on them from straight behind the observer. This is consistent with most phase functions that describe the dependence of scattered light versus the incident radiation phase angle.

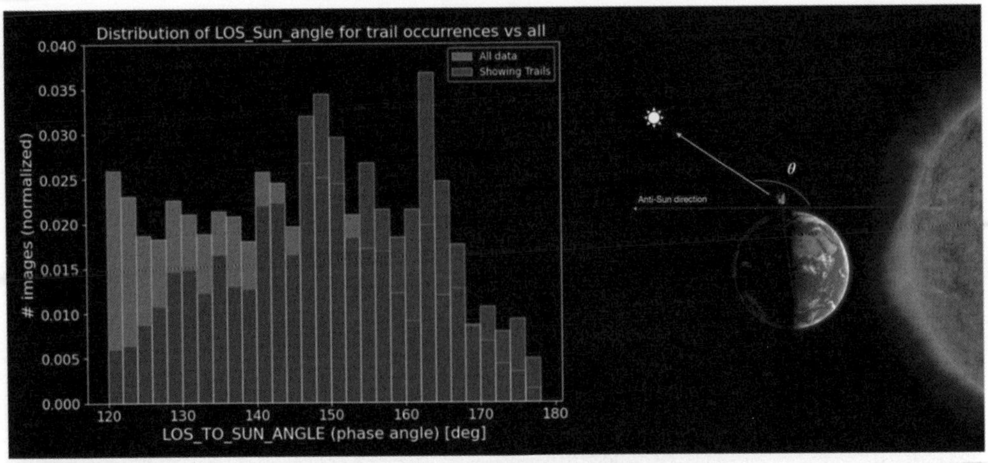

Figure 3. Distribution of angular distances between line-of-sight (LOS) and Sun direction.

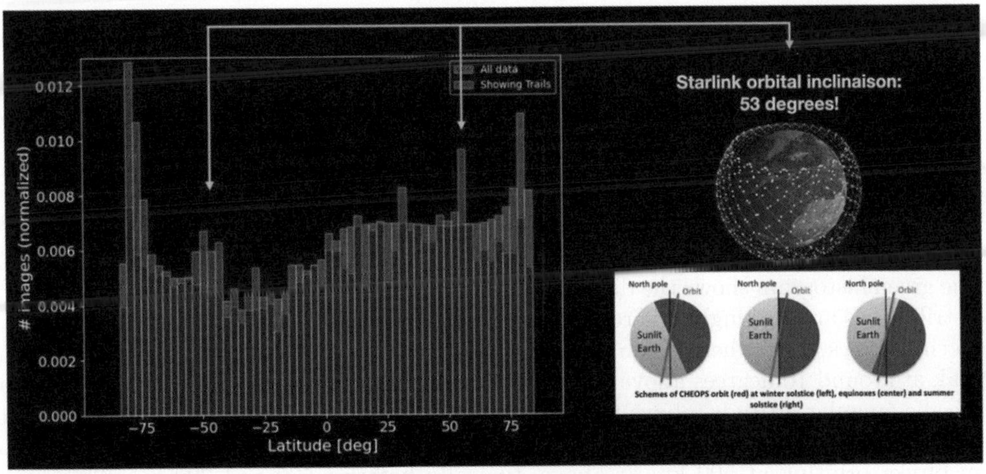

Figure 4. Distribution of latitudes at time of observations.

4.3. *Starlink imprint in geographic coordinates*

We now consider the geographic position of CHEOPS at the time of observation to search for trends and features. Figure 4 shows the distribution of latitudes at which CHEOPS was located when it was observing.

The blue histogram shows the distribution for the entire dataset of 1.25 million images. Note that, with an orbital inclination of 98.7 degrees, CHEOPS always remains below a latitude of 81.3 degrees. The asymmetry close to the poles is also due to the inclination of CHEOPS' orbit as the spacecraft spends on average more time on the day side of the Earth when close to the South Pole than close to the North Pole. As CHEOPS is very sensitive to stray light, images acquired close to the South Pole are often affected by stray light and therefore rejected on-board, thus leading to a deficit of images at very low latitudes. In addition, the dip in the blue distribution between latitudes 0 and -50 degrees is due to the South Atlantic Anomaly (SAA). When CHEOPS crosses the SAA, data are not recorded on-board because of a too-high level of cosmic ray hits on the detector.

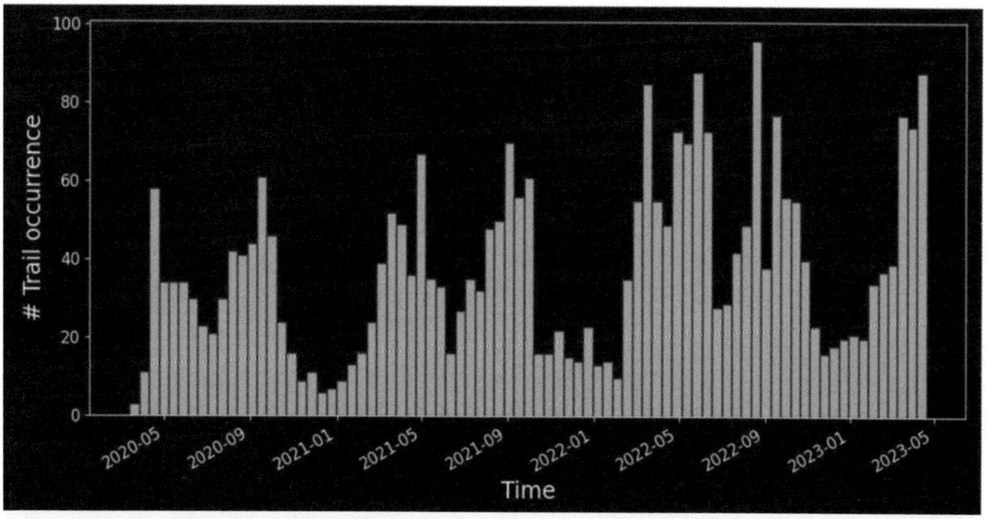

Figure 5. Trail detection number count in bi-weekly bins.

The green histogram shows the distribution of the same quantity but exclusively for images affected by trails. Both distributions are again quite different. The asymmetry between South and North Poles, also caused by CHEOPS' orbital inclination, goes in the opposite direction now. Close to the North Pole, CHEOPS spends a fraction of its time in the shadow of the Earth during winter months, in the so-called eclipse season. During this period, a fraction of the objects crossing the field-of-view of CHEOPS will go unnoticed as they would also be in the shadow of the Earth during observations. This explains the deficit of trail detections close to the North Pole. Conversely, close to the South Pole, objects crossing the field-of-view spend comparatively more time illuminated by the Sun, thus increasing the number of detections at those low latitudes. The green distribution also shows a couple of peaks at around -50 and +50 degrees latitude. Coincidentally a large fraction of the Starlink constellation have an inclination of 53 degrees, which means that Starlink satellites spends comparatively more time at latitudes ± 53 degrees than at other latitudes thus increasing the likelihood of being observed by CHEOPS at those latitudes. We therefore interpret the couple of features at ± 53 degrees latitude as being the imprint of Starlink.

4.4. *Increased trail occurrence rate*

With the injection of thousands of Starlink satellites in low-Earth orbit over the past couple of years, we expect to see an increase in the number of trails detected with CHEOPS. This is indeed what is observed and presented in Figure 5 where we plot the bi-weekly number count of trail detections. The histogram shows strong modulations due to seasonal effects, with a severe deficit of detections during winter months. This is related to CHEOPS entering the eclipse season, as mentioned in Section 4.3, where the spacecraft spends a fraction of its time in the Earth's shadow. During these winter months, many objects crossing the CHEOPS field-of-view will also be in the shadow of the Earth thus making them too faint to be detected in the images, and consequently reducing the number of detections. Summer months suffer from a similar effect, but to a lesser extent.

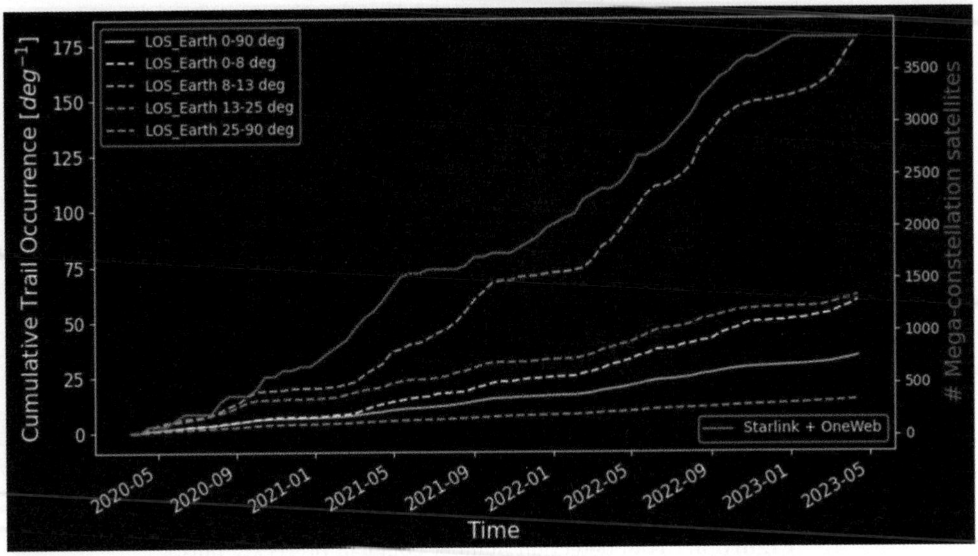

Figure 6. Cumulative distribution of trail occurrences for various angular distances between line-of-sight and Earth horizon, along with the number of Starlink and OneWeb satellites injected into low Earth orbit over the same period.

Independently of these seasonal effects, the overall trend seen in Figure 5 is indeed an increase in the detection rate. This trend becomes even clearer when considering specific 'angular altitudes' above the Earth horizon.

Figure 6 presents the cumulative distribution of trail occurrences for various bins of angular distances between the line-of-sight and the Earth horizon, as described in Section 4.1. The distribution is normalized by the bin size to represent a proxy of the debris/satellite density in each bin. When considering only trails detected far above the horizon, e.g., between 25 and 90 degrees, Figure 6 shows that the cumulative distribution is nearly linear. This means that the number of detections, per degree above the horizon, remains relatively constant, i.e., the population of objects detected in this zone has changed very little, if at all, over the monitored period. Conversely, the densest zone probed between 8 and 13 degrees above the horizon, corresponding to the peak in figure 2, shows the strongest increase in the number of trail detections. As this zone traces mostly the population of resident space objects in low-Earth orbit, we interpret this steep rise as a significant increase in the number of objects now populating this zone. This is consistent with the fact that Starlink and OneWeb have launched thousands of satellites into low-Earth orbit over the past couple of years, as shown in Figure 6.

5. Object identification

5.1. *TLE projection*

For a detailed analysis of each individual detection, the observational circumstances, such as observer-target distance, phase angle and angular velocity are required. To obtain this information, we linked each observed trail to the catalog of known space objects provided by the U.S. 18th Space Defense Squadron, published on the space-track website. Orbital information on space objects is usually provided by Two-Line Element sets (TLEs), that correspond to the mean keplerian orbital elements at a given epoch. This results in the accuracy of positions computed from propagating the TLEs to decrease with increasing distance from the TLE epoch. Since the CHEOPS data was acquired

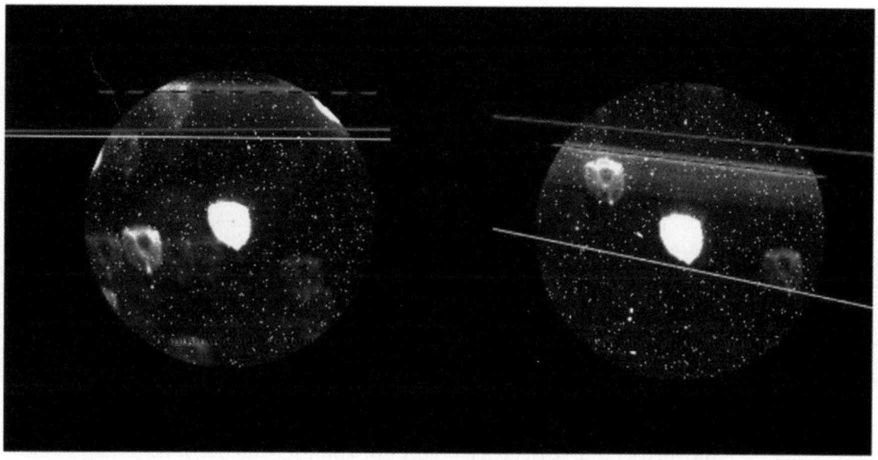

Figure 7. Trail detections (dashed red lines), projections of all cataloged objects that cross the field of view (solid lines) and the assigned known object (solid red line).

over the past 3 years, historic TLEs are required to identify the detected trails with cataloged space objects. The space-track API provides access to historic orbital information but only limited functionality to constrain the search. Further, the number of requests is limited which made it impractical to use the space-track API since we needed to link thousands of trails. This forced us to download the entire catalog of TLEs over the period of CHEOPS operations (March 2020 to June 2023), which contained 49 280 118 individual TLEs in total.

5.2. *Identification process*

For each CHEOPS image that contains a trail, we selected the TLE of CHEOPS that was closest to the observation time to compute the observer location, and for every cataloged space object, we selected its TLE closest to the observation time from an interval of 2 days around the observation epoch. This resulted in about 18 000 to 25 000 TLEs for each observation. Using the precise observation time and the astrometric solution stored along with the CHEOPS data, we were able to compute which of those objects were crossing the observed field during the exposure and projected their track into the image. This projection is of limited precision due to several reasons. Firstly, because the observer location and the positions of the cataloged space objects were computed from TLEs which are inherently imprecise, and the uncertainty of positions, computed by propagating the orbit, grows with increasing distance to the TLE epoch. Secondly because CHEOPS is nearly nadir-locked, i.e., slowly rotates in order keep its radiators always away from Earth, and this rotation is not accounted for in the astrometric solution stored along with the data. However, by defining margins for the identification, it was still possible to assign the majority of the trails to cataloged objects.

Figure 7 shows two examples of the trail identification. The dashed red line corresponds to the detected trail and the solid lines to cataloged objects crossing the field during the exposure. The red solid line corresponds to the cataloged object that was identified with the detected trail. In order to account for the inaccuracy of the projection, we used margins of 350 px for the distance and 4 degrees for position angle between trail and projected object for the assignment. The margins were chosen to account for an expected uncertainty in the positions computed from propagating the TLEs of about 1 km at 600 km distance and a rotation rate of CHEOPS of 6 degrees/minute on average. In case

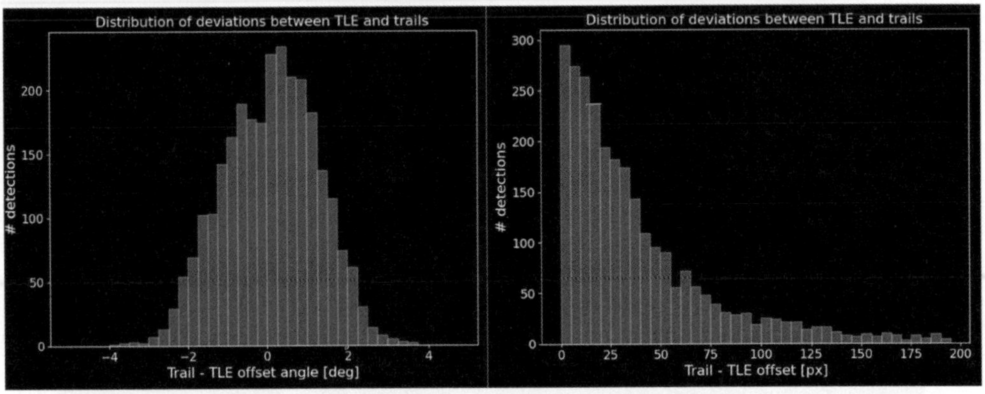

Figure 8. Distributions of the offset in position angle (left) and the distance between trail and projected trajectory (right) of the identified trails.

there were multiple known objects within these margins, the one closest to the detected trail was selected. The distribution of pixel offset and difference of the position angles of the identified objects is shown in Figure 8. The majority of identifications are actually much closer to the detected track than the accepted limits, suggesting that the TLEs selected for identification generally have good accuracy.

As already explained in section 4.4, we expect that a significant number of trails were caused by satellites from the Starlink constellation. Indeed, out of the 2742 trails that could be identified with known objects, 805 were caused by Starlinks, making the Starlink constellation responsible for about one third of the detected trails. Considering only trails caused by payloads (1799), Starlink satellites are responsible for almost 50% of the detected trails. We further identified 785 trails belonging to rocket bodies and 140 to orbital debris.

6. Conclusions

Astronomical observations have been increasingly affected by resident space objects over the past few years. Even sending telescopes into low-Earth orbit does not make them immune to intruding objects crossing their field-of-view. The current projection for the next decade, including all planned private and governmental mega-constellations, is that the low Earth orbit could host hundreds of thousands of satellites. This is two orders of magnitude larger than it is today. If CHEOPS were to still be in operations then, the fraction of affected images would go from a fraction of a percent to nearly a quarter of the collected images, which would seriously impair the science case of CHEOPS.

Despite its detrimental impact on observational astronomy, serendipitous observations of space debris by astronomical facilities can be turned into useful information. The dataset presented in this article is indeed very valuable for the Space Situational Awareness community because it provides in-situ observations of satellites and other space debris over a period of 3 years that covers the start of the mega-constellations era. Besides, the CHEOPS dataset is complementary to ground observations, and in a sense ahead of its time when we consider that several projects dedicated to in-situ observations of space debris are currently in development (Rainbow 2022; Cirillo, Michel & Utzmann 2023). We aim at making our dataset public via the International Astronomical Union's CPS† (Centre for the Protection of the Dark and Quiet Sky from Satellite

† `https://cps.iau.org`

Constellation Interference). This dataset could be used to refine observing strategies and mission concepts for these future endeavors.

References

Benz, W., Broeg, C., Fortier, A. et al. 2021, Exper Astro, 51(1), 109

Fortier, A., Simon, A. E., Broeg, C. et al. 2024, A&A, 687, A302

Kruk, S., García-Martín, P., Popescu, M. et al. 2023, Nat Astro, 7, 262

Rainbow J. 2022, Getting SSA off the ground, SpaceNews. `https://spacenews.com/getting-ssa-off-the-ground/`

Cirillo, G., Michel, M. & Utzmann, J. 2023, Airbus Robotic Telescope and SPOOK as test-bed for space-based space surveillance, 2nd NEO and Debris Detection Conference `https://conference.sdo.esoc.esa.int/proceedings/neosst2/paper/54`

Astronomy and Satellite Constellations: Pathways Forward
Proceedings IAU Symposium No. 385, 2026
P. Grimley, J. Barentine, D. Galadí-Enríquez & C. Walker
doi:10.1017/S1743921324000887

The impact and mitigation of satellite constellations on astronomical laser propagation

Julian C. Christou

Gemini Observatory / NSF's NOIRLab
email: julian.christou@noirlab.edu

Abstract. U.S. adaptive optics telescopes using laser guide stars for an artificial reference source are impacted by the presence of satellite constellations (SATCONs) because of the requirement to follow Department of Defense laser deconfliction protocol as administered by the U.S. Space Command's Laser Clearinghouse. The Laser Deconfliction protocol is designed to protect damage to satellites and their operation due to inadvertent illumination. As SATCONs begin to dominate the number of satellites in orbit they contribute highly to the number of closure windows when the astronomical lasers cannot be propagated. If left unchecked, the proposed number of SATCONs over the next decade will severely impact the laser open-shutter time, especially for the U.S. extremely large telescopes (US-ELTs) reducing the competitiveness of U.S. astronomical adaptive optics. Mitigation of the closure windows is therefore important.

Keywords. Adaptive Optics, Laser Guide Stars, Satellite Constellations

1. Overview

Laser guide star (LGS) adaptive optics (AO) is commonplace at AO-enabled telescopes around the world. The U.S. astronomical laser propagating community is required to follow the laser deconfliction protocols set up by the U.S. Department of Defense (DoD) and administered by the US Space Command's Laser Clearinghouse (LCH) (CJCSI 2023). The purpose of the laser deconfliction protocol is to protect potentially vulnerable satellites from inadvertent illumination. Observatories register their lasers with the LCH and the LCH creates a list of potentially vulnerable satellites for each observatory based on the laser power, wavelength, beam width, and pointing accuracy. Prior to an observing run using LGS AO, the observatories present a list of astronomical targets, identified by their on-sky coordinates (right ascension and declination), and the LCH provides each observatory with closure windows when the lasers are required to be shuttered.

Prior to the launch of the satellite constellations (SATCONs), beginning with SpaceX's Starlink in 2019, the impact of the closure windows on observational efficiency and productivity was considered to be minimal based on a 2010 study prepared by the Institute for Defense Analyses Science and Technology Policy Institute for the National Science Foundation (NSF) (Kramer 2010). However, a 2020 study prepared on behalf of the NSF and the U.S. Department of Energy (DoE) by the JASONs, an independent group of scientists which advises the U.S. government on matters of science and technology, on the impact of the SATCONs on observational astronomy estimated that the number of closure windows could increase by up to two orders of magnitude (Long 2020). This would significantly impact LGS operations of current and future telescopes systems including both U.S. extremely large telescope (US-ELT) candidates, the Thirty Meter Telescope and the Giant Magellan Telescope, which are planning to use LGS AO on a regular basis.

In 2015 there were an estimated 1400 active satellites in Low Earth Orbit. As of August 2022 there were 5311 active satellites, with that number growing consistently as more SATCONs are placed in orbit. This increase has seen the number of times optical astronomy observatories have had to pause observations increase by a factor of at least two. SpaceX, working with the NSF, has already waived the DoD LCH Laser protection, which has reduced the number of closure windows. While this waiver does not extend to other satellite constellation operators, especially those that don't require approval from U.S. regulation authorities, the Federal Communications Commission has recently added required coordination with NSF for optical astronomy to several satellite authorizations. OneWeb recently signed a coordination agreement with NSF and has agreed to also consider requesting a waiver from the LCH after an analysis of astronomical lasers on their system is completed.

To address this growing problem, representatives from the LCH, the US Air Force Research Laboratory's (AFRL's) Satellite Assessment Center (SatAC), the NSF, NSF's National Optical-Infrared Astronomy Research Laboratory (NOIRLab), NASA and the satellite industry including SpaceX, convened for a workshop at Vandenberg Space Force Base on 12–13 April 2023. The workshop concluded that the best solution would likely be to reclassify astronomical lasers from Category III to a lower category, although other potential avenues were also discussed. Category II requires the coordinates of the point of origin of the lasers, and the laser's parameters including power and wavelength, to be made available to satellite constellation operators so they can request mitigation if necessary. Category I declares a laser to be safe and not to pose a threat to satellites.

To determine which category is appropriate for which laser, the workshop concluded that the best way forward was for a Probabilistic Risk Assessment (PRA) to be conducted to determine the vulnerability of satellites from specific astronomical lasers. This risk assessment would then be submitted to the LCH for coordination. The PRA will be carried out by SatAC and the NSF are currently discussing logistics for this study with the AFRL. This study should begin in early 2024 with the technical details being coordinated between NOIRLab and SatAC.

To ensure that this process runs smoothly to the satisfaction of all parties, and to allow ground-based astronomy to continue to thrive, close collaboration between the NSF, satellite owner-operators, the Department of Defense, the Department of Commerce's Office of Space Commerce, and current and future observatories is essential.

A detailed discussion of the LGS lasers, the affected observatories, and DoD Laser Deconfliction can be found in the "Workshop on Mitigating the impact of Satellite Constellations on Astronomical Laser Propagation" Report published by NOIRLab to which the reader is directed (Christou 2023).

References

CJCSI 2023, Procedures for Management of Illumination of Objects in Space, CJCSI 3225.01B

Kramer, S. D. 2010, The Impact of Predictive Avoidance Restrictions on Astronomical Observatories, IDA Document: d-4202

Long, G. 2020, The Impacts of Large Constellations of Satellites, The MITRE Corporation: JSR-20-2H

Christou, J. C. 2023, Workshop on Mitigating the impact of Satellite Constellations on Astronomical Laser Propagation, NOIRLab: NOIR-03-DEV-RPT-000648

Astronomy and Satellite Constellations: Pathways Forward
Proceedings IAU Symposium No. 385, 2026
P. Grimley, J. Barentine, D. Galadí-Enríquez & C. Walker
doi:10.1017/S1743921324001042

Starlink in your backyard: A DIY approach to satellite monitoring and data analysis

Robert A. Watson

JBCA, Department of Physics and Astronomy,
The University of Manchester,
Manchester, M13 9PL, United Kingdom
email: bob.watson@manchester.ac.uk

Abstract. This work provides a detailed walk through the design, construction, and operation of a DIY monitoring station, emphasising its affordability, accessibility, and ease of use. The station leverages readily available components, including software defined radio (SDR) receivers and a Raspberry Pi. These tools allow for constant 24/7 monitoring and automatic storage of band usage and the detection of Doppler-shifts in the satellite beacons, providing valuable data for analysis.

Using this DIY setup can lead to indications of power levels and beam shapes of the Starlink satellites and user link strategy. Additionally, an experimental interferometer mode is being implemented, further expanding the capabilities of this home-based monitoring station with direct positional information.

Keywords. Software defined radio, radio astronomy, radio receivers, radio noise

1. Introduction

Some of visible impacts of Starlink mega-constellation satellites can be seen by any observers using their unaided eyes at dusk and dawn, and seeing sunlight reflecting off these satellites, especially with the long trains that occur in the days after launch. In the radio bands the impact is potentially greater, with satellites transmitting all the time over the whole sky, but it is less obvious. So the idea behind this project was to make an accessible and relatively inexpensive device to detect Starlink and OneWeb downlinks and so gain some idea of the cadence of satellites over an observation site. The Starlink and OneWeb down-link band is the same 10.7-12.7 GHz band as used by satellite TV, but unlike those fixed geostationary satellites, Starlink and OneWeb satellites are in low earth orbit (LEO), and can rapidly cross almost any part of the sky. Both these sets of satellites are as bright as the Sun as seen in their down-link bands, hence only a small antenna is needed to receive them. Given the same band, cheap low noise blocks (LNB) for satellite TV can be used to pick up and amplify their signals, while the rise of inexpensive software defined radio (SDR) devices can be used to tune in on those signals. Finally the development of affordable single board computers (SBC) such as the Raspberry Pi and their clones allows for the easy interfacing and data logging from those devices.

2. 3D printed antenna

One could just use a TV satellite dish, especially if one obtained a complete system second hand or a discarded one, given the shift from direct broadcast to streamed content via the internet. The disadvantage of such a dish is that it is cumbersome to mount and point for a beginner and has a 4° beam and so will miss a large part of the sky. The design here uses a cylindrical paraboloid, which gives a wide beam $\approx 40°$, the same size as the LNB feed, in one direction. In this way, it can be used as a transit instrument with the long axis set north-south and so it will pick up most of the satellites transiting overhead. Also it is easier to point, as it just needs to be set on a flat level surface. Furthermore, it is relatively easy to make by bending a flat metal plate or mesh into a parabola.

The first prototype was made from fibre board with a stiff plastic sheet bent into a parabola and kitchen foil stuck to it on the reflective surface. This made for a very inexpensive antenna, which survived a few months outside. Other materials such as plywood or laser-cut acrylic, which with aluminium plate or mesh would make a nearly permanent solution. Nevertheless, a 3D printed design was chosen, as hobbyist printers are common place now and the required parabola shape and focus can be built in. It is more robust, accurate, weatherproof and reproducible and the 3D STL model files are easily distributed.

The antenna was designed in Fusion 360 in several parts to be able to be printed on a small 22×22 cm printer, which can be glued together to make the full antenna and focus mount. The files and build instructions are on the Instructables website Watson (2023a) and should be at a suitable level for high school projects and interested hobbyists and therefore good for public engagement.

3. Software defined radios - SDR

With the increase in power and speed of modern digital circuitry, the analogue part of a standard radio receiver can now be implemented in digital electronics, although a good quality low noise front-end amplifier is still needed to boost the power level to feed the digital receiver. While SDRs did exist before, they were specialised and expensive, it was the amateur use of common low-cost DVB-T USB dongles with the Realtek controller and tuner which made them affordable and accessible. Such dongles can be bought for 10 to 20 pounds, with better versions for general use for 20-40 pounds. These have about 1 MHz of bandwidth and can be used to see the 11.325 GHz beacons on Starlink satellites and with kHz resolution they can easily see the change in their Doppler shifts as they go over. More expensive versions such as the HackRF or ADALM-PLUTO have 20 MHz bandpass, but are capable of fast scanning over the whole 2 GHz down-link band and cost in the region of 200 to 300 pounds. Further there are devices like the Kerberos and Kraken SDR, which have 4 and 5-port inputs for direction finding and interferometry that can locate a satellite within the large beam and are comparable in price to the other devices.

4. Single board computer and monitoring

A single board computer, as the name implies, is a complete computer built on single (usually credit card sized) circuit board, and used for development and educational systems, or for use as embedded controllers in industrial systems. At the moment the most popular and best supported is the Raspberry Pi, which can run full a Linux distribution such as Ubuntu server. Although the default Raspberry Pi OS, being a Debian-based Linux distribution, will also work and it uses the standard RTL-SDR drivers, while giving a more user-friendly desktop.

When coupled with a 512GB USB SSD storage it allows for a relatively cheap platform for 24/7 monitoring. RTL-SDR software provides continuous scanning such as rtl_power and hackrf_sweep which can be piped into python routines to partially process and package up in hourly datasets. Crontab schedules downloading of TLE from Celestrak, hourly calculations of expected transits for the next hour and plots everly 10 minutes which are placed on a light weight webserver. All the scripts, software and install guide are available at Github Watson (2023b).

Data are produced by piping the output of the command line scanning application "hackrf_sweep" into a python script. With this IF band is scanned from 0.25 to 2.35GHz 4 times a second at 0.1MHz resolution, but this gets averaged down to 5MHz (420 frequency bins). Meanwhile the rtl_sdr 1MHz bandwidth device (Kerberos in this case) scans using the "rtl_power" command to cover the 1 MHz band centred on 11.325 GHz at 1 kHz resolution (1024 frequency bins) every second, piped into another python script. These python scripts slice up and redirect the data stream to different directories per hour and per day. The Linux facility "crontab" downloads TLEs, creates directories and produces plots every 10 minutes which are shown on a simple website

To promote the mining of such data, all the data from the month of March 2023 have been made available on Zenodo (Watson 2023c). This includes zip files of all the hourly HackRF and Kerberos data files, 10 minutes plot, orbital TLE files from the day and csv files of expected satellite transits. This could be useful for trying automated identification of satellites via machine learning for example.

5. Summary

The hope of this article is to provide reference material to people interested in being able to make a simple Starlink monitoring station for a few hundred pounds, and so increase the awareness of the impact of mega-constellations of satellites on radio astronomy. Rather than being a definitive guide and design for such a station, it is a starting point for the various parts that are needed; antenna, RF amplifier, receiver, logging computer, and software. It is hoped that interested groups and individuals will come up with their own projects and ideas. Indeed, with a group of such stations spread over many kilometres, one could start to reconstruct the size and shape of the downlink beam.

References

Watson, R. A. 2023a, "Simple 3D Printed Starlink Antenna for Monitoring", Instructables, `https://www.instructables.com/Simple-3D-Printed-Starlink-Antenna-for-Monitoring`

Watson, R. A. 2023b, "Starlink_Monitoring_station", Github, `https://github.com/DrWhatson/Starlink_Monitoring_station`

Watson, R. A. 2023c, "Raw Starlink/OneWeb monitoring data from Altrincham UK for March 2023 - Test for IAU 385 (v1.0) [Data set]", Zenodo, `https://doi.org/10.5281/zenodo.8400658`

Astronomy and Satellite Constellations: Pathways Forward
Proceedings IAU Symposium No. 385, 2026
P. Grimley, J. Barentine, D. Galadí-Enríquez & C. Walker
doi:10.1017/S1743921324000693

The effects of satellite constellations on amateur astronomy

Paul A Daniels

Federation of Astronomical Societies, UK RAS Megaconstellation Working Group (Optical), UK
email: paul@qsoft.ltd.uk

Abstract. This paper presents some results from a survey carried out by the UK's Federation of Astronomical Societies to determine the effect of satellite constellations on (principally) amateur astronomers' observations, their mitigation methods and the effectiveness of those methods.

Keywords. FAS, Survey, Satellites, Constellations, Megaconstellations

1. Introduction

The survey described here was originally conceived as a UK-based follow-on from the survey conducted by Arion & Larsen (2021) for the SATCON2 workshop but with some additional questions to account for the larger number of satellites and the observing experience gained over the additional couple of years, and to exploit the author's access to a large number of amateur astronomers in the UK. The survey was originally intended to be only for the approximately 12 000 individual members of the around 200 member societies of the Federation of Astronomical Societies (FAS) but, following encouragement from Meredith Rawls and Olivier Hainaut, the scope of the survey was expanded to be global and thanks are due to them and to Federico Di Vruno for helpful suggestions on improving the questions asked.

1.1. *The survey, setup and publication.*

The JotForm system, https://www.jotform.com, was used to create, publish and host the survey. The survey form was developed and tested using the free 'Starter' option that permits access to all of the features but is limited to 500 form submissions in total and no more than 100 per month - ample for development and testing.

The survey comprised 32 questions of different types:

- Single line free text input,
- Multi-line free text input,
- Single-choice from multiple options and
- Multiple-choice from multiple options

Single- and multiple-choice questions could also have an 'Other' option allowing the responder to enter a free-text response. In addition, some question sections could be made conditional on earlier responses to avoid the responder being asked irrelevant questions.

Figure 1 shows the JotForm form builder used to create the form, including setting up the conditional question logic, the form when published and the data from the form submissions being deposited as rows in a spreadsheet in the FAS' Google Drive.

Following development and testing, a one-month 'Bronze' basic subscription was taken out and the form was published on 26 August 2023. The survey officially closed on

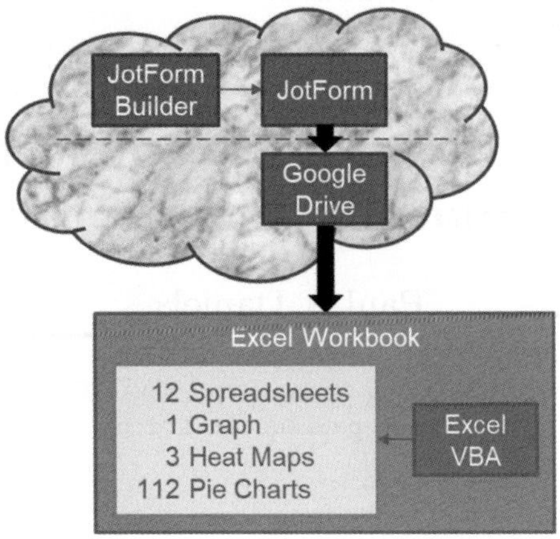

Figure 1. The JotForm system

22 September 2023 (but five valid responses after the closing date were accepted) for a total of 547 received responses. Even after the expiry of the JotForm subscription, the Google Drive data remained accessible.

1.2. *The data*

An Excel workbook was created to download the data from the Google Drive spreadsheet (read-only). Each submitted form occupies a single spreadsheet row with one or more columns representing responses to questions. The single- or multi-line text responses occupy a single spreadsheet cell within a row. For single- or multi-selection multiple-choice questions, the selected option labels (in full as shown on the survey form) are concatenated with a carriage-return, CHAR(10), as separator; free-text given as a response to an 'Other' option is similarly concatenated with any other question selections.

1.3. *The analysis*

Some of the analysis was performed by formulae in the spreadsheet cells but most of the analysis was performed using Visual Basic for Applications (VBA) code which was developed even whilst the survey was in progress. The correctness of the VBA code was checked by carrying out a parallel manual analysis of the data and comparing it with the results from the VBA coded analysis. The ability to swiftly re-load and re-analyse the survey results meant that, when the survey closed, the analysis results were immediately available.

One of the principal goals was to compare results from different latitudes. Survey question 3 (Q3) required the responder to provide their latitude in the form, *e.g.*, 51 N or 35 S by using an entry template to constrain the final letter to only be either 'N' or 'S'. The intention was to convert these to, respectively, +51 or −35 in the source data once loaded into the analysis workbook.

Unfortunately, it was subsequently discovered that the entry template also permitted the N/S letter to be omitted leaving only the placeholder underscore character. In total, 26 of the latitudes were ambiguous in this manner and two other latitudes were considered to be incorrect (one was for 00 N and the other 01 N). Fortunately, either (a) the responder was known to the author, (b) the responder confirmed their hemisphere by email, (c) the responder's email domain was btinternet.com (a UK provider) or, in one case, the email address was clearly Greek or (d) in just six cases, the numeric part of the provided latitude was in the range for UK latitudes and so was assumed to be North.

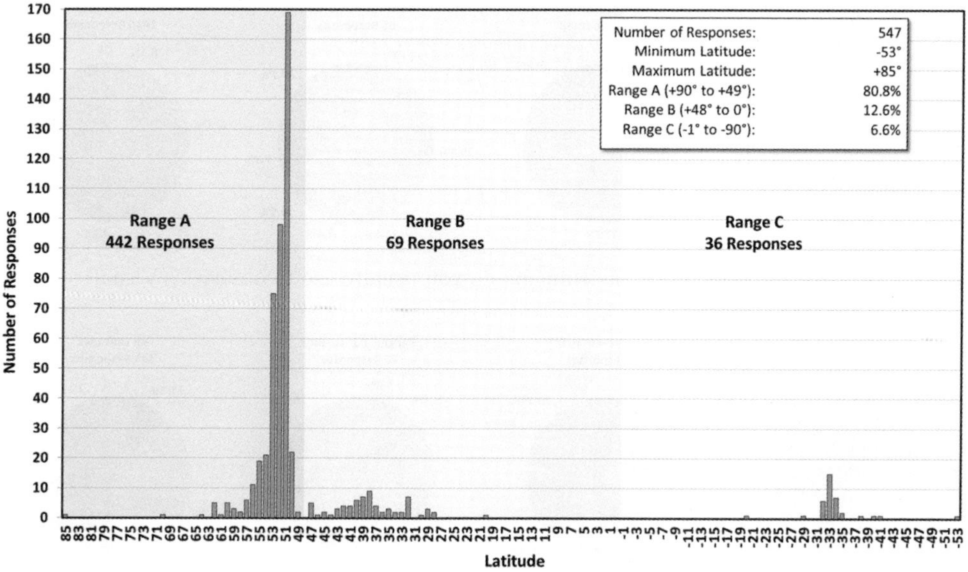

Figure 2. Latitude Distribution of Responders

A lookup table was created with the correct latitudes and was used in a post-loading, preparatory step to adjust the supplied latitudes in those 28 instances.

Figure 2 shows the distribution of responder's latitudes. There is a convenient division in the data between the latitudes for the UK, northern Europe and Canada and more southerly latitudes for the USA and southern Europe down to the equator. There is a single, distinct data group in the southern hemisphere (principally the Astronomical Society of New South Wales and the Northern Sydney Astronomical Society, both in Australia). These latitude groups have been labelled Range A (+90° to +49°), Range B (+48° to 0°) and Range C (−1° to −90°) respectively. The analysis divides the responses to questions between these latitude ranges and also, where reasonable, aggregates them into an 'All' range.

Unfortunately, owing to the quantity of results and limited space for publication, only a small selection of the results can be presented in this paper. Some of the questions also have free-text 'Other' responses but there is no room to list those here.

The analysis produced:

- 12 Spreadsheets,
- 1 Graph,
- 3 Heat Maps and
- 112 Pie charts

A full description of the survey, all data (names and email addresses removed) and VBA code will be made available by the FAS and can also be obtained from the author at fas@qsoft.ltd.uk.

2. Results

There were generally good responses to most of the questions with the exception of a disappointing shortage of radio observer responses. In consequence, the results from the radio responders are mostly inconclusive and are not reported in this paper.

In a few cases the number of responses was low and it may be that the conclusions that can be drawn are less certain. Some of the questions may also have biased responses and these are discussed in the relevant question sections.

NB: for all of the following pie charts, the order of the items in the accompanying legend matches the pie slices in a clockwise direction from the top.

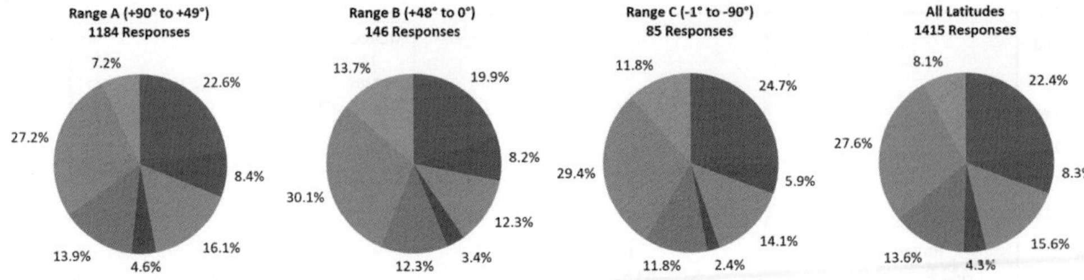

Figure 3. Q5a: What instruments do you usually use to observe? (Visual)

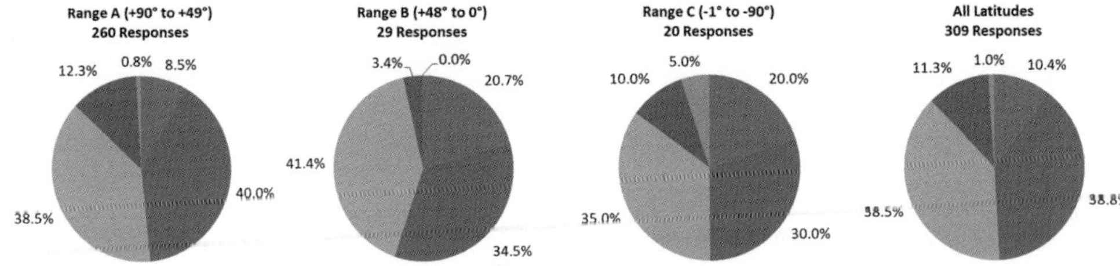

Figure 4. Q6: How significant are the effects of satellites and space debris, to your current naked eye observation?

2.1. *Q5a: What instruments do you usually use to observe? (Visual)*

This question asks the responders which instruments they use for visual observations either by naked eye or with the aid of binoculars or a telescope; Figure 3 shows relatively little difference between the latitude ranges. Examples of other instruments reported were Hα telescopes for solar work, meteor cameras, photometers for variable star work, radio meteor detection and spectroscopes.

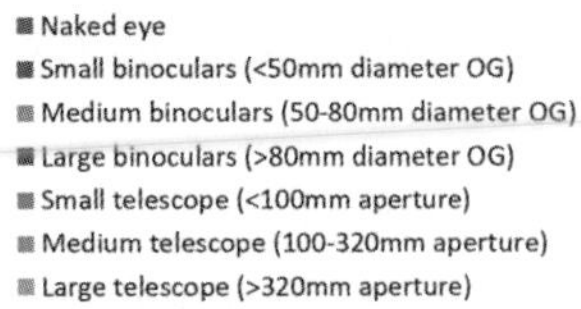

2.2. *Q6: How significant are the effects of satellites and space debris, to your current naked eye observation?*

It was not expected that there would be a significant reported impact to naked eye observation because of the limiting magnitude of about 5th or 6th magnitude for the naked eye, most satellites being just fainter than this limit, the ephemeral nature of such observing and the ability of the observer to simply ignore satellite intrusions.

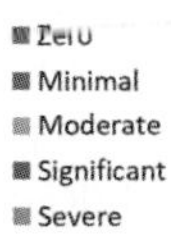

Figure 4 shows that 49.2% of the responses stated minimal or less impact, 87.7% stated moderate or less impact and only 12.3% stated significant impact or more. Some responders stated that satellites were not necessarily unwelcome and made the night sky more 'interesting'.

2.3. *Q8: How significant are the effects of satellites and space debris on your current binocular or visual telescopic observations?*

Compared to the responses for Q6, figure 5 shows that only 37.4% of the responses stated minimal or less impact, 82.4% stated moderate or less impact and 17.6% stated significant impact or more. This expected shift towards greater reported impact is probably due to the use of optical equipment permitting observation of the greater number of fainter satellites.

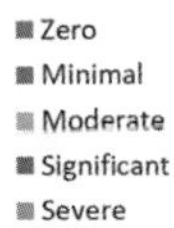

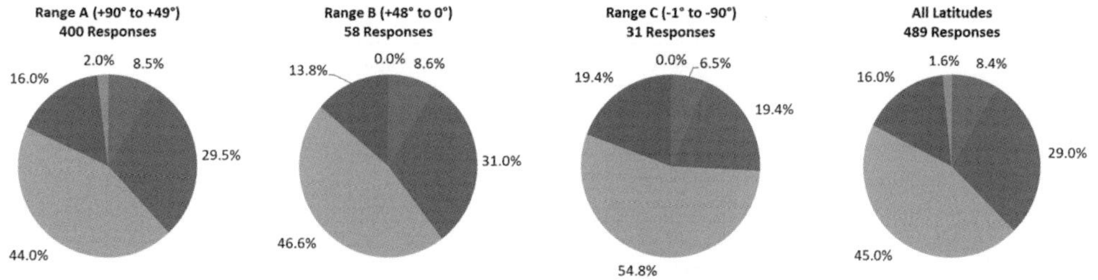

Figure 5. Q8: How significant are the effects of satellites and space debris on your current binocular or visual telescopic observations?

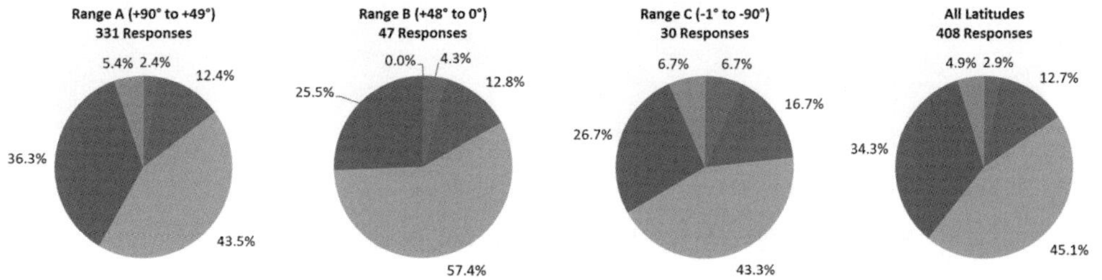

Figure 6. Q10: How significant are the effects of satellites and space debris on your current astrophotography?

2.4. *Q10: How significant are the effects of satellites and space debris on your current astrophotography?*

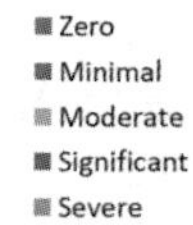

Compared to the responses for Q6 & Q8, figure 6 shows that just 15.6% of the responses stated minimal or less impact, 60.7% stated moderate or less impact and 39.2% stated significant impact or more. Again, this is a shift towards greater reported impact with a combination of image exposures and the use of optical equipment capturing fainter satellites over a period of time. Also, the observations are no longer ephemeral and captured images may reveal satellite trails that might have been missed by a visual observer.

There's a potential trend in response bias for both Q8 and Q10 in that those who have invested in optical equipment (and especially those who perform astrophotography) take their observing 'more seriously' and so are more likely to report a greater impact.

2.5. *Q11: What types of your images are most affected?*

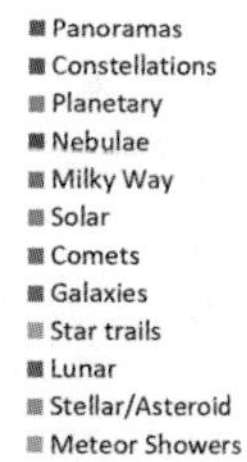

This asks which types of imaging are most affected by satellite trails. An inherent problem with this question is that the responses are a conflation of both the effects of satellites and the observing preferences of the responders and so doesn't simply reflect the problem posed for different observing targets. Figure 7 shows little variation between the latitude ranges. though there is a, perhaps insignificant, larger number of southern hemisphere observers responding with 'Lunar'. The author wonders what satellite problems were encountered by the twelve solar observers.

Table 1 groups celestial targets into, broadly, 'faint, fuzzy & extended', 'wide FOV' and 'bright or point source'. It shows a generally greater impact from satellites for extended objects requiring longer exposures.

Table 1. Q11: Distribution of responses for categories of observing target.

Category	%age
Nebulae, Milky Way, Comets & Galaxies	57%
Panoramas, Constellations, Star Trails & Meteor Showers	32%
Planetary, Solar, Lunar & Stellar/Asteroid	11%

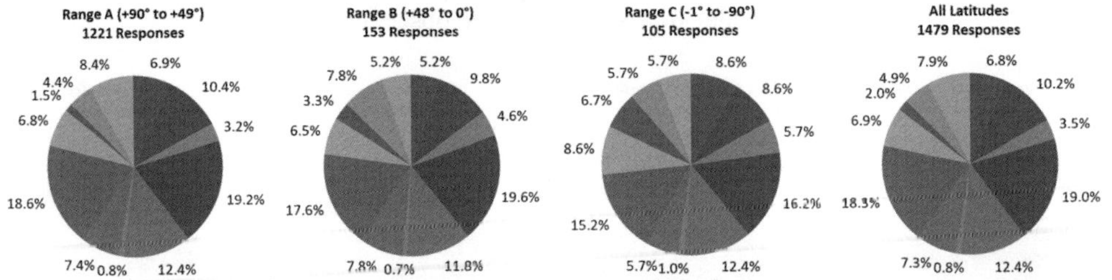

Figure 7. Q11: What types of your images are most affected?

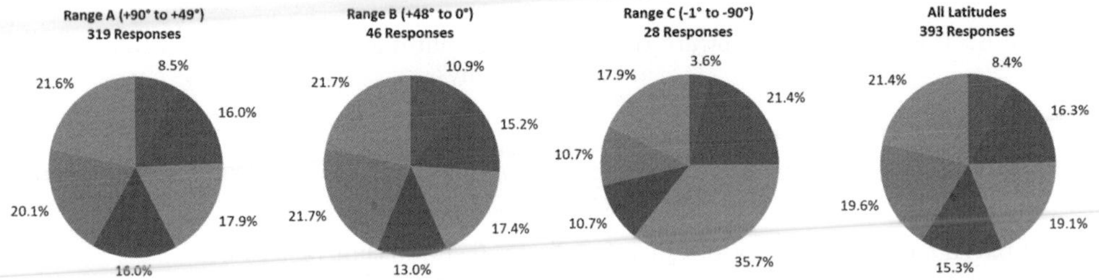

Figure 8. Q12: What is your estimate of the change in those effects compared to five years ago?

2.6. *Q12: What is your estimate of the change in those effects compared to five years ago?*

This question asked responders to subjectively compare their experience of satellites' impact with that of five years ago. In retrospect, perhaps this question should have asked for a comparison with three years ago as several of the responses pointed out they "weren't doing astronomy five years ago", "didn't take any notice of satellites back then" or they've "simply forgotten so can't make a comparison".

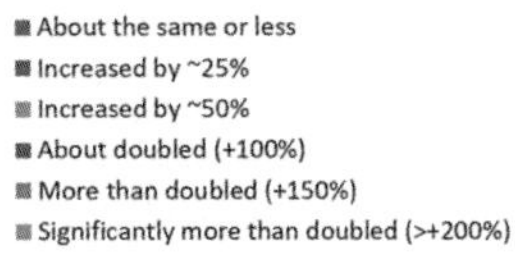

Figure 8 shows that the number of responses for Range C (and, to some extent, Range B) are too small to be conclusive but it seems odd that 42.4% of Range A and 43.5% of Range B responders report an increase of 50% or less but that 60.7% of Range C report same.

2.7. *Q13: What techniques do you use to mitigate the effects of satellite trails in your images?*

This asks how the responders mitigate the effects of satellite trails in their imaging. Figure 9 shows little difference between the latitude ranges, that image stacking is the principle mitigation (58.1% with imagers discarding unusable images) and trail masking

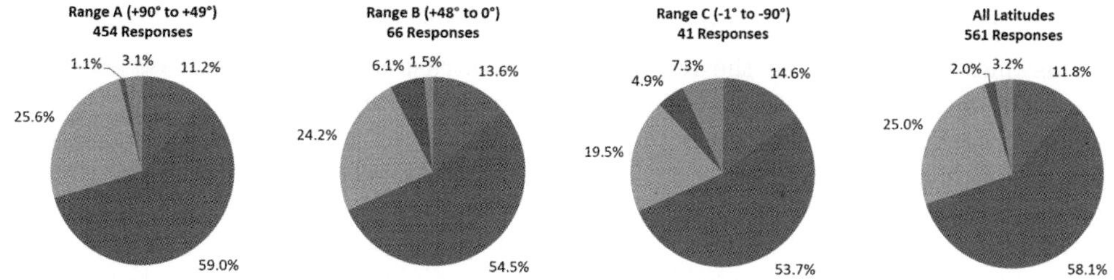

Figure 9. Q13: What techniques do you use to mitigate the effects of satellite trails in your images?

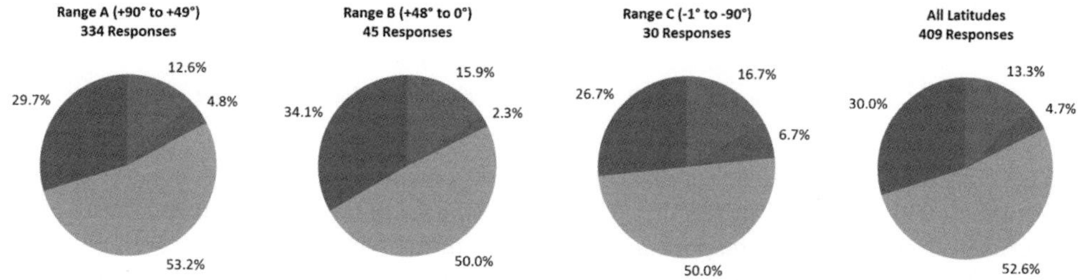

Figure 10. Q15: How effective are those mitigation techniques?

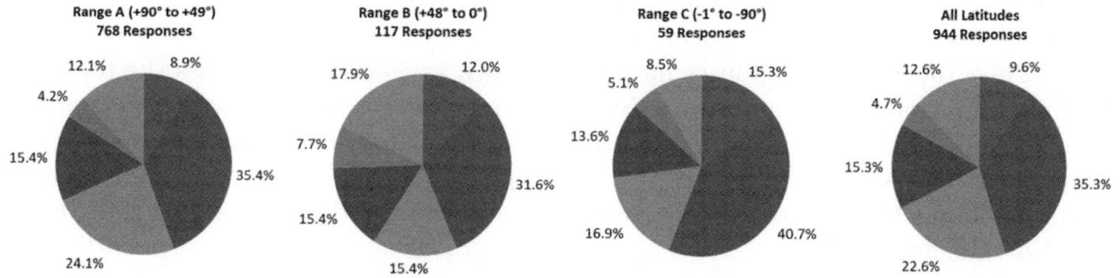

Figure 11. Q23: What time(s) of night do you find the satellites most intrusive?

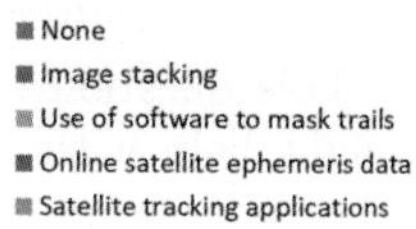

software came second (25%). Those using online satellite ephemeris data or satellite tracking applications formed a relatively small proportion of the responses (5.2% in total). This question allowed multiple options to be selected so some responders may be using more than one mitigation.

2.8. *Q15: How effective are those mitigation techniques?*

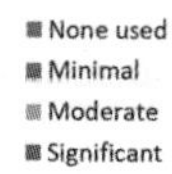

It's not detailed here but Q14 asks about the effort involved in applying the mitigation techniques in Q13 with 36.6% reporting minimal or less, 69.5% reporting moderate or less and 30.5% reporting significant or more. This question, Q15, asks about the effectiveness of the mitigation techniques reported in Q13. Figure 10 shows that 52.6% of the responders thought the mitigations moderately successful and another 30% significantly successful. This is heartening though it's difficult to judge whether the, often qualitative, improvements in images sought by amateur astronomers mean that they are more easily satisfied with the results of the mitigations than professional astronomers with their more demanding quantitative requirements.

2.9. *Q23: What time(s) of night do you find the satellites most intrusive?*

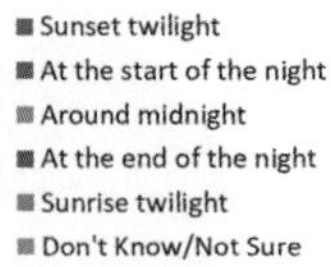

This question asks when the responder finds intrusion of satellites into their imaging to be worst. Figure 11 shows that nearly 57.9% of the responses reported those times to be between the start of the night and around midnight. There is some selection bias here because many amateurs rarely observe very much beyond midnight and so may be unaware of any problems between midnight and sunrise twilight. This may explain why the 9.6% affected before the start of night is about twice that of the 4.7% affected after the end of night. Range C responses were more biased towards 'before midnight' than Ranges A or B but, without knowing more about those responders' observing locations and conditions, it may be speculative to suggest that more accessible or clearer skies have produced this bias.

2.10. *Q24: What parts of the sky (Azimuth-Elevation) are most affected by the intrusion of satellites?*

This question asked responders to report which parts of the sky they considered were most affected by satellite intrusion. The question was presented as an 8-column by 3-row grid of check-boxes corresponding to eight azimuth compass points North clockwise through North West and three elevations High, Medium and Low. A further two check-boxes were available for 'Near the Zenith' and 'Don't Know/Not Sure'.

Figure 12 shows 'heat maps' for the three latitude ranges (an 'all-latitudes' version makes no sense). The dark hues represent the relatively most affected and the lighter hues represent relatively least affected parts of the sky.

The 'Near the Zenith' responses have been shown as a single bar across the top of the eight columns and the 'Don't Know/Not Sure' results are not included. For symmetry the left-hand column has been duplicated on the right and, to allow more direct visual comparison between the northern and southern hemisphere heat maps, the compass points for Range C (southern hemisphere) have been 'rotated' to be South-facing rather than North-facing.

2.11. *Q27: Were you aware of the availability of the following resources to help you mitigate the problem of satellite constellations on your observations?*

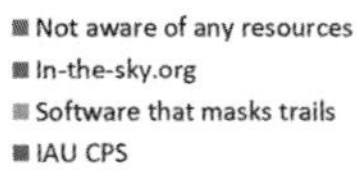

This question asks what resources the responder is aware of that might help with mitigation. Figure 13 shows a disappointing 40.6% were not aware of any resources (with an even greater proportion, 59.5%, in the Range C latitudes). 41.1% were aware of software for masking trails but only 7.1% were aware of the IAU CPS (but none in the Range C latitudes!).

2.12. *Q28: The introduction to this survey suggests that there could be a more than 40-fold increase in the number of LEO satellites. How aware were you of the plans to launch that many satellites into LEO?*

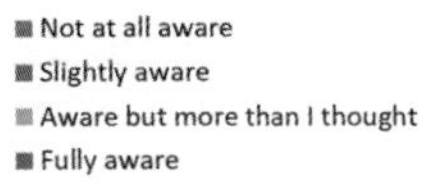

This asks whether responders were aware of the large number of proposed LEO satellites. Figure 14 shows a disappointingly low awareness of plans for such numbers, overall 27% for slightly aware or less (but 47.2% for Range C latitudes). 46.4% of respondents were aware of the growing problem but didn't realise how bad it could be.

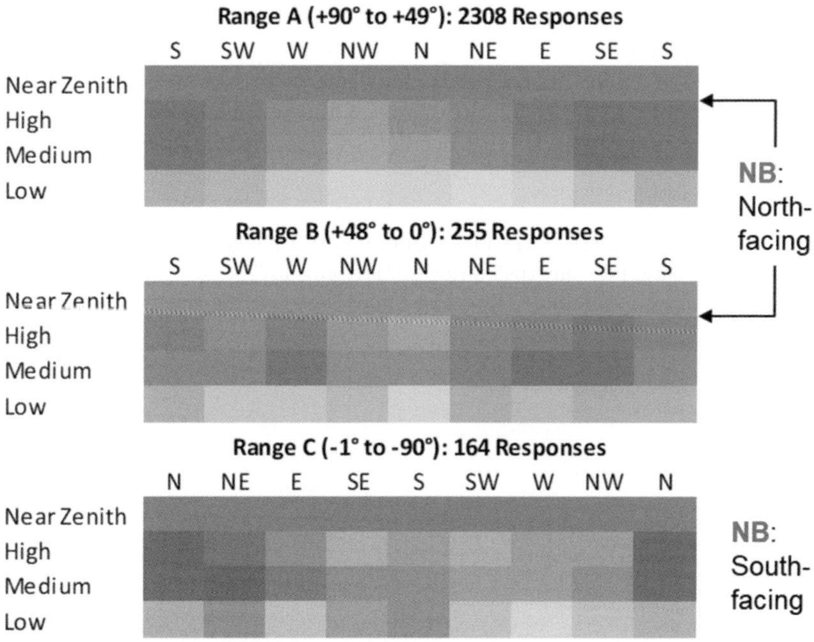

Figure 12. Q24: What parts of the sky (Azimuth-Elevation) are most affected by the intrusion of satellites?

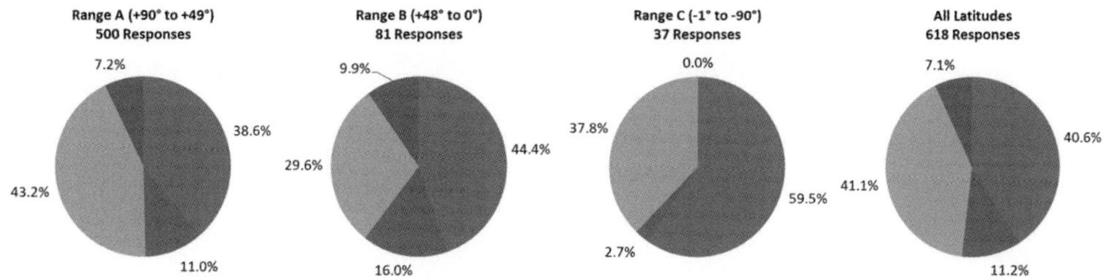

Figure 13. Q27: Were you aware of the availability of the following resources to help you mitigate the problem of satellite constellations on your observations?

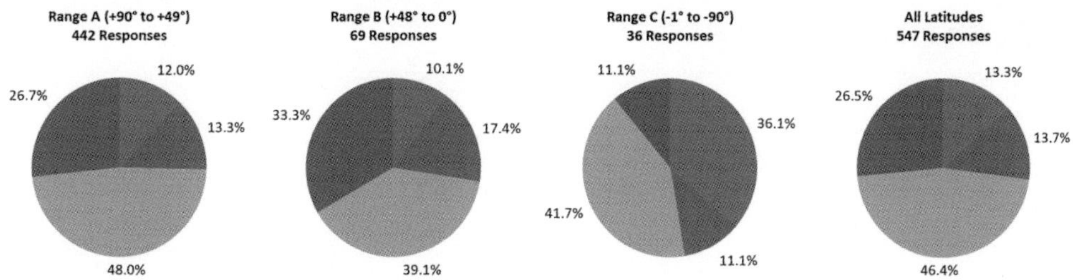

Figure 14. Q28: The introduction to this survey suggests that there could be a more than 40-fold increase in the number of LEO satellites. How aware were you of the plans to launch that many satellites into LEO?

3. Summary

Overall the survey was successful with good support from observers around the world. It would be interesting to repeat the exercise in a couple of years but to run the survey for longer and attempt to capture a more even dispersion of latitudes. In particular, the amount of data for radio observers was low and statistically inconclusive so it would be good to improve on that.

What has also come from this survey is that there is less awareness of the problem in southern latitudes both in terms of the scope of the problem and also the mitigations available: it may be that the IAU CPS needs to do more work there.

Reference

Arion, D. N. & Larsen, K. 2021, SATCON2: Community Engagement Working Group Report, https://baas.aas.org/pub/enwrps6k/release/1?from=48753&to=48758#survey-of-the-amateur-astronomy-community-regarding-impacts-of-satellite-constellations

Astronomy and Satellite Constellations: Pathways Forward
Proceedings IAU Symposium No. 385, 2026
P. Grimley, J. Barentine, D. Galadí-Enríquez & C. Walker
doi:10.1017/S1743921324001467

PASO - astronomy and space situational awareness in a dark sky destination

Domingos Barbosa[1], Bruno Coelho[2,5], Miguel Bergano[2,4], Constança Alves[1,3], Alexandre C. M. Correia[5,8], Luís Cupido[12], José Freitas[9], Luís Gonçalves[1,5], Bruce Grossan[14], Anna Guerman[7], Allan K. de Almeida Jr[2,4], Dalmiro Maia[4], Bruno Morgado[4], João Pandeirada[1,6], Valério Ribeiro[10], Gonçalo Rosa[11], George Smoot[14,15], Timothée Vaillant[2,4,5], Thyrso Villela[13] and Carlos Alexandre Wuensche[13]

[1]Instituto de Telecomunicações, Universidade de Aveiro,
Campus Universitário de Santiago, 3810-193, Aveiro, Portugal
email: dbarbosa@av.it.pt

[2]ATLAR, Ed. Multiusos, Rua Rangel de Lima, 3320-229 Pampilhosa da Serra, Portugal

[3]University of Aveiro, Campus Universitário de Santiago, 3810-193, Aveiro, Portugal

[4]CICGE, Faculdade de Ciências da Universidade do Porto, 4169-007 Porto, Portugal

[5]CFisUC, Departamento de Física, Universidade de Coimbra, 3004-516 Coimbra, Portugal

[6]Instituto Superior Técnico, Avenida Rovisco Pais 1, 1049-001 Lisboa, Portugal

[7]C-MAST, Center for Mechanical and Aerospace Science and Technology,
University Beira Interior, 6201-001 Covilhã, Portugal

[8]IMCCE, UMR8028 CNRS, Observatoire de Paris, PSL Université,
77 Avenue Denfert-Rochereau, 75014 Paris, France

[9]Dep. of Space, Armed Forces General Staff, Avenida Ilha da Madeira,
1449-004 Lisboa, Portugal

[10]KLM, Data & Technology - Platform Ground,
PO Box 7700, 1117 ZL Schiphol, the Netherlands

[11]Ministry of Defense, Avenida Ilha da Madeira, 1449-004 Lisboa, Portugal

[12]LC Technologies, Rua da Vila Verde 39, 3800-810 Eixo / Aveiro, Portugal
[13] Instituto Nacional de Pesquisas Espaciais, Divisão de Astrofísica, Av. dos Astronautas,
1758, 12227-010 - São José dos Campos, SP, Brazil

[14]Lawrence Berkeley National Laboratory, 1 Cyclotron Rd, Berkeley, CA 94720, United States

[15]University of California, 366 Physics North, MC 7300, Berkeley, CA 94720-7300, United States

Abstract. The Pampilhosa da Serra Space Observatory (PASO) is located in the centre of the continental Portuguese territory, in the heart of a dark-sky destination certified by the Starlight Foundation (Aldeias do Xisto) and has been an instrumental asset to advance science, education and astrotourism certifications. PASO hosts astronomy and Space Situational Awareness (SSA) activities including a node of the Portuguese Space Surveillance & Tracking (SST) infrastructure network, such as a space radar currently in the test phase using the GEM radiotelescope, a double Wide Field of View (WFOV) Telescope system, and a European Union SST optical sensor telescope. These instruments allow surveillance of satellite and space debris in LEO, MEO and GEO orbits. The WFOV telescope offers spectroscopy capabilities enabling light curve analysis and cosmic sources monitoring. Instruments for space weather are being considered for installation to monitor solar activities and expand the range of SSA services.

Keywords. Dark Sky, radioastronomy, space, space debris, space situational awareness

---

1. Early days: from GEM to SSA

The number of space activities, including the launch of constellations of satellites and the associated space debris population, has been steadily increasing, raising concerns over the impact on the night sky quality for astronomical and space science purposes. Astronomy and Situational Awareness (SSA) services have been developed at the Pampilhosa da Serra Space Observatory (PASO), in Pampilhosa da Serra municipality in the Central Region of Portugal since 2011. This space observatory has been instrumental to foster astronomy and space science, citizen science and astro-tourism, contributing to showcase the quality of the Dark Sky "Aldeias do Xisto" ("Schale villages") tourist destination. This dark-sky destination is named after its stone (shale) widely used in traditional construction across the region.

PASO took its origin after a site selection to install a radiotelescope for the Galactic Emission Mapping project in Portugal (GEM-P) (Fonseca et al. 2006), with a careful site survey on climate, low radio-frequency interference (RFI) levels, and infrastructure capabilities. The GEM project was an international collaborative effort in the fields of radio astronomy and cosmology. The project goal was to accurately determine the spatial distribution and absolute intensity in the radio and microwave spectrum of the large-scale synchrotron radiation emitted by the Milky Way and by the unresolved blend of external galaxies; see Torres et al. (1996).

The GEM experiment concept was pioneered by the Berkeley team to address the problems of cosmic microwave background (CMB) foreground cartography at the time of the detection of temperature anisotropies in the CMB radiation by the Differential Microwave Radiometers (DMR) onboard the Cosmic Background Explorer (COBE) satellite (Smoot et al. 1992), a major scientific breakthrough for cosmology. COBE DMR all-sky maps at 30 GHz, 53 GHz, and 90 GHz have shown the need to extend our knowledge of diffuse Galactic microwave emissions (Bennett et al. 1992), and eliminate foreground contamination from pristine cosmic signals. Since COBE, high resolution map-making made a huge leap forward from the data returned from more recent experiments like, MAXIMA (Hanany et al. 2000; Stompor et al. 2001), Boomerang (de Bernardis et al. 2000), DASI (Leitch et al. 2002), NASA's Wilkinson Microwave Anisotropy Probe (WMAP; Bennett et al. (2013), and the European Space Agency's Planck Surveyor satellite (Tauber et al. 2010; Planck Collaboration et al. 2020), that offer a direct glimpse into the physics at the cosmic surface of last scattering, providing constraints on cosmological parameters and tests of theories of large-scale structure formation, and favoring the inflationary paradigm†.

GEM was first designed as a portable, double-shielded 5.5-m radiotelescope on a constantly rotating ground platform in an alt-azimuth configuration to map 60-degree-wide declination bands from different observing sites with a slow circular scan pattern of the sky at zenith angles of 30°. This GEM design characteristic proved to be of great interest for space surveillance and tracking of satellites and space debris, as is discussed in Section 2. GEM made observations from Bishop, California; Leyva in Colombia, which is close

† For a comprehensive list of CMB and Foreground experiments: NASA Legacy Archive for Microwave Background Data Analysis (LAMBDA); `https://lambda.gsfc.nasa.gov/`

to the equator; and Teide, Tenerife, Spain; before settling in Cachoeira Paulista, Brazil, to maximize homogeneous sky observations.

Since the beginning of this project, total power maps of diffuse Galactic synchrotron emission at 408, 1465 and 2300 MHz were accomplished with total power receivers, with the later addition of custom-developed 5 GHz polarimeters to provide I, Q and U Stokes parameter maps (Bergano et al. 2011; Ferreira 2008). These frequencies were important to cover the frequency gap at centimetre wavelengths on synchrotron radiation. Meanwhile, other projects like C-BASS (Harper et al. 2022) followed and embraced similar concepts to improve our knowledge on diffuse Galactic emissions. To integrate for large sky areas, and since sensitivity is more important than resolution, the GEM scanning strategy consists of an azimuthal dish rotation until the required sensitivity is attained, thus allowing integration over large sky areas and providing useful foreground templates to CMB teams. The Moon was used to calibrate the antenna temperature scale, and the preparation of the map required direct subtraction and destriping algorithms to remove ground contamination as the most significant source of systematic error (Tello et al. 2013). To add the northern hemisphere coverage, and thus produce a template of most of the sky, Portugal's PASO location was assessed as a suitable location to host a 9-metre antenna to complement the original GEM southern hemisphere maps and enable future expansion towards observation of polarized Galactic diffuse emission. The GEM working frequencies were chosen because below 20 GHz, atmosphere contribution and polarization contamination are negligible. Improved mapping of the spectral and large-scale properties of Galactic synchrotron radiation would result from an observational strategy aimed to minimize ground contamination and long-term systematics to guarantee baseline uniformity. However, we acknowledge concerns that spectral occupancy of satellite communications in C and X bands could severely impact the map-making capability of future sensitive Galactic surveys.

2. Space surveillance & tracking

The GEM-P team used a 9-metre diameter Cassegrain Vertex antenna with alt-az mounting. This antenna was originally a telecommunications antenna for C-band operations with a King Post pedestal mounting. The new observational strategy and mechanical modifications allowed a steady, constant, slow azimuthal rotation of 0.3-1 rpm at a fixed elevation angle of 30° from zenith while allowing for travel speeds of up to 6 deg/sec in azimuth. The operational requirements for the GEM-P antenna with precise pointing at constant speed implied the need for key mechanical and structural changes and enabled novel observation strategies for space debris surveillance (Pandeirada et al. 2024). Further refurbishments mean the antenna is able to cope with fast angular speeds and responds to the operational requirements of LEO surveillance. Therefore, the equipment became included in the Portuguese node of the European Space Surveillance & Tracking program (EUSST)† under contract with the Portuguese Ministry of Defense (PT MoD). The system operates at 5.56 GHz with a beamwidth of 0.73° and aims to provide range and range-rate measurements of objects in LEO with RCS above 10 cm^2 at 1000 km (Table 1).

Additionally, PASO got the installation of a double wide field of view telescope system with a maximum FoV of 4.3° × 2.3° (see Fig. 1). The sensor consists of two small aperture (30 cm) telescopes. The system can observe in white light, in BVRI bands, and in the narrow bands Hα and O[III]. The equatorial mount has very fast speeds (max. slewing speed of 40°/s), enabling tracking capability of LEO objects. This design is meant to support science and space surveillance services with its own scheduling tools. Besides

† `https://www.eusst.eu/`

Table 1. PASO SSA instruments.

Radio Instruments	Frequency	Size	Mount	Type	Speed	FoV[3]
L-Band radiotelescope (Science, Outreach)	1.42GHz	5-m	Alt-Az	Front-fed parabolic		2.9°
C-Band space radar[1] (LEO (< 1000 Km))	5.56GHz	9-m	Alt-Az	Cassegrain	$\leqslant 10°$/sec	0.5°
Solar Radiospectrograph[2] (Space Weather)	0.1-1GHz	5-m	Alt-Az	Front-fed parabolic		2° (2GHz)
Optical Instruments	**Bands**	**Size**	**Mount**	**Type**	**Speed**	**FoV**
Twin WFOV Telescope[4] (LEO - GEO)	BVRI,Hα OIII	2×0.3 m	Equ	Cassegrain	$\leqslant 40°$/sec	$4.3° \times 2.3°$
WFOV Optical Telescope[5] (MEO/GEO)	I	0.28 m	Equ	RASA 11"	$\leqslant 6°$/sec	$\sim 2.6 \times 2.6°$

Notes:
[1] Former GEM radiotelescope.
[2] In commissioning by FCUP. Expected start of operations for mid-2024, pending Concept of Operations (CONOPS) protocol and final site agreement with PT MoD.
[3] FoV is taken as beamwidth at -3dB.
[4] It-UC-MDN partnership.
[5] EUSST MDN telescope.

Figure 1. Left: ATLAS, the PASO 5.56 GHz monostatic radar tracking sensor; Right: the PASO WFOV Double telescope System.

EUSST, the WFOV system was designed to enable observations of transient sources and provide regular monitoring of novae and other powerful cosmic nuclear explosions (Chomiuk et al. 2021) and contribute in the near future to surveillance networks of Time-domain Astronomy and gravitational-wave sources like the Global Rapid Advanced Network Devoted to the Multi-messenger Addicts (GRANDMA, Agayeva et al. 2022) pending a Concept of Operations (CONOPS) final agreement. The PASO location has excellent all-sky clearance on a hilltop at an elevation of 840 m above sea level and is surrounded by mountain ranges that allow protection from the light pollution of the coastal regions of the country. The conditions are among the best in continental Portugal for optical use in SST operations, where the sky background reaches 21 magnitudes per square arcsecond or darker with more than 200 clear nights per year on average (Coelho et al. 2021). Near-future sensor fusion concepts using space radar and very wide-field tracking telescopes for Space Surveillance and Tracking (SST), such as the PASO telescopes (Coelho et al. 2022), will provide additional information that is much needed to accurately model the orbital parameters and attitudes of satellites and debris (de Almeida et al. 2024).

Using several types of sensors in the same place as PASO has many benefits: radiotelescope and associated space radar measurements give precise radial velocity and distances

Table 2. Overview of current satellite Constellations major parameters

Parameter	GEO	MEO	LEO
Altitude	36,000 Km	5,000 - 20,000 Km	500 - 1,200 Km
Coverage area	Vast	Medium	Narrow
Downlink and uplink rate (signal speed)	Slow	Medium	Fast
Ground station spacing	Distant	Regional	Local

to the objects, while optical telescopes give better sky-coordinate measurements. With the installation of radar and optical sensors, PASO can extend the observation time of satellite megaconstellations and space debris and correlate information from optical and radar sources in real time. During twilight periods both sensors can be used simultaneously to rapidly provide orbital parameters and enable computing of new two-line elements (TLEs) for LEO objects, eliminating the time delays involved in the data exchange between sites in a large SSA/SST network. This concept will not replace the need for an SST network with sensors in multiple locations around the globe, but it will provide a more complete set of measurements on object passages and therefore increases the added value for initial orbit determination or monitoring of atmospheric reentry campaigns.

2.1. *Space weather*

PASO is also a suitable location to address space weather observations. Candidate projects include solar observation telescopes dedicated to monitoring and outreach in partnership with the local Pampilhosa da Serra Town Hall and the Faculty of Sciences of University of Porto (FCUP) solar radiospectrograph for radio observations (Table 1). This solar radio instrument is a fully refurbished and upgraded radiospectrograph, an instrument dedicated to monitor radio emissions produced during space weather events that awaits site commissioning due to happen by late-2024. This fast instrument will sweep the entire band from 100 MHz up to 1 GHz, with a bandwidth of 2 MHz and 5 msec integration time.

An earlier version of the FCUP radiospectrograph (Maia et al. 1999), had a large history of collaboration with major solar instruments, namely namely with the Nançay Radioheliograph and also with the Large Angle and Spectrometric Coronagraph (LASCO) onboard the Solar and Heliospheric Observatory (SOHO) for the detection and monitoring of several powerful and energetic solar events.

The refurbished and upgraded FCUP radiospectrograph will benefit from newly available instruments, in particular the new radio imaging instruments with great sensitivity and large spectral coverage like the Square Kilometre Array (SKA; Nakariakov et al. 2015). Potential contributions from the SKA include interplanetary scintillation (IPS), radio-burst tracking, and solar spectral radio imaging with superior sensitivity. These will provide new insights and results in topics of fundamental importance, such as the physics of impulsive energy releases; magnetohydrodynamic oscillations and turbulence; the dynamics of post-eruptive processes; energetic particle acceleration; the structure of the solar wind; and the development and evolution of solar wind transients at distances up to and beyond the orbit of the Earth. The FCUP radiospectrograph design is adequate for monitoring a wide range of various heliospheric phenomena of fundamental interest, in particular relating to flare physics; coronal mass ejection (CME) initiation; shock propagation; and energetic particle acceleration. CMEs are recognized as primary drivers of

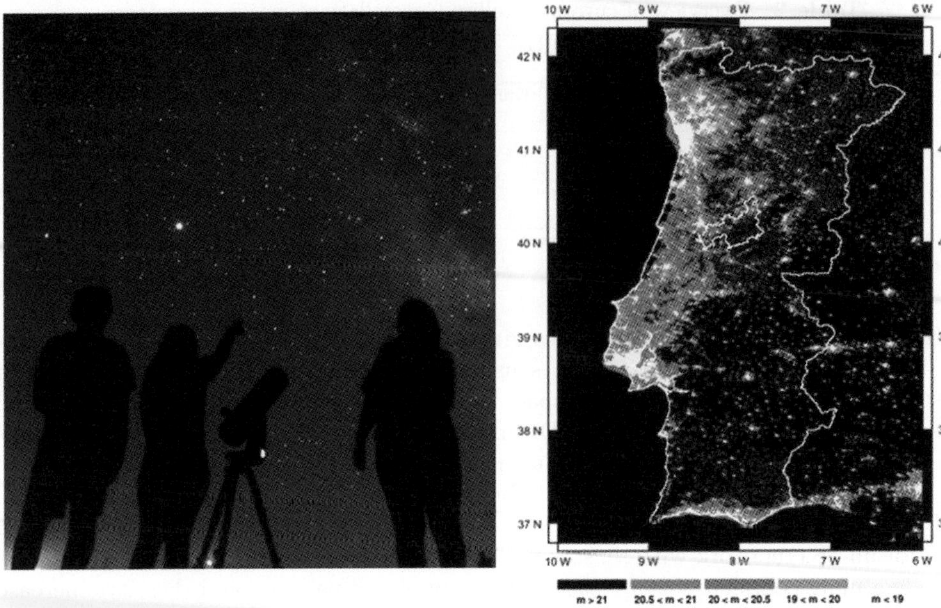

Figure 2. Left: Cibercosmos activities with a new digital Unistellar eVscope: a summer Milky Way, from (Barbosa et al. 2022); Right: A map of the brightness of the night sky over the Portuguese mainland. The false colors indicate sky brightness in units of V magnitudes per square arcsecond. The contours of the area defined by the seven councils of reference are superposed on the map (Maia et al. 2018).

disturbances in the interplanetary medium. As such, they can have a profound impact on the near-Earth space environment. A longstanding problem has been understanding the origin of CMEs and the details of their relationship to a number of associated phenomena, including solar flares, coronal and interplanetary shocks, and solar energetic particle (SEP) events. Radio observations can be of paramount importance in capturing the formation stages of CMEs, and in following CME evolution in the corona and heliosphere; see Vourlidas et al. (2020).

3. Astrotourism

Astrotourism and related citizen-science activities are becoming a major trend of a sustainable, high-quality tourism segment, core elements to the protection of dark skies in many countries. Indeed, dark skies are a natural asset that is becoming rarer due to light pollution. Therefore, it became recognized that preserving dark skies and developing a plethora of astronomical activities for the common citizen can be integrated into sustainable and inclusive tourism experiences, contributing to the conservation of natural resources and cultural heritage. Moreover, astrotourism within dark sky locations requires minimal infrastructure and therefore presents the potential for good return on investment. As well described in Dalgleish et al. (2021) and Dunn and Edensor (2023), the United Nations Societal Development Goals (SDGs) benefiting from astrotourism and related science education activities for development through education, sustainable job creation, reduced inequalities, and the preservation of local tradition and indigenous knowledge impact at least SDGs 4, 5, 8, 9, 10, and 11: quality education; gender equality; decent work and economic growth; industry, innovation, and infrastructure; reduced inequalities; and sustainable communities. Good examples of this awareness were built through capacity-building projects such as the National Astrophysics and Space Science Programme (Buckley 2019), the SKA Human Capital Development Programme (HCDP), the DevelOpment of PaloP knowLEdge in Radioastronomy (DOPPLER) and the development in Africa with Radio Astronomy (DARA; Hoare 2018). Being an active observatory in expansion, PASO's dark sky was thought of as the centre of a new dark-sky destination. The GEM Portugal and FCUP teams started site testing campaigns and analysis of satellite data performed within the classical scheme of optical seeing properties, air transparency, sky brightness, and cloudiness. For the purpose of Starlight Tourist Destination certification those properties were checked to determine compliance with the following criteria:

- Seeing better than 3 arcsec.
- A cloud free sky ($> 50\%$ of nights per year)
- Visual magnitude of the sky higher than 21 mag/arsec^2 at zenith.
- Sky transparency such that the Naked Eye Limiting Magnitude (NELM) in the visible is higher than 6.

A methodology using semi-empirical models (Cinzano et al. 2001) inferred the area and the seven municipalities that fulfilled at least two of the certification criteria. Using satellite observations we were able to infer the average values and time variability of cloud cover and aerosol content over the area of interest. By combining nighttime satellite imagery with field testing we were also able to infer the sky brightness in magnitudes per square arcsecond in the area of interest. The analysis revealed that an area corresponding to most of the Lousã and Açor Mountains fulfills the criteria for certification regarding cloud cover, atmospheric transparency and sky brightness. The seeing measurements made as part of field work confirmed that the most promising locations, in the mountain ranges at heights above 500 metres, fulfilled the criteria for certification regarding cloud cover, atmospheric transparency and sky brightness. Finally, by 2018, an area encompassing 33 parishes in 14 municipalities was certified as part of the Dark Sky Tourist Destination "Aldeias do Xisto" by the Starlight Foundation, endorsed by the UNWTO and UNESCO. Local outreach activities boomed and in the summer of 2020. In the middle of the COVID-19 pandemic, we started an initiative to train young students - "Cyber-Cosmos" - using a Unistellar eVscope, a smart, compact and user-friendly digital telescope that offers unprecedented opportunities for deep-sky observation and citizen-science campaigns. Sponsored by the Ciência Viva Summer program, Pampilhosa da Serra, a region at the heart of this certified dark-sky destination, was the chosen location for this project where we expect astrotourism and citizen science to flourish and contribute to astronomy and space sciences education (Barbosa et al. 2022). This was probably the first continuous application of this digital equipment in a pedagogical, citizen-science and pandemic context.

4. Conclusions

With its new capabilities, PASO science and space service programs like the EUSST can provide invaluable data to its SSA activities and help understand the evolution of satellite megaconstellations and the space debris field orbiting our planet and contribute to astronomical science through the monitoring of transient cosmic sources. PASO constitutes a perfect site for the development and testing of new radar and optical data fusion algorithms and techniques for space debris monitoring and for regular night sky quality assessment. From its early days, GEM Portugal paved the way for a plethora of scientific, education, outreach and astronomy activities that flourished into a new observatory within a certified Dark Sky Destination. The local PASO SSA studies may be of future interest to contribute to the International Astronomical Union Centre for the Protection of the Dark and Quiet Sky from Satellite Constellation Interference (IAU CPS). Although concerns exist with mounting evidence of the rise of light pollution trends that threatens this small dark and quiet spot in the middle of Portugal, PASO has already contributed to public awareness to help protect the local night sky.

Acknowledgements

We would like to remember the memory of the late Ivan Ferreira, from INPE and UNB, Brazil. DB acknowledges partial support by the European Regional Development Fund (FEDER), through the Competitiveness and Internationalization Operational Programme (COMPETE 2020) of the Portugal 2020 framework [Project SmartGlow with Nr. 069733] (POCI-01-0247-FEDER-069733)]; IT, UC and FCUP acknowledge further support from ENGAGESKA Research Infrastructure, ref. POCI-01-0145-FEDER-022217, funded by COMPETE 2020 and Fundacção para a Ciência e a Tecnologia (FCT), Portugal; IT team members acknowledge support from Projecto Lab. Associado UID/EEA/50008/2019. DM, BM, AKAJ and TV were partially supported by the project Centro de Investigação em Ciências Geo-Espaciais, reference UIDB/00190/2020, funded by COMPETE 2020 and FCT, Portugal. AG acknowledges support from the Center for Mechanical and Aerospace Science and Technologies (C-MAST) funded by FCT through project UIDB/00151/2020. AC was supported by CFisUC (UIDB/04564/2020 and UIDP/04564/2020). JP acknowledges support by FCT through PhD grant No-2022-12341-BDANA. AKAJ acknowledges support from ATLAB Innovation and from Instituto Nacional de Pesquisas Espaciais (INPE)-Brazil. The team acknowledges support by the European Commission H2020 Programme under the grant agreement 2-3SST2018-20, additional funding from 1-SST2018-20 and the partnerships led my MoD and EMGFA. We are grateful for the EUSST / MoD support that enabled the planning, deployment and operation of new SSA instruments. The Team acknowledges support from the Town Hall of Pampilhosa da Serra and the Parish Council of Fajão-Vidual. We warmly thank George Smoot for its pioneering support and guidance without whom the installation of the GEM-P radiotelescope that later turned into a space radar would not be possible. The GEM project has been supported in part by the US Department of Energy, by the US National Science Foundation, by the Progetto Italia Antartide, by INPE-Brazil, and by Colciencias of Colombia. DB would like to warmly thank the Starlight Foundation and Instituto de Astrofísica de Canarias for the hospitality and an excellent meeting.

References

Agayeva, S., Alvazyan, V., Alishov, S., et al. 2022, Proc SPIE, 12186, 121861H

Barbosa, D., Coelho, B., Antón, S., et al. 2021, Adv Space Res, 68, 3064

Barbosa, D., Coelho, B., Bergano, M., et al. 2022, Acta Astron, 200, 612

Bergano, M., Fernandes, F., Cupido, L., et al. 2011, Exper Astro, 30, 23

Bennett, C. L., Smoot, G. F., Hinshaw, G., et al. 1992, ApJL, 396, L7

Bennett, C. L., Larson, D., Weiland, J. L., et al. 2013, ApJS, 208, 20

Buckley, D. A. H., 2019, Nat Astro, 3, 369

de Bernardis, P., Ade, P. A. R., Bock, J. J., et al. 2000, Nature, 404, 955

Cinzano, P., Falchi, F., & Elvidge, C. D., 2001, MNRAS, 323, 34

Chomiuk, L., Linford, J. D., Aydi, E., et al. 2021, ApJS, 257, 49

Coelho, B., Barbosa, D., Bergano, M., et al. 2021, New SST Optical Sensor of Pampilhosa da Serra: studies on image processing algorithms and multi-filter characterization of Space Debris., arXiv:2107.02315

Coelho, B., Barbosa, D., Bergano, M., et al. 2022, International Astronautical Congress (IAC), International Astronautical Federation, Paris, France, 18-22 September, IAC-22-72306

Dalgleish, H. & Bjelajac, D., 2022, Encyclopedia of Tourism Management and Marketing, Cheltenham, 1

Dalgleish, H., Mengistie, G., Backes, M., et al. 2021, ASP2020: Embracing the Future: Astron. Teaching and Public Engagement, 531, 17

Dunn, N., Edensor, T. (Eds.), 2023, Dark Skies: Places, Practices, Communities (1st ed.), Routledge, ISBN 9781032528021

Ferreira, I.S. 2008, PhD Thesis, INPE, Brazil, `http://www.inpe.br/posgraduacao/ast/arquivos/teses/tese_ivan.pdf`

Fonseca, R., Barbosa, D., Cupido, L., et al. 2006, New Astro, 11, 551

Hanany, S., Ade, P., Balbi, A., et al. 2000, ApJL, 545, L5

Harper, S. E., Dickinson, C., Barr, A., et al. 2022, MNRAS, 513, 5900

Hoare, M. G., 2018, Nat. Astron., 2, 505

de Almeida, A. K., Aljbaae, S., Vaillant, T., et al. 2024, Acta Astron, 215, 548

Leitch, E. M., Kovac, J. M., Pryke, C., et al. 2002, Nature, 420, 763

Maia, D., Vourlidas, A., Pick, M., et al. 1999, JGR, 104(A6), 12507-12513

Maia, D., Morgado, B., Barbosa, D., et al. V. A. R. M., 2018, Service Provision Contract AD 21/2017 Final Report, FCUP.

Nakariakov, V., Bisi, M. M., Browning, P. K., et al. 2015, Advancing Astrophysics with the Square Kilometre Array (AASKA14), 169

Pandeirada, J., Bergano, M., Marques, P., et al. 2024, Rem Sens, 16, 704

Planck Collaboration, Aghanim, N., Akrami, Y., et al. 2020, A&A, 641, A6

Smoot, G. F., Bennett, C. L., Kogut, A., et al. 1992, ApJL, 396, L1

Stompor, R., Abroe, M., Ade, P., et al. 2001, ApJL, 561, L7

Tauber, J. A., Mandolesi, N., Puget, J.-L., et al. 2010, A&A, 520, A1

Tello, C., Villela, T., Torres, S., et al. 2013, A&A, 556, A1

Torres, S., Cañon, V., Casas, R., et al. 1996, Astroph Space Sci, 240, 225

Vourlidas, A., Carley, E. P., & Vilmer, N., 2020, Front Astro Space Sci, 7, 43

Welch, D., Dick, R., Treviño, K., et al. 2024, The world at night: Preserving natural darkness for heritage conservation and night sky appreciation., IUCN WCPA Good Practice Guidelines Series No. 33, Gland, Switzerland: IUCN, ISBN: 978-2-8317-2260-3

Astronomy and Satellite Constellations: Pathways Forward
Proceedings IAU Symposium No. 385, 2026
P. Grimley, J. Barentine, D. Galadí-Enríquez & C. Walker
doi:10.1017/S1743921324000632

Analytic vs. Montecarlo simulations of satellite visibility

David Galadí-Enríquez

Departamento de Física, Universidad de Córdoba,
Edificio Einstein, planta baja, Campus Rabanales, ES-14071-Córdoba, Spain
email: david@galadi.net

Abstract. The predicted impact of large satellite constellations on optical astronomy has been worked out, from the beginning, using Montecarlo methods: individual satellite orbits inside each shell are randomly initialised and satellite tracks are computed one by one to check whether or not they hit the field of view, and under what conditions. However, it is known that the density of satellites inside each shell follows the analytic double-sine probability density distribution, which offers new ways to get similar results at a much lower computational cost. We compare results obtained applying the two approaches and we derive some conclusions on the advantages and drawbacks of each method. The original FORTRAN code of the software is available at https://galadi.net/satsim/.

Keywords. Artificial satellites, probability density functions, numeric methods

1. Discrete simulations of satellite visibility

Performing simulations to quantify the number of illuminated satellites that may affect some optical observation was among the first actions undertaken by the astronomical community to address the risk posed by large satellite constellations. See, for instance, Chapter 6 in Walker et al. (2021).

The first attempts tried to quantify the number of satellites that would be above the horizon (or above a specific angular elevation from the horizon) and illuminated by the Sun, as a function of different variables, such as solar time, observatory latitude and season of the year. The techniques later moved to making predictions for observations performed in specific directions and with realistic fields of view, that always are significantly smaller than a whole hemisphere. See another good example of all this at Appendix B.1. in NOIRLab (2020).

With few exceptions (one of them being Hainaut & Williams 2020), the most realistic of all these simulations followed the Montecarlo, or discrete, approach. That implies considering all the satellites of each shell of the constellation, and moving them along their circular, Keplerian orbits in steps of, for instance, 1 s or 0.1 s. At each step the algorithm checks which illuminated satellites are present inside the field of view for the given pointing direction.

From the computational point of view, this simulation scheme is very different from the computation of real orbits of true satellites because, for this approach, all orbits are perfectly circular and there is no need to take into account effects such as atmospheric drag, manoeuvres, departures from a central force problem, etc. Under these conditions, the Montecarlo approach is conceptually simple, but requires significant amounts of computer time.

Introducing some prospective considerations allows making the computations not for all satellites, but only for those with some chance of coming into the field of view. Refining these criteria saves a lot of computing time. However, even with some sort of prospective analysis, in the end there remain many satellites that have to be followed. The number of satellites to be tracked grows with integration time, what leads to a significant and non linear increase of computer time.

The discrete nature of the problem requires many iterations, with random initialisation of the initial conditions, in order to average out statistical noise. This is especially true for simulated observations affected by a small quantity of satellite traces, because a really large number of iterations are needed in order to get a significant number of traces, avoiding small number statistics.

The extreme case occurs when there are no illuminated satellites at all in the field of view. Although some intelligence may be introduced to avoid this, in a normal blind simulation the algorithm has to run for quite a long time to get some certainty that zero is a good approximation to the result.

2. The analytic approach

Bassa et al. (2022, hereafter BHG22) showed that the density of satellites on each shell of a satellite constellation is described by a simple analytic function.

It is well known that large satellite constellations are made up of shells. Each shell consists of a number of orbits, with some number of satellites in each of them. Ideally, the most uniform distribution of N satellites in one shell would result from setting up $\sqrt{N}$ orbits, each of them with $\sqrt{N}$ satellites, but this is not always the case because of other engineering constraints.

For our purposes, a shell is characterised by its altitude h above sea level (or the orbit radius $r = h + r_\oplus$, where $r_\oplus$ is Earth's radius), orbit inclination i and number of satellites N. The details of the distribution of satellites inside the shell are irrelevant.

BHG22 showed that one single satellite in circular orbit induces a probability density function that varies with latitude φ following this simple analytic form:

$$P(\varphi, i, h) = \frac{1}{2\pi^2 r^2 \sqrt{\cos^2\varphi - \cos^2 i}}. \tag{2.1}$$

We may add, for completeness, that the probability density function is zero for latitudes beyond the north and south boundaries of the shell ($P = 0$ for $|\varphi| > i$).

For a shell with N satellites, the satellite density results from simply multiplying the probability density function by N, to give the so called frequency function.

Celestial mechanics make impossible a really uniform distribution of the satellites in a shell. Equation 2.1 implies that the probability density grows significantly towards the north and south boundaries of each shell, and reaches a minimum at the equator, as shown in Figure 1. In fact, the probability density diverges positively at the extreme latitudes, which means that there are two parallels per shell where we find an infinite density of satellites. The latitude of those parallels is $\varphi = \pm i$.

We do not count satellites by their density, but by their number: we need to integrate the Eq. 2.1 over the field of view to get the fraction of the sample contained in the area, or to integrate the frequency function under the same conditions to get the number of satellites present.

In BHG22 some results are presented for areas of the sky where the probability density distribution is smooth. Under these conditions, and if the field of view is not very large, we can assume that the density is constant and simply multiply its value by the shell area covered by the field of view.

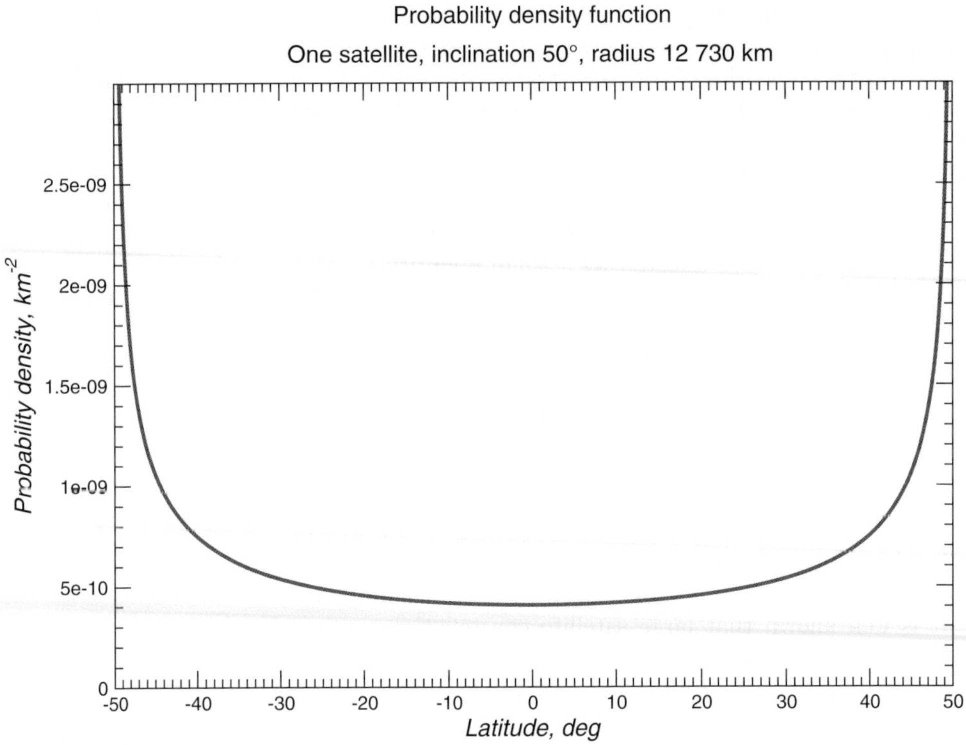

Figure 1. The probability density function (Eq. 2.1) for one satellite following an orbit with radius equal to twice that of the Earth, and an inclination of 50°

However, we know that the density is not uniform in sky areas close to or containing shell boundaries. Also, the density cannot be considered uniform for large fields of view, independently of their location on the sky. So, for an accurate result, we should go one step further than BHG22 and integrate Eq. 2.1 over the field of view. The function is not difficult to integrate on a region A of the shell between latitudes φ_0 and φ_1 and longitudes θ_0 and θ_1, with this result:

$$\int_A \frac{\mathrm{d}S}{2\pi^2 r^2 \sqrt{\cos^2\varphi - \cos^2 i}} = \int_{\theta_0}^{\theta_1}\int_{\varphi_0}^{\varphi_1} \frac{r^2 \cos\varphi \; \mathrm{d}\varphi \, \mathrm{d}\theta}{2\pi^2 r^2 \sqrt{\cos^2\varphi - \cos^2 i}}$$
$$= \frac{\theta_1 - \theta_0}{2\pi^2}\left(\arcsin\frac{\sin\varphi_1}{\sin i} - \arcsin\frac{\sin\varphi_0}{\sin i}\right) \qquad (2.2)$$

This integral can be performed only on areas limited by meridians and parallels on the shell, and our fields of view never have that shape. So, a part of the procedure will approximate our region by means of integrable pieces of that kind, with enough accuracy.

3. Description of the analytic algorithm

Let us discuss in some detail the steps necessary to implement the analytic approach to counting satellites in the field of view.

From the longitude and latitude of the observatory we compute the corresponding geocentric cartesian position vector. We also have the properties of the shell, r, i, N. The pointing direction is normally specified in terms of azimuth and elevation, that can be transformed to a topocenric vector that defines a line that intersects the shell at some

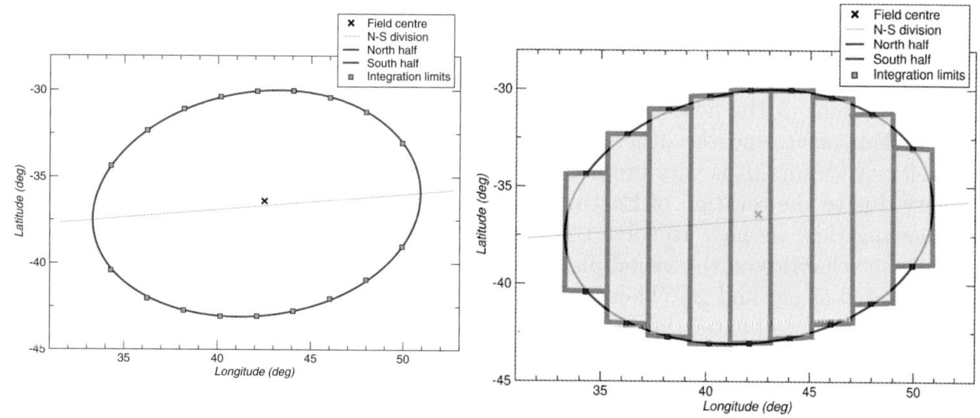

Figure 2. The intersection of the shell with the conic bundle that defines the field of view may differ significantly form an elliptic shape. The left figure shows an example in which a field of view of 15° intersects the shell of Figure 1 on an elongated and deformed area. The line separates the north and south halves of the perimeter, defined in this case by 5000 points. The squares mark the limits for the nine strips shown on the right, used to compute the integral according to Eq. 2.2

point. A quadratic equation is involved in the computation of the intersection point and the right solution has to be selected.

Having the position of the centre of the field of view projected onto the shell is not enough. We have to specify the extent of the field and find the projection of its outline as well. For the moment we deal only with circular fields of view, but the procedure could be easily generalised to other shapes. So, for now, one number suffices to define the field: the angular width of the view cone.

From the pointing direction and the angular width we build a conic bundle of lines whose intersection with the shell defines the region in which we will perform the integral of the frequency function. An example is shown in Figure 2, left.

The field of view is circular, but it is easy to show that its projection on the shell is not even an ellipse, but, in general, something more complicated that we manage by simply outlining it with a sufficiently large number of individual points. Those points are classified into two sets, the northern and the southern boundaries, that later are used to set the starting and ending values of the latitudes for the integrals.

As indicated at the end of Section 2, we now have to approximate the integration region by means of meridians and parallels. We do that by defining vertical strips and integrating each of them according to Equation 2.2. We set up an iterative procedure that starts from four strips and increases their number one by one. If I_n is the value of the integral obtained with n strips, in our implementation the iteration ends when $|I_n - I_{n-1}|/I_n < 10^{-3}$. Figure 2, right, illustrates an example.

The iterative integration procedure yields the instantaneous number of satellites present in the field of view at the beginning of the observation, N_0, but we have to add the satellites that will enter the field of view during the integration time t. The way to evaluate this requires some more geometry that is totally absent in the Montecarlo approach.

The orbital speed is the same for all satellites in circular orbits with radius r, and is equal to $\sqrt{GM_\oplus/r}$, where G is the constant of universal gravitation and $M_\oplus$ represents the mass of the Earth. The velocity vector of each satellite is horizontal, tangent to the orbit. This means that we have all the information necessary to build the velocity vector of the satellites in the field.

To do this for one shell, we have to take into account that half of the N_0 satellites that are present in the field at the start of the integration time are in the Northwards part of their orbit, and the other half are in the southwards part. So, we have to build two velocity vectors. It is very important to subtract from both vectors the observatory's velocity due to the rotation of Earth.

Following this, we have to solve the non straightforward geometrical problem of projecting the velocities on the visual plane and deducing two apparent angular speeds, that we may label as ω_N and ω_S. Then, adapting some results from BHG22, the number N_ω of satellites that will enter the field of view during the integration time t is given by

$$N_\omega = \frac{N_0 t}{\pi(1 - \cos\zeta)}(\omega_N + \omega_S), \tag{3.3}$$

where ζ indicates the angular radius of the field of view. Solving in detail the projection of the velocities on the field of view, one can deduce not only the apparent angular speeds already mentioned, but also their position angle. Given an observatory, there is a pair of apparent angular velocities (apparent angular speeds plus position angles) for each shell at each point of the sky, and they are deterministically fixed.

The approach to counting the satellites that enter the field implicit in Eq. 3.3 is only approximate, in the sense that the angular velocities vary over the field, and we are taking their values at the centre of the image as representative of all. This approximation is very good, as we will show later, but it evidently will get poorer as the size of the field of view grows, which widens the velocity distribution of the satellites that cross the image.

If we have some way of computing the apparent brightness of the satellites, it can be applied at this step from the information available. Much work is being done to improve the photometric predictions, but this is not the main purpose of our algorithm and, so, for the moment we are using a simplistic approximation to satellite brightess drawn from Hainaut & Williams (2020).

4. Some intricacies of the method

The procedure explained up to now is geometrically non straightforward, but clear. However, there are some practical intricacies that we have to keep in mind to produce a working code.

We define the poles of a satellite shell of radius r as the points of the Earth's rotation axis that are at distance r from the centre of the planet. The observed area may include one of the poles of the shell. These configurations have to be foreseen and the integration routine has to be adapted to deal with them.

The field of view may include one of the north or south shell boundaries and this has to be kept into account to perform the integral and, also, to compute the apparent angular speeds ω_N and ω_S.

The part of the shell in the field of view may be inside the Earth's shadow, totally or partially. It is very important to integrate the probability density function only on those parts of each shell that are illuminated by the Sun. Our algorithm currently scans each of the integration strips in latitude, at small steps of 10 arcseconds, checking which fraction of each strip is in sunlight. The parts of the strips that are in shadow are omitted from the integral. As Figure 3 shows, it may happen that some strips have to be divided into several parts (two or three), if they are only partially illuminated.

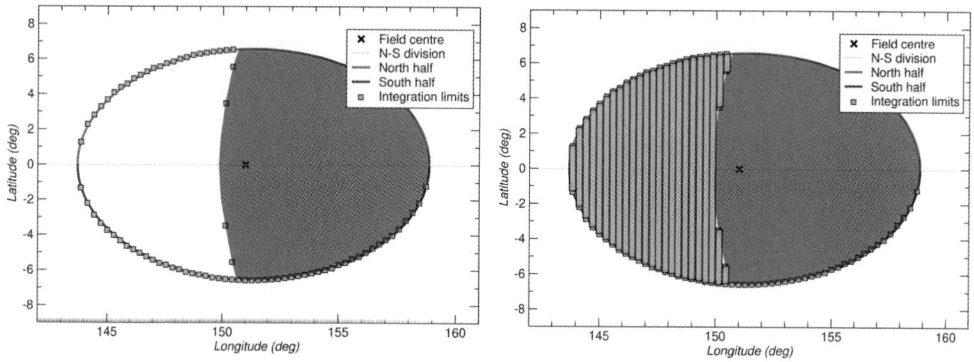

Figure 3. Example of an intersection of a field of view that spans 15° with a shell (again, that of Figure 1) that is only partially illuminated by the Sun. For the strips in shadow, the algorithm sets the latitudinal integration limits (Equation 2.2) to the same value ($\varphi_1 = \varphi_0$), causing the result to vanish. The figure on the right overlays the integration strips in the same way as in Figure 2. Note that there are two strips that had do be divided into three sections, because only their central segment is in shadow.

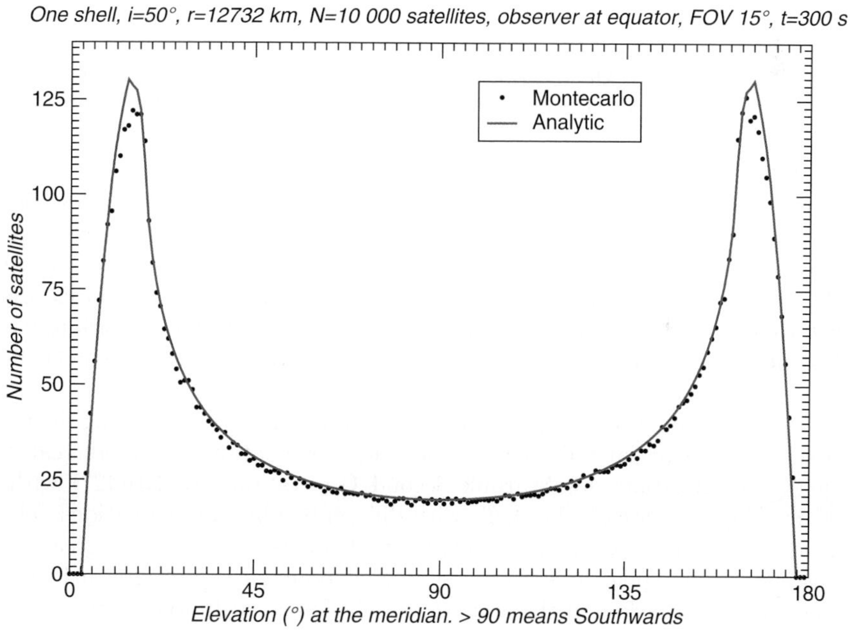

Figure 4. Comparison of Montecarlo (dots) and analytic (line) simulations. Only one shell at an altitude above sea level equal to the radius of Earth, $i = 50°$, 10 000 satellites, field of view diameter 15°, observatory at the equator. Observations along the meridian from zero elevation due north to the zenith and to the south (pointing south is indicated by elevations larger than 90°).

5. Some results

As a first example, in Figure 4 we show a comparison of Montercarlo and analytic results for a fictitious constellation made up of one single shell of 10 000 satellites with a radius of 2 Earth radii. The inclination of the shell is 50°. The observatory is at the equator. The simulation is performed in daylight and observing at the meridian, covering

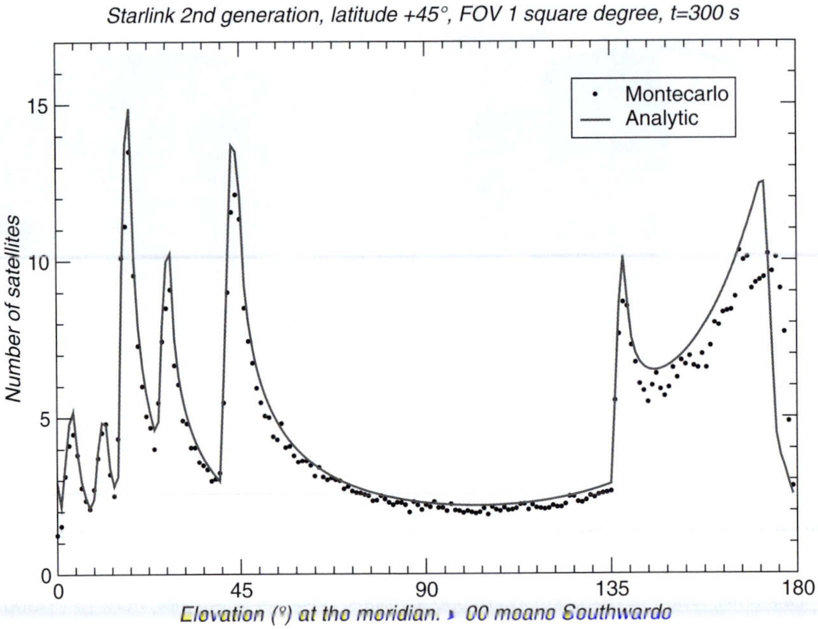

Figure 5. Comparison of Montecarlo (dots) and analytic (line) simulations. Constellation Starlink Second Generation, see the text and the caption of Figure 4 for more details.

an arc from north to south. We use a generous field of view 15° wide. Most often, the Montecarlo results fall slightly below the analytic ones, a situation that we attribute to the random stochastic nature of Montecarlo simulations: when filling a circle with dots randomly generated, the size of the cloud is always smaller than the circle diameter. The greater excess of the analytic results around shell boundaries is attributable to the approximation of the same angular speeds for all satellites in a heterogeneous field and deserves further consideration.

Figure 5 shows a similar experiment, observing from north to south along the meridian, from an observatory at latitude +45°. Now the constellation is a more realistic one, following the specifications for Starlink Second Generation (see BHG22) with 11 927 satellites distributed among eight shells with different radii and inclinations. The width of the field of view is now 213 arcminutes, what spans one square degree inside our circular window. The spikes due to crossing shell boundaries are very evident in this graph, and the Montecarlo and the analytic results follow each other very closely.

6. Summary and conclusions

The probability density function from BHG22 allows a radical change of philosophy for simulations of the impact of large satellite constellations on optical astronomy.

With this approach, the computational weight of Montecarlo simulations is transferred to the geometric complexity of the problem.

Our implementation implies two iterative processes: one to check the areas in shade, and another one to decide the width of the integration strips.

The algorithm solves some intricacies, none of them straightforward, with a non negligible amount of geometrical complexity. It can be said that the computational weight of the Montecarlo methods is transformed, in the analytic approach, into a more complex algorithm. In our case, the analytic code is more than 25% longer than our version of the Montecarlo program. However, the gain in computer time is overwhelming, reducing the execution by a factor of hundreds, some thousands in some cases.

Some excess in satellite counts derived from the analytic method, compared to the Montecarlo approach, may arise from the way of counting the traces that enter the field of view during the integration time, and deserve further consideration. The algorithm may be easily modified to accept fields of view with square or rectangular shapes and different orientations. The FORTRAN code of the software is available at `https://galadi.net/satsim/`.

References

Bassa, C. G., Hainaut, O. R., & Galadí-Enríquez, D. 2022, A&A, 657, A75 (BHG22)

Hainaut, O. R. & Williams, A. P. 2020, A&A, 636, A121

NOIRLab 2020, Technical document: Appendices to "Impact of Satellite COnstellations on Optical Astronomy and Recommendations Toward Mitigations, https://www.noirlab.edu/public/products/techdocs/techdoc004/

Walker, C., et al. 2021, Dark and Quiet Skies for Science and Society: Report and recommendations, Zenodo, `https://doi.org/10.5281/ZENODO.5898785`

Astronomy and Satellite Constellations: Pathways Forward
Proceedings IAU Symposium No. 385, 2026
P. Grimley, J. Barentine, D. Galadí-Enríquez & C. Walker
doi:10.1017/S1743921324000826

Satellite and space debris photometric phase functions used for astronomical images contamination prediction

Jiří Šilha, Adam Rigan, Peter Jevčák, Matej Zigo, Tomáš Hrobár and Roman Ďurikovič

Faculty of Mathematics, Physics and Informatics, Comenius University in Bratislava
email: jiri.silha@fmph.uniba.sk

Abstract. The increasing amount of space debris and the number of mega-constellations are creating a serious problems for terrestrial astronomy. Light trails caused by these objects contaminate astronomical images, rendering them ultimately worthless. Knowing where in the sky these objects will be and how bright they will be are essential to optimising observation planning and avoiding such contamination. Photometric phase functions are a powerful tool for understanding asteroids and resident space objects (RSOs) such as satellites, upper stages, and space debris. However, they also provide a means of predicting how these objects will reflect sunlight toward the observer. We consider a model that assumes a spherical object with diffuse and specular components. We defined the phase functions of more than 600 rotating RSOs using more than one million photometric measurements of more than 10 000 objects from the Mini-MegaTORTORA (MMT) light curve catalogue. We discuss in detail the photometric properties of space debris and satellites extracted from the MMT photometric catalogue using the model. We introduce our freely available online service that predicts the brightness of RSOs, including space debris and mega-constellations, for the purpose of supporting astronomers' decision-making.

Keywords. space debris, photometric phase function, online ephemeris service

---

1. Introduction

Inoperative satellites and space debris, artificial space objects that no longer have any purpose, are in general a threat to the space environment and to its long-term sustainability. Their high relative speeds and kinetic energies are making them a potential cause of a satellite's catastrophic fragmentation. In the recent years, new populations of satellites, the so-called mega-constellations, have created a new question regarding the sunlight reflected toward Earth causing light pollution. There are several studies investigating the phase functions of mega-constellations (e.g. Mallama 2021, 2022). However, there are few studies focusing on effect of space debris on light pollution.

2. Photometric catalogues

For our analysis we used the Mini-MegaTORTORA database (MMT) of Kazan Federal University (Karpov, Katkova, Beskin, et al. 2016) which contains more than 400 000 light curves of more than 10 000 objects. Each light curve is processed and analysed. For rotating objects the apparent rotation periods are extracted by the MMT team and are reported on the official website (Karpov, Katkova, Beskin, et al. 2016). For comparison, in Figure 1 are shown the distributions of apparent rotation periods extracted from the MMT and the Space Debris Light Curve Database (SDLCD) catalogue created and

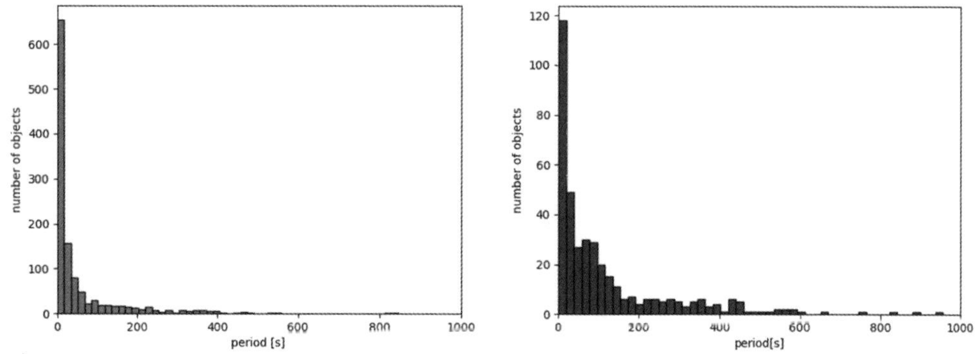

Figure 1. Distribution of obtained measured periods for 1211 objects observed by MMT system in the years 2014 to 2023 (left) and 406 rotating objects observed by AGO70 telescope in the years 2017 to 2022 (right).

maintained by the Comenius University in Bratislava (Šilha, Krajčovič, Zigo, et al. 2020; `sdlcd.space-debris.sk`). From all the available objects, for 12.9% in MMT and 39.4% in the SDLCD catalogue the apparent rotation periods could be extracted.

When the apparent period is known, the lightcurve can be split into rotational curves when the length of one rotational curve is the apparent rotational period. Each rotational curve is then fitted with the Fourier series to extract the Fourier coefficient a_0 (mean value) used for further analysis.

3. Photometric phase function

Phase functions help to understand asteroid surface properties such as porosity, roughness, particle size, complex refractive index etc. In the asteroid community, the phase function is defined by several parameters, including the absolute magnitude H, which is the visual brightness of an object in the V filter when viewed at 1 AU by an observer at a phase angle of 0°. The parameter H defines the intrinsic brightness of the asteroid and its dependence on the size of the object and the albedo. Another parameter that defines the phase function is the so-called steepness factor G, which describes the shape of the phase function (Bowell, Hapke, Domingue, et al. 1989). The inclination factor G showed a strong dependence on asteroid taxonomy. In 2010, this model was improved by Muiknonen, Belskaya, Cellino, et al. (2010) when the G_1 and G_2 coefficients were introduced and basis functions were replaced by cubic-spline basis functions. This model was introduced to capture the opposition effect that Bowell's model did not cover.

In the last two decades, extensive research has been carried out on the phase functions of man-made objects. It included laboratory measurements of representative samples (Murtazov 2014; Peltoniemi, Zubko, Virkki, et al. 2022; Hostetler & Cowardin 2019) simulation and modelling (Hejduk 2011), and observation acquisition and data fitting (Hejduk 2011; Africano, Kervin, Hall, et al. 2005; Hostetler & Cowardin 2019).

Hejduk (2011) states that the best way to model any type of satellite or artificial space object is to assume a diffuse sphere as a first approximation. This is because there is no variation in the photometric return with satellite aspect, and the brightness is a straightforward deterministic calculation based on the solar phase angle, satellite albedo, and satellite size (spherical diameter). Hejduk (2011) introduced the phase function which contains mixture of diffuse and specular component defined as follows:

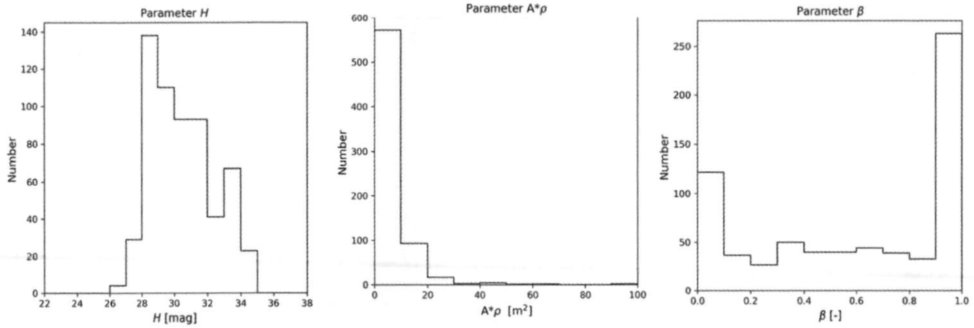

Figure 2. Extracted parameters absolute magnitude H (left), $A\rho$ (middle), β (right) analyzing 598 objects from the MMT database.

$$m_V(R,\phi) = -26.74 - 2.5log(A\rho[\beta F_1(\phi) + (1-\beta)F_2(\phi)]) + 5log(R) \quad (1)$$

Where ϕ (deg) is the phase angle, A (m) is the sphere's cross-sectional area, ρ is the unite-less bond albedo, which is defined as the fraction of incident light scattered in all directions by the surface, R (m) is the satellite-observer range, β is a unit-less mixing coefficient representing portion of each component in the resulting brightness, and $F_1(\alpha)$ and $F_2(\alpha)$ are diffuse and specular sphere phase functions (Hejduk 2011).

We fit more than 100 000 folded light curves for 598 individual objects to extract the a_0 coefficient. For each object we constructed a phase function diagram, phase angle versus apparent magnitude V_{app}, where V_{app} is the coefficient a_0 reduced to 1 AU. For detailed data reduction see Šilha et al. (in preparation).

We processed 190 satellites, 307 rocket upper stages and 101 space debris objects. In Figure 2 are plotted histograms of obtained absolute magnitudes, $A\rho$ and β. Satellites showed equally redistributed β with small maxima for fully diffuse and specular reflections. Rocket upper stages showed highly diffuse reflections, where more than 50% had β larger than 0.9. Space debris objects showed two maxima for the β parameter, fully diffuse or fully specular reflections.

Examples of four phase curves constructed for four different object types are shown in Figure 3. Shown are phase curves for two satellites, the box-wing type of satellite Globalstar, the irregularly shaped satellite Iridium, the rocket upper stage Atlas 2AS Centaur and fragmentation debris from the CZ-4 rocket upper stage. Both satellites have small β which indicates the presence of the specular reflection. For Atlas upper stage the reflectance seems to be predominantly diffuse. Debris objects also show signs of specular reflection.

All the high quality phase curves for more than 400 objects will be published in publicly available Space Debris Phase Curve Database (SDPCD; Šilha et al., in preparation).

4. Online prediction service

We are developing an online space debris forecasting service for professional and amateur astronomers as well as astrophotography enthusiasts. With this service, the user can define a planned line-of-sight (LoS) to be observed at a specific time, and the service will provide a list of objects that theoretically contaminate FITS frames. It is built on the long-term software solution SatEph (Šilha, Krajčovič, Zigo, et al. 2020), which has been used by the Comenius University for more than a decade to observe space debris

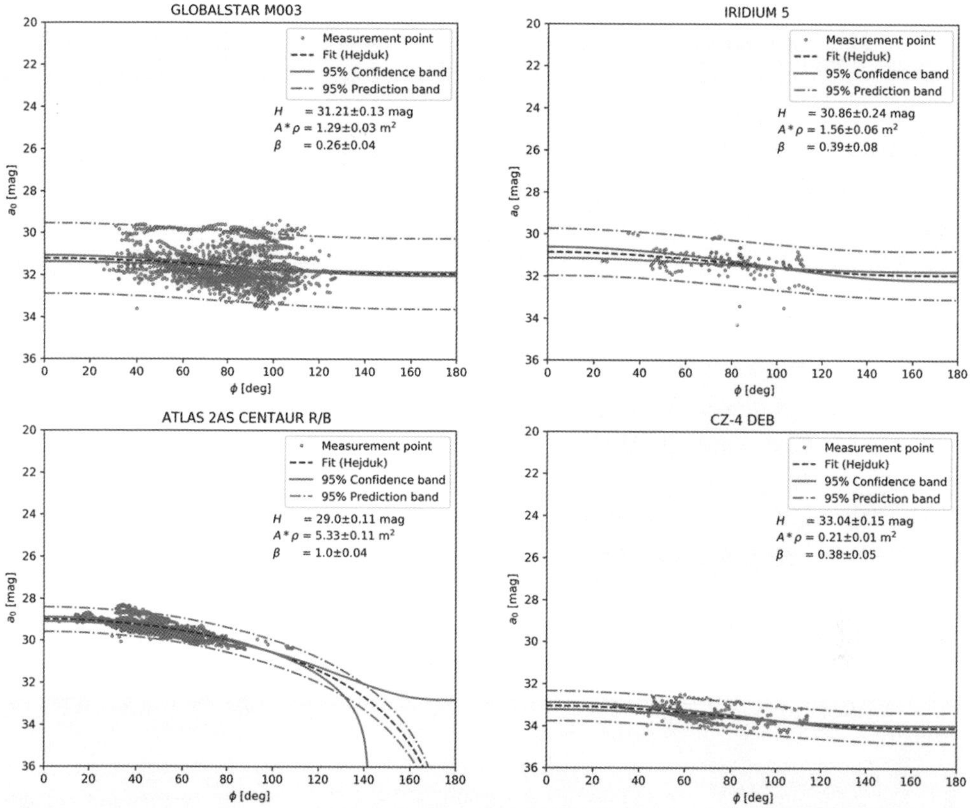

Figure 3. Photometric phase functions of four individual objects Globalstar, Iridium, Atlas 2AS Centaur and fragmentation debris from CZ-4 upper stage.

(Figure 4). It is a fast, accurate and reliable algorithm built on the well-known simplified general perturbations model SGP4 Vallado, Crawford, Hujsak, & Kelso (2006).

We used the SatEph engine and developed around it a web interface to create an online service for satellite and space debris sky position and brightness prediction. The first functionality, the calculation of the position for the specific objects, has been implemented, including the frontend. Figure 5 shows the front-end interface of the developed online service. Plotted are all four steps necessary to get ephemeris of artificial object including setting up time (Step 1), defining the observation site location (Step 2), adding the object identification number (Step 3) and providing the results window which contains ephemerides in topocentric horizontal and equatorial coordinates (Step 4). Currently, the service uses as inputs the Minor Planet Center (MPC) database of observation sites and orbital elements provided by the online service USSTRATCOM (2023).

We will continue to develop the service, continuing with the implementation of a service where a list of all objects visible within the defined LoS during a defined period of time will be provided. The interface with the SDPCD will be developed to enable the brightness of the object to be calculated and the interface with the www.space-track.org service to secure up-to-date orbital elements of the objects.

5. Conclusions

We have presented phase functions of artificial objects such as satellites and space debris. Phase functions are defined by two major parameters, $A\rho$ which defines the mean cross section multiplied by geometric albedo, and β defining the dullness of the object's

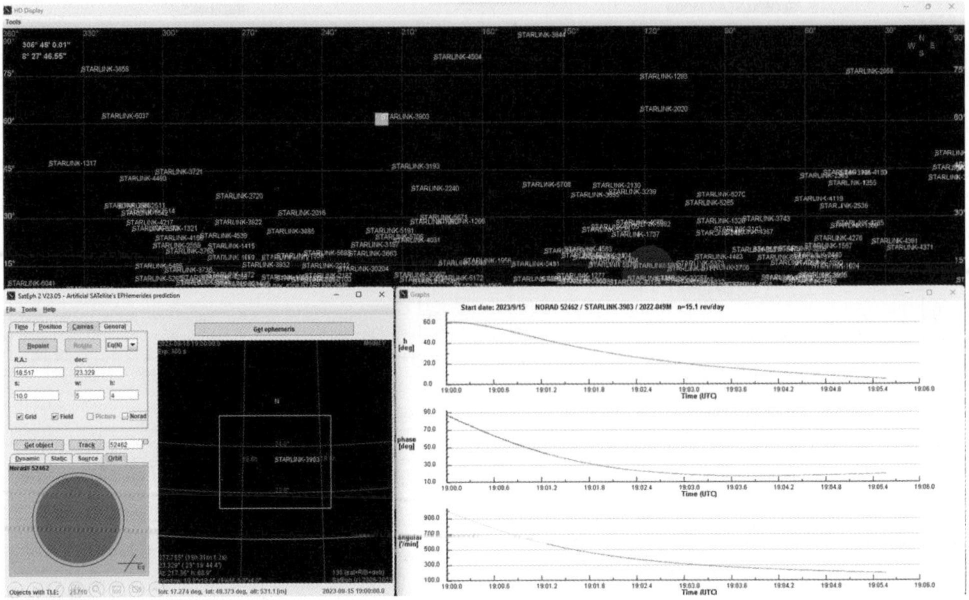

Figure 4. Screenshot of Satellite Ephemerides (SatEph) S/W, main window, graphs and horizontal coordinates.

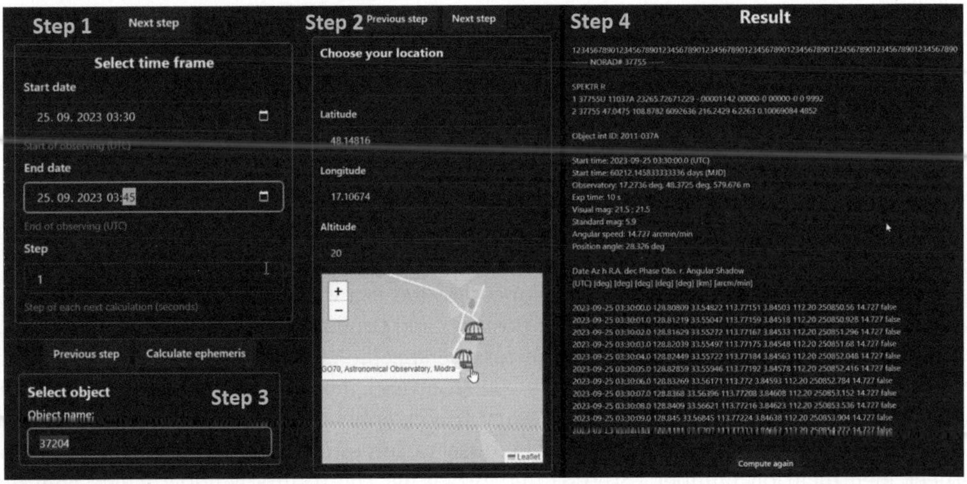

Figure 5. Screenshot of the developed front-end web interface for the Online contamination prediction service.

surface. Investigation showed that for satellites there are typical either specular or diffuse reflections. Rocket upper stages show in the main strong diffuse reflection which indicates a relatively homogeneous surface with few highly reflective components. Space debris mostly showed highly specular reflections. All the phase curves will be publicly available in the SDPCD.

We have developed the first module for an online contamination prediction service which uses a Simplified General Perturbations model to predict the position of a selected object and given observer location on the sky. The frontend has been developed for easier work with the engine and allows the service to be used from any type of operating system. In the near future the service will be improved by adding the line of sight of the observer and generating a list of objects which will be crossing the LoS during specific time period. Additionally, interfaces will be developed with SDPCD to also implement calculation. Once the service has been properly tested it will be deployed on our server and publicly accessible to any user.

Acknowledgements

This work has been supported by ESA project under the seventh call of the Plan for European Cooperating States in Slovakia with Contract no. *4000139802/22/NL/SC - Modelling the night sky brightness produced by space objects.*

References

Africano, J., Kervin, P., Hall, D., et al. 2005, in Danesy, D. (ed.) Proceedings of the 4th European Conference on Space Debris (ESA SP-587), 141

Bowell, E., Hapke, B., Domingue, D., et al. 1989, in Proceedings of the Conference Asteroids II (Tucson, AZ, University of Arizona Press), 524

Hejduk, M. 2011, in Ryan, S. (ed.) Proceedings of the Advanced Maui Optical and Space Surveillance Technologies Conference, E15

Hostetler, J. & Cowardin, H. 2019, International Orbital Debris Conference, IOC 6148

Karpov, S., Katkova, E., Beskin, G., et al. 2016, Revista Mexicana de Astronomía y Astrofísica (Serie de Conferencias), 48, 112

Mallama A. 2021, arXiv:2111.09735

Mallama A. 2022, arXiv:2203.05513

Muinonen, K., Belsakaya, I. N., Cellino, A., et al. 2010, Icarus, 209, 542

Murtazov, A. 2014, in Muinonen, K. (ed.), Proceedings of the conference Asteroids, Comets, Meteors 2014, M379

Peltoniemi, J. I., Zubko, N., Virkki, A. K., et al. 2022, Adv Space Res, 70, 10

Šilha, J., Krajčovič, S., Zigo, M., et al., 2020, Adv Space Res, 65, 8.

USSTRATCOM 2023, space-track.org

Vallado, D. A., Crawford, P., Hujsak, R., & Kelso, T. S. 2006, AIAA 2006-6753-Rev1

Astronomy and Satellite Constellations: Pathways Forward
Proceedings IAU Symposium No. 385, 2026
P. Grimley, J. Barentine, D. Galadí-Enríquez & C. Walker
doi:10.1017/S1743921324000814

Precise optical observations of orbital objects in severely trailed images: application to LEO satellites, extremely faint objects, and space debris

Alex Drago-González[1,2] and **Jorge Núñez**[1,3]

[1]Departament de Física Quàntica i Astrofísica, Institut de Ciències del Cosmos (ICCUB), Universitat de Barcelona (IEEC-UB),Martí i Franquès 1, E08028 Barcelona, Spain
email: jorge@fqa.ub.edu

[2]Medical Ultrasonics Laboratory (MEDUSA), Department of Neuroscience and Biomedical Engineering, Aalto University, 02150 Espoo, Finland
email: alex.dragogonzalez@aalto.fi

[3]Observatori Fabra, Reial Acadèmia de Ciències i Arts de Barcelona, Rambla dels Estudis, 115, 08002 Barcelona, Spain

Abstract. Optical observation is used to track objects in orbit, but it's challenging for objects in Low-Earth Orbit (LEOs) due to high-speed satellites leaving long trails. We present a study that uses a multilayer neural network software called Source-Extractor (SExtractor) to accurately determine the position of stars and orbital objects. By modifying input parameters, we can draw ellipses around star trails, including very long ones. Using images from the TFRM telescope and Fabra Observatory, we accurately determine the position of objects in all types of orbits, including of 6000 km-high 1U CubeSats and up to magnitude 18.0 on 10-second exposure images following the object showing star trails over 400 pixels long. This technique can recognise all types of objects, including other satellites or debris, if bright enough. This technique can help study the negative impact on astronomical observations caused by megaconstellations of satellites like Starlink or OneWeb, for professional and amateur observatories.

Keywords. Low earth orbit satellites, Space debris, Satellite constellations, Satellite tracking

1. Introduction

Since ancient times, humans have looked to the night sky for answers. Today, those same stars help us to guide ourselves and develop knowledge about the Universe. However, since 1957, the sky started to look different with new bright and moving points in sight: artificial satellites. They reflect sunlight to Earth, especially during the early and late hours of the night, and sometimes show up in our optical observations through telescopes and cameras (Leslie 2020). However, the lack of an efficient international regulatory commission in space affairs makes space debris (Nazarenko & Usovik 2022) and mega-constellations of satellites, such as Starlink or OneWeb, a source of concern that can interfere with night sky observation of astronomers, both professional and amateur.

In this paper we propose an on-demand, fast and compact way to solve the issues with the optical recognition of long star trails from FITS images by adapting an open-source software (Bertin & Arnouts 1996), which will enable amateur and professional astronomers to optically track and to study the brightness evolution of satellite mega-constellations, even those in Low-Earth Orbit (LEO). We aim to adapt the values of the input parameters of the software to draw non-fragmented ellipses of large eccentricity to represent the trails left by the stars during an observation following satellites, with high accuracy of their geometrical center and length.

To show the effectiveness of the technique, we show the astrometric reduction of images of geostationary orbit (GEO; Hispasat constellation), Medium-Earth Orbit (MEO; Navstar and Cubesats) and LEO (Soil Moisture and Ocean Salinity SMOS satellites, Starlink and OneWeb) satellite trackings. The optical images were taken with the Fabra-ROA Telescope Montsec (TFRM). The telescope (Fors, Núñez, Muiños et al. 2013) is the transformation for CCD observation of a Baker-Nunn telescope of 50 cm aperture made by Boller & Chivens for Smithsonian and U.S. Air Force satellite-tracking programs, which was designed to track artificial satellites. The TFRM, with its CCD camera of 4096×4096 pixels, is ideal for observing space debris thanks to its large Field of View ($4.4° \times 4.4°$) and its focal ratio (focal length / telescope aperture) of 0.96, i.e., extremely fast, which allows shorter exposures and for operating in fully robotic mode. Each TFRM image is in FITS format (Wells, Greisen & Harten 1981) and contains several thousand images of stars. Given its huge field of view, the telescope can register, near the GEO ring, up to some tens of satellites and space debris in a single image. In addition, we use images tracking on the orbital object, either using its Two-Line Elements (TLE) or, in the case of GEO objects, by stopping the telescope. Therefore, the satellites present a circular shape while the stars appear as elongated trails. The technique allows us to detect faint satellites since the entire exposure time is applied to the object. However, when doing sidereal tracking, the fast movement of the satellite makes the effective exposure time (on a specific pixel) very short, so only bright satellites can be observed.

2. Software

Optical observation is commonly used to track satellites in orbit and is very efficient for high orbits because, due to the low speed relative to the stars, when we make observations following the object, the stars leave short trails on the detector. However, for low orbits, the satellites present high speeds with respect to the stars and the trails become longer, which complicates their recognition and treatment. This is especially notable in the case of LEO orbits as trails can be of hundreds of pixels long even for very short exposure times of 1 or 2 seconds. Figure 1A shows an image of a MEO satellite of the Navstar constellation. It can be seen that the reference stars used to obtain the position of the satellite (circular image at the bottom outlined in red) have a notable length and inclination that makes it difficult to accurately determine its position. In addition, in Figure 1A it can be seen that faint stars leave trails that tend to divide into several "sub-trails", which makes almost impossible the astrometric reduction of the image.

There are various ways to attack this problem of long trails left by stars or orbiting objects (Schildknecht 2007). In this work we have opted for a simple and compact solution consisting of determining ellipses that fit optimally to the geometry of the trails. To do this, we will use the Source Extractor (SExtractor) software (Bertin & Arnouts 1996) with the appropriate input parameters for each of the cases under study (GEO, MEO, LEO) for either bright, faint or very faint objects. This is not the first time in which SExtractor has been used for star or satellite trail processing or as comparison software (see, for example Park, Keum, Lee et al. 2013, Sun & Zhao 2013, Petit et al. 2021, Yang, Zhao, Liu & Liu 2021), but to our knowledge it has not been used for very long trails.

By default, SExtractor uses as input parameters like those listed in Figure 2, but, for recognizing the trails left by stars or satellites, these default values do not work well. This is because the default parameters are intended for sidereally-tracked images where stars appear as circular images, but not as long trails. Galaxies and other

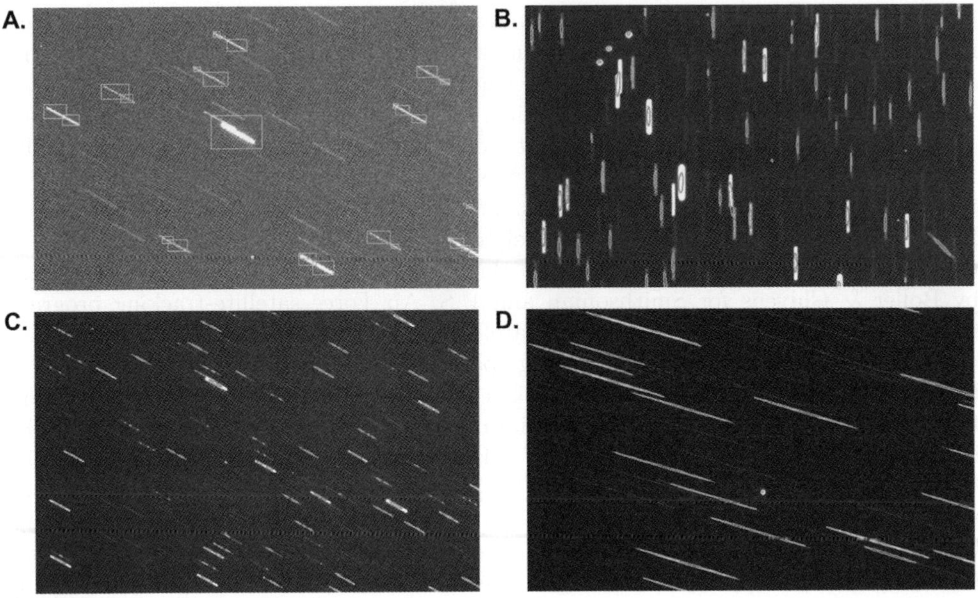

Figure 1. *A.* Recognition without adaptation of trails left by the stars during a Navstar satellite tracking (red: Satellite, green: stars). *B.* GEO example, Hispasat (top-left) surveillance with 10 seconds of exposure time and unidentified object (bottom right), potential candidate of space debris or satellite, showing a clearly different angle of motion. *C.* MEO example, Navstar (center) tracking with 5 seconds of exposure time, with long and sloping trails left by the stars. *D.* LEO example, SMOS tracking with 1 second of exposure time, with trails left by the stars that have an approzimate length of 450 pixels.

```
#---------------- Catalog ----------------
CATALOG_NAME      Test.cat
CATALOG_TYPE      ASCII
PARAMETERS_NAME   default.param

#---------------- Extraction ----------------
DETECT_TYPE       CCD         # CCD (linear) or PHOTO (gamma cor.)
DETECT_MINAREA    5
DETECT_THRESH     1.5
ANALYSIS_THRESH   1.5
FILTER            Y
FILTER_NAME       gauss_1.0_7x7.conv
DEBLEND_NTHRESH   32
DEBLEND_MINCONT   0.005
```

Figure 2. Default configuration file for SExtractor. The list of variables is much longer (see `http://tdc-www.harvard.edu/wcstools/sextractor/default.sex`).

diffuse objects appear as ellipses, but of limited eccentricity. The main input parameters that were key for a successful adaptation of the software to detect satellites at different orbits are DETECT_MINAREA, DETECT_THRESH, DEBLEND_NTHRESH and DEBLEND_MINCONT. To help locate the tracked satellites with a circular shape we also add an additional parameter called DETECT_MAXAREA.

Since our aim is to represent the long trails left with the stars by an ellipse of large eccentricity, among the large number of output parameters given by the SExtractor software, we select those from Figure 3 that refer to the position of the center of the ellipse, the semi-major and semi-minor axes, the inclination with respect to the x-direction of the detector pixels, the light flux and the flags found during the reduction.

```
X_IMAGE
Y_IMAGE
A_IMAGE
B_IMAGE
THETA_IMAGE
FLAGS
FLUX_ISOCOR
```

Figure 3. Output parameters for SExtractor.

3. Results

For each of the referred cases, we adjust the parameters to obtain the star trails fitted by ellipses that allow an accurate astrometric reduction of the image, while also tracking the satellite and other orbital objects appearing in the image. In this section we share the process of adapting the software for each example: with relatively short trails such as GEO objects (Hispasat, at an altitude of 36000 km), images with long star trails such as MEO objects (Navstar, at an altitude of 20200 km) and images with very long star trails such as LEO objects (SMOS, Starlink and OneWeb, at heights of 755, 550 and 1200 km, respectively) or very faint objects in low MEO that need long time exposures giving, also, very long star trails (Cubesats in MEO, at an altitude of 6000 km).

3.1. *Sidereal tracking*

We started with an image, taken in sidereal tracking, of the Virgo Cluster of galaxies to understand how SExtractor works. In this case there are no trailed star images. We were able to check that the outputs were correct and coincident with the star catalogs and the well-known angular sizes of the galaxies of the cluster. In this example we used the default parameters listed, but we noted that the software detected several hot pixels as real objects. To solve this issue we changed the default value of DETECT_MINAREA, which is used to determine the minimum number of pixels to detect an object. By default this value was set to 5, but we found that the best value for this image was 10 pixels. Another problem we faced was that the detection of objects gets messy on the edges of the image. The geometry of the objects breaks and the ellipses stop being reliable.

3.2. *Images with relatively short star trails: GEO object survey*

For short trails, as happens with satellites in the GEO ring, we used a series of images of the Hispasat satellite constellation obtained in GEO survey mode (telescope stopped). Figure 1B shows one of the images that has been reduced. This set of images also featured an object moving in a different direction, which created a trail with a different position angle than the rest of reference star trails. If SExtractor is working properly, an ellipse should be detected in the direction of such trail, meaning that we have a row in the output file that has a different THETA_ANGLE that corresponds to a satellite or a space debris candidate. By working with the Hispasat set of images and several similar others, we adjusted the input parameters to obtain optimal results. The DEBLEND_MINCONT parameter was adjusted to 0.06, which is discussed in the next section. The modified parameters are listed in the second row (i.e. *Hispasat*) in Table 1.

3.3. *Images with relatively long star trails: MEO object observation*

For the case of images with relatively long trails, we used a series of images of some Navstar satellites obtained in tracking mode. The tracking was made using the Two Line

Table 1. Optimal values for the SExtractor input parameters for the presented examples. "IN1" is DETECT_MINAREA. "IN2" is DETECT_THRESH, "IN3" is DEBLEND_NTHRESH. "IN4" is DEBLEND_MINCONT. "Avg. ellipses" shows the mean number of ellipses detected. The last column shows the mean accuracy of the position in arc seconds.

Image	Orbit	IN1	IN2	IN3	IN4	Avg. ellipses	Acc. (")
Virgo Cluster	-	5	1.5	32	0.005	15165	0.5
Hispasat	GEO	5	1.5	32	0.06	2065	0.7
Navstar	MEO	5	1.5	32	0.06	1226	0.9
SMOS (sat.)/(field)	LEO	30/1000	2.5/0.9	2/2	0.06/0.7	5/172	–/1.4
1U CubeSat (field)	Low MEO	1000	0.9	2	0.7	215	1.4

Elements (TLEs) of the satellite. Figure 1C shows one of the images. Despite the shorter exposure time of 5 seconds compared to the GEO survey case (10 seconds), the trails left by the stars appear longer and sloping.

In Figure 1C we notice that, when computing the ellipses to fit the trails, several of them (especially the faint ones) present "sub-trails" consisting of two or more ellipses in a row along the trail that, if interpreted as individual stars, can cause the astrometric software to fail. The reason for this error is that trails have multiple peaks of flux along the directions of their orientations, and SExtractor tends to separate the trails in 2 or more "sub-trails". SExtractor's default parameters are set to deblend images of galaxies and other extended objects that are blended in astronomical images but, in our case, we need just the opposite; that is, not to deblend the trails of the stars. SExtractor controls the deblending using the DEBLEND_MINCONT parameter that gives the threshold contrast for deblending and is set by default to 0.005. If in a trail the contrast between two peaks and the corresponding valley is higher than the threshold set by default (0.5%), the software automatically interprets two different trails instead of one. Therefore, we changed the default value of the DEBLEND_MINCONT parameter from 0.005 to 0.06 in order to join close "sub-trails" that had similar fluxes per pixel and directions. The optimal parameters for the case of long trails are listed in the third row (i.e., *Navstar*) in Table 1.

3.4. *Images with very long star trails*

In our opinion, the most interesting result of this paper is to handle the very challenging problem of trails left by the stars that are very long, of even hundreds of pixels. This situation is typical in the case of 1) Tracking of LEO objects in which, due to the high speed of the object crossing the sky, the trails can be very long even for short exposures of 1 second and 2) The case of very faint satellites or objects in higher orbits (as low MEO) that, due to their faintness, long exposures are needed and entail very long trails in the image.

For the case of LEO objects, we used a set of images of the SMOS satellite. Figure 1D shows an image from the SMOS satellite of 1 second of exposure time with star trails of about 450 pixels. With this image, the default values of the parameters leave multiple recognitions of objects in one single trail, with multiple little ellipses. To solve this problem, we separated the analysis in two steps, one directed to detect only the object or objects and the other to detect the long trailed reference stars. The first one consisted of limiting the maximum number of pixels (DETECT_MAXAREA), and we raised the DETECT_THRESH to a very high value of 2.5, resulting in a single ellipse that corresponded to the tracked SMOS satellite (see center of Figure 1D). For this step, other centroid algorithms, such as Synthetic Aperture or Gaussian PSF fitting, can be used to detect the tracked object. The second step aimed to detect only the trails left

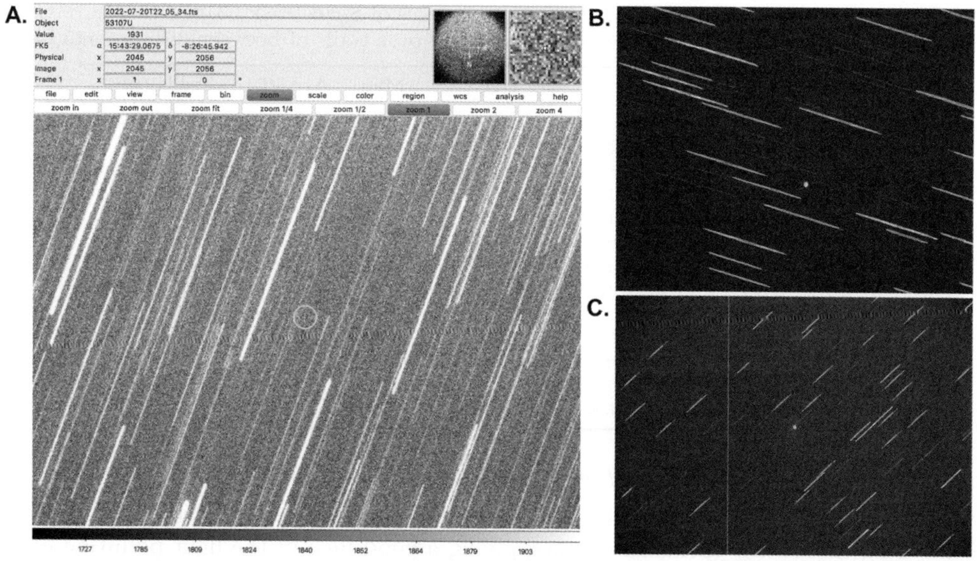

Figure 4. *A.* Observation of a very faint low MEO object (1U CubeSat at 6000 km altitude, Spacetrack 53107 "OBJECT C") tracked using TLEs with 10 seconds of exposure time. The satellite is at the center of the green circle with magnitude 17.6; the trails left by the stars have lengths of about 370 pixels. *B.* OneWeb tracking image with the software applied. *C.* Starlink tracking image with the software applied.

by the stars. The DEBLEND_NTHRESH was raised to 0.7 and the minimum number of pixels (DETECT_MINAREA) was also increased to 1000 to avoid detecting "sub-trails". This approach allowed us to detect more ellipses perfectly positioned with optimal length and inclination. Then, the two output files containing the parameters were mixed. The optimal parameters for the case of very long trails are listed in the fourth row (i.e., *SMOS* (satellite run) and *SMOS* (field stars run)) in Table 1.

For the case of very faint satellites or objects in higher orbits, such as Cubesats, we have used a set of images made with the TFRM of a 1U Cubesat located in a low MEO orbit at an altitude of 5850 km. Specifically, the satellite observed was Spacetrack object number 53107 "OBJECT C" (COSPAR 2022-080-C). Figure 4A shows an image tracking "Object C", with the satellite in the center of a green circle with a magnitude of 17.6. Since Object C is a 1U Cubesat of $10 \times 10 \times 10$ cm located at an altitude of almost 6000 km, its observation by radar is almost impossible due to the distance. The same happens in case of short-exposure optical observations due to the faintness of the object. To have optical detection possibilities, the exposure time must be increased, which implies very long trails of up to 450 pixels. In this example, due to the faintness of the object, we used the combined method of synthetic aperture (mentioned in the previous section) and, to solve astrometrically the images, we used the values that are listed in the eighth row (i.e., 1U Cubesat (field stars run)) in Table 1. Therefore, using this combined method, we have been able to astrometrically solve the images and obtain astrometric positions of a fast moving object, with magnitude up to V=18 and with a mean error of 1.38 arcsec (1/3 of a pixel). This is a remarkable result giving that the trails left by the reference stars are between 365 and 445 pixels. To show the performance of the method on such a faint moving object we present the astrometric results for some of the images of Spacetrack 53107 "OBJECT C" in Table 2.

Table 2. Astrometric results for some of the images of Spacetrack 53107 "OBJECT C". Mean astrometric error is 1.38 arcsec (1/3 of a pixel) over trails of between 365 and 445 pixels.

Date and time	R.A.	Dec.	Mag	Acc. (")	Trails (pix.)
2022 07 20.911887	14 44 47.061	+21 29 17.17	17.7	2.02	445
2022 07 20.912442	14 49 43.021	+19 31 18.15	17.6	1.12	440
2022 07 20.912627	14 51 18.808	+18 51 55.55	16.8	1.25	440
2022 07 20.912812	14 52 53.174	+18 12 29.18	17.1	1.06	435
2022 07 20.915220	15 11 27.141	+09 40 46.10	17.4	1.32	420
2022 07 20.917257	15 24 52.923	+02 36 08.74	17.6	1.42	405
2022 07 20.917627	15 27 04.419	+01 20 15.34	17.7	1.52	395
2022 07 20.917812	15 28 15.455	+00 42 29.26	16.8	1.14	400
2022 07 20.918368	15 31 30.739	−01 09 47.69	17.5	1.38	395
2022 07 20.918553	15 32 34.511	−01 47 00.04	15.7	1.64	390
2022 07 20.920220	15 41 39.918	−07 15 19.50	17.4	1.50	375
2022 07 20.920590	15 43 29.124	−08 26 47.91	17.6	1.18	370
2022 07 20.920775	15 44 31.596	−09 02 02.48	16.2	1.37	365

3.5. *Application for megaconstellation satellites*

The results of this work allows us to observe the magnitudes of satellites in LEO and comparing those with background stars, giving a good idea of the light pollution of the brightest satellites. The main issue that can be targeted is the effects of the Starlink and OneWeb megaconstellations, as shown in Figure 4B and C. As of March 2024, SpaceX has over 5500 functioning Starlink satellites in orbit and the effects caused by the constellation are beginning to be noticed by the astronomy community.

The software presented in this work gives a tool to professional and amateur astronomers to analyze how the megaconstellations of satellites are reflecting light (see, column 4 of Table 2).

4. Conclusion

In this work, we present a new technique that allows an integration and an operation for a complex system for space (i.e., optical satellite tracking), and it can be useful in helping to control the space debris problem. We have shown that in optical observations of objects like satellites and space debris with severely trailed images, with trails of more than 400 pixels, it is possible to perform an astrometric reduction and obtain precise positions of orbital objects. To do this, we have developed a technique based on the modification of the input parameters of free software (SExtractor) that allows us the recognition and precise astrometric positioning of trails of any length left by the reference stars and of the objects of study. We also saw that this technique allows us to detect anything that orbits in a different direction than the satellite we are tracking, which can help us to recognise all kinds of objects if those are bright enough for detection.

In Table 2 we summarise the software optimised parameters for the examples presented in this work. The first image that we analyzed was in sidereal tracking (the Virgo Cluster of galaxies) with a mean astrometric accuracy of 0.5". For the second set of images, with short trails, we used a set of 3 images of the geostationary Hispasat satellite constellation taken in GEO survey mode (telescope stopped), recognizing 2065 star trails on average, the 4 satellites and other GEO objects moving in a different direction with a mean astrometric accuracy of 0.7". For the third set of images, with relatively long trails as star images, we used Navstar satellite tracking, recognizing 1226 star trails on average and the satellite, with a mean astrometric accuracy of 0.9". For the fourth one, that presents very long trails, we used 7 sets of images of the SMOS satellite recognizing 172 star trails on average, and we correctly positioned the satellite in 6 of them, with a mean astrometric accuracy of 1.4". Finally, for the fifth set of images, also presenting very long trails and a very faint satellite, we used a set of images of a 1U CubeSat in a low MEO orbit at about 6000 km of altitude. The mean number of star trails found was 215 and we correctly positioned the faint satellite using the synthetic aperture method to obtain the centroid. Again, the mean astrometric accuracy was 1.4". We also tested the method with other sets of images for trails short (GEO), long (MEO) and very long (LEO or faint satellites in low LEO) with similar results.

The obtained results can have applications in many fields of astronomy and, by the determination of the precise magnitude of the orbital objects, to advise on the negative impact on astronomy of new megaconstellations of LEO satellites.

Acknowledgements

The authors acknowledge Dr. Daniel del Ser for his fruitful advice to use the SExtractor input parameters and all the team of TFRM for providing all the images presented in this paper. This work was supported, in part, by the Spanish National Research Agency (AEI) under grant n. PID2019-105510GB-C31/AEI/10.13039/501100011033.

Competing Interests

The authors declare that they have no conflict of interest.

References

Bertin, E. & Arnouts, S. 1996, A&AS, 117, 393
Fors, O., Núñez, J., Muiños, J. L. et al. 2013, PASP, 957, 522
Leslie, M. 2020, Engineering, 6(8), 849
Nazarenko, A. I. & Usovik, I. V. 2022, Acta Astron, 194, 383
Park, S.-Y., Keum, K.-H., Lee, S.-W. et al. 2013, J Astro Space Sci, 30, 193
Petit, A., et al. 2021, Procedings 8th European Conference on Space Debris
Schildknecht, T. 2007, A&AR, 14, 41
Sun, R.-Y. & Zhao, C.-Y. 2013, Res Astro Astroph, 13(5), 604
Wells, D. C., Greisen, E. W. & Harten, R. H. et al. 1981, 44, 363
Yang, W., Zhao, Y., Liu, M. & Liu, D. 2021, Opt Expr, 29(22), 35348

Astronomy and Satellite Constellations: Pathways Forward
Proceedings IAU Symposium No. 385, 2026
P. Grimley, J. Barentine, D. Galadí-Enríquez & C. Walker
doi:10.1017/S1743921324001108

Report of the International Institute of Space Law on light pollution of the night sky from a space law perspective

Rafael Moro Aguilar

Department of Politics and International Relations
Florida International University
11200 SW 8th Street, Miami, Florida 33199, United States of America
email: rmoroagu@fiu.edu

Abstract. The International Institute of Space Law (IISL) is the largest worldwide association of space lawyers. In 2021, IISL created a Working Group (WG) on light pollution of the night sky caused by space objects, particularly by satellite constellations. The WG completed its research early in 2023 and issued a report that was adopted by IISL in June 2023. The report shows that light pollution generated by spacecraft in the optical range is a new problem that is currently unregulated. However, a number of existing and emerging rules in international law and in the domestic laws of many countries support the protection of astronomy and the night sky. The eventual adoption of standards limiting the brightness of satellites as well as international agreements protecting dark skies was also explored by the WG. The author, in his capacity as coordinator of the WG, summarizes the contents of the report and explores its implications.

Keywords. Satellite constellations, light pollution, Space Law, International Law

1. Introduction

Since May 2019, we are in the age of large constellations of satellites (satcons). Several companies are building networks of hundreds or even thousands of satellites each. Their main goal is to provide low-latency broadband internet access from space to virtually any place on the planet. Other companies and nations have plans for setting up their own constellations in low Earth orbit (LEO) with the same purpose, or to provide other communications or remote sensing services. These developments take advantage of the decreasing launch and satellite costs.

Providing worldwide internet connectivity and helping bridge the digital divide is a beneficial application of space technology. However, the surge in the number of satellites in LEO implies technical and regulatory challenges, and it is having an environmental effect: a steady growth in the light pollution of the night sky.

These thousands of new space objects in LEO are changing the appearance of the night sky. Astronomers are concerned with the impact that a night sky increasingly filled up with shiny artificial objects will have on scientific research. Other impacts of this contamination can be expected as well: on the culture and ceremonies of Indigenous communities, which may suffer an alteration; on many animal wildlife species, which may see their daily life cycles and their orientation capacities altered; and on the ability of the general public to view and enjoy a natural and pristine night sky.

In terms of numbers of people affected, potential economic damage, and scientific repercussions, the disruption of astronomical research is particularly worrisome. Studies suggest that some astronomical observations will be particularly affected by large satcons, such as detection of objects (including near-Earth objects) and observations conducted during twilight, wide-angle surveys, and radio astronomical observations (AAS 2020).

There is no need to recall the many long-term scientific and cultural contributions of astronomy to humanity (see, for instance, UNGA 2007a, which declared 2009 as the International Year of Astronomy, particularly preamble paragraphs 2 and 3). However, should all the planned satellite constellations become a reality, the situation may become unsustainable for ground-based astronomy, even if all currently proposed mitigation measures are adopted. It may become unsustainable for much space-based astronomy as well, as telescopes orbiting in LEO such as the Hubble Space Telescope are also affected by other satellites, and in such cases, mitigations are more challenging to implement[1].

This emerging problem has been the object of several international conferences. It has prompted statements from the International Astronomical Union (IAU) and from other institutions, such as the European Southern Observatory (ESO) and the International Dark Sky Association (IDA). It has also attracted the attention of the public at large, both inside and outside the space community.

These concerns have reached the United Nations and its Committee on the Peaceful Uses of Outer Space (COPUOS), where some national delegations and observer organisations have presented the issue. At the 58th session of the COPUOS Scientific and Technical Subcommittee (STSC), in February 2021, a group of countries and the IAU submitted the conference room paper "Recommendations to Keep Dark and Quiet Skies for Science and Society" (COPUOS 2021). At the same session, the STSC decided to add to its agenda an item entitled "General exchange of views regarding satellite system effects upon terrestrial-based astronomy", with the purpose of allowing delegations to provide in the future their views on both the impact of satcons on astronomy and its relevance to the work of COPUOS (COPUOS 2022a).

Discussions were held during the sessions of the STSC in 2022 and 2023. The issue of the preservation of dark and quiet skies was included again on the agenda of the STSC in February 2024, under a different name.

The plenary of COPUOS (65th session, June 2022) recognised that "astronomical observations for both optical and radio astronomy [a]re an essential aspect of space activities and should be protected from interference," as reflected in its annual report[2].

2. The IISL Working Group on Light Pollution of the Night Sky from a Space Law Perspective

Founded in 1960, the International Institute of Space Law (IISL) is the most important global association of space lawyers[3]. It is composed of individuals and institutions from more than 50 countries, comprising academic, governmental, industrial, and legal practice sectors. The IISL's key mission is the promotion of further development of space law and the expansion of the rule of law in the exploration and use of outer space. It includes the studies of legal and social science aspects of the exploration and use of outer space and holding of meetings, colloquia, and competitions on juridical aspects of space activities.

At its spring meeting of 2021, the Board of Directors of IISL approved the creation of the study group "IISL Working Group on Light Pollution of the Night Sky from a Space Law Perspective"[4]. The Working Group (hereinafter, WG) started operating in

[1] See https://www.iau.org/static/publications/uncopuos-stsc-crp-8jan2021.pdf at paragraph 18. For more information on the HST, see e.g. https://www.nature.com/articles/s41550-023-01903-3. On the impact on other space telescopes, see the statement from India at the COPUOS STSC about India's AstroSat at https://www.unoosa.org/documents/pdf/copuos/stsc/2023/Statements/16_PM/17_India_16_Feb_PM_1.pdf.

[2] See https://www.unoosa.org/oosa/en/ourwork/copuos/2022/index.html

[3] https://iisl.space/

[4] https://iisl.space/index.php/2023/11/07/iisl-general-assembly-successfully-held-on-6-november-2021/

September 2021. The purposes of this WG were to collect information about the light pollution of the night sky caused by satellite constellations, analyse this problem from a space law perspective, raise awareness of the topic among all IISL members, and provide the IISL Board of Directors with an up-to-date report on the current situation.

The WG completed its report early in 2023, with 10 contributors and 110 pages. The main conclusions were made public during the Symposium held in the 2023 session of COPUOS LSC in late March. The report was then reviewed by the IISL Directorate of Studies and adopted by the IISL Board of Directors in June 2023. The report was presented to the IISL community in June 2023 at a webinar which was well attended and which provided some additional input. The report (IISL 2023) will be shared with all the delegations of COPUOS in 2024[5].

The subject matter of the report is the light pollution caused by space objects in the night sky, and in particular the optical impact of the large satcons, which are being placed in LEO. These satcons are also informally known as "mega-constellations". The WG agreed to use all three names in their deliberations, although the term mega-constellation, which, it has been pointed out, is technically inaccurate, has been used only occasionally in the report.

Another terminological issue discussed by the WG was whether this is a case of light pollution, or a case of visual or optical interference with the work of astronomers. Actually, it can be considered both. Besides the individual interferences caused in the images and on other data by the trails of satellites, the sky as a whole is becoming increasingly polluted, as the combined brightness of all orbiting spacecraft increases the diffuse skyglow (see, for instance, Kocifaj, Kundraci, Barentine & Bará 2021).

Several members supported the WG's consideration of the work already done by other bodies on the topic of light pollution of the night sky and its impact on astronomy and society, in particular the reports of the IAU/UNOOSA/Spain Dark & Quiet Skies for Science and Society I and II meetings that were held in October 2020 and October 2021 (Walker et al. 2021; Walker & Benvenuti 2022).

The WG agreed that the aforementioned reports were a significant source of information, and that they would be used as a basis for the discussions. The recommendations addressed to all stakeholders concerning mitigation measures contained in those reports are noteworthy, and they should be taken into account by everyone who approaches this issue. Several members of the IISL WG, including the present author, contributed to the 2021 Dark & Quiet Skies II Conference and its associated reports. The D & Q Skies II reports can be found at Walker & Benvenuti (2022). A summary of the reports was presented to COPUOS in 2022 as UN Document A/AC.105/1255 (COPUOS 2022b).The WG underlined that it should proceed independently in its examination of the topic, and that it would produce its own report on behalf of IISL.

3. Light pollution of the night sky as a legal problem

This is first a highly technical problem that requires technical solutions. However, it can also be seen as a legal problem, as there are two conflicting human activities: large constellations of satellites, and astronomical observations (together with general stargazing and other affected groups).

The interference posed by these satellites is still moderate, since there are not too many satellites in LEO yet, and most of them fall below the threshold of naked-eye visibility as soon as twilight ends[6]. At the same time, this is a real problem that is already affecting professional and amateur astronomers, posing a serious risk to their research.

[5] The Report is available at https://iisl.space/index.php/2023/07/05/iisl-publishes-report-of-iisl-light-pollution-wg/.

[6] Not all satellites do, however: https://www.space.com/bluewalker-3-prototype-satellite-brightest-objects-sky

This issue should be addressed promptly, before it becomes much worse. Note the similarity with the space debris problem, which was moderate initially, but grew over the years, owing to an almost total lack of regulation and the slow adoption of measures. Space debris is also, similar to light pollution stemming from satcons, an unintended negative effect that results largely from activities that are considered beneficial.

Satcons providing global telecommunication services are a lawful activity with practical benefits. Ground-based astronomical observation is also a lawful activity that provides benefits (if more intangible) for mankind. It is desirable to achieve a coordination of activities so that they can both be developed in a sustainable manner.

International space law can play a role in achieving that consensus. There is currently a lack of any regulation of optical interferences (unlike interferences in the radio range) in international law. However, a number of international space law rules, general international law rules, and pieces of national legislation around the world either can apply or at least be relevant to this situation, and they can help provide solutions.

4. Applicable rules of international space law: the 1967 Outer Space Treaty (OST)

First, the WG examined the potential application of international rules in force for outer space activities, particularly the 1967 Outer Space Treaty (OST). This Treaty is considered the cornerstone of contemporary international space law (UNGA 1966b).

Some rules contained in the OST that may apply were identified by the WG and by other legal authors, most notably Article I (which establishes freedom of access to and exploration and use of outer space), Article III (which opens the way to the application of international law in outer space), Article VI (which brings the duty to authorise and supervise private operators of space objects), and Article IX (with its "due regard," "harmful interference," and "duty of consultation" provisions).

Article I OST establishes freedom for all states to conduct exploration of and scientific investigations in outer space. It establishes the free utilisation of outer space, including by way of satcons. Therefore, both astronomy and satcons constitute legitimate and lawful activities according to the OST. The Treaty does not establish any priority between competing space activities. Then, both activities could be considered to be on equal footing, while having in mind that it is large satcons that are interfering with astronomy, and not the other way around (see p. 109 of Boley & Byers 2023).

Article III OST establishes that all human activities in space must be conducted in accordance with international law. This opens the way to the application of other branches of international law in outer space, such as international environmental law.

Important principles of international environmental law that might serve as an inspiration to solve this problem include the prevention of transboundary harm, the precautionary principle, and the principles of equitable use and sustainable development.

Principle 21 of the 1972 Stockholm Declaration of the United Nations on the Human Environment and Development is particularly relevant: States have the responsibility to ensure that activities within their jurisdiction or control do not damage the environment outside their jurisdiction (UNEP 1972).

Article VI OST establishes:

(i) State responsibility for all national activities, including those activities carried out by non-governmental (i.e. private) entities.
(ii) As a result, the requirement for private space activities to be subject to authorisation and continuing supervision.

Article VI means that States are internationally responsible for ensuring that all space actors conduct their activities in conformity with the OST and with general international law. As a part of the compliance by private operators, including operators of satcons, of applicable international law, States must guarantee a responsible and sustainable use of the Earth's orbital space environment. The responsible and sustainable use of outer space by States should include limiting the light pollution that may be caused by their space objects.

Article IX constitutes another limitation on the freedom of exploration and use of outer space established in Article I. From this provision, States Parties to the OST must conduct all of their activities in outer space: 1) paying due regard to the corresponding interests of all other States Parties to the Treaty; 2) avoiding harmful interference with the activities of others; 3) avoiding the harmful contamination of outer space and celestial bodies, as well as adverse changes in Earth's environment; and 4) engaging in international consultations with affected States Parties in case of an ongoing or potential harmful interference.

Article IX OST seems suitable to our problem, provided that 1) the term "harmful interference" includes visual interference on astronomical activities, and that 2) ground-based astronomy may be considered as an activity contained within its scope. The latter is not clear from the text, but there are strong arguments in favor of considering astronomy, ground-based or space-based, as an activity of exploration of outer space, and therefore under the remit of the Outer Space Treaty (pp. 98 ff. of Boley & Byers 2023).

Historical interpretation of Article IX OST favours the view that essential activities conducted by States on the ground must be protected under its provisions. The analysis of the events which led to formation of Principle 6 of the United Nations General Assembly Resolution 1962 (XVIII), (UNGA 1963), a direct predecessor of Article IX OST, suggests that the intention of the drafters was to protect the interests and activities of States carried out on Earth as well as in outer space.

As explained in detail in Section 4 of IISL (2023) and elsewhere (Moro Aguilar 2024, also available at pp. 114 ff. -Subchapter 4.4- of Walker & Benvenuti 2022), the origin of these two provisions can be traced to certain large-scale experiments conducted in near-Earth orbits by the space powers during the early 1960s: Project West Ford (The Harvard Crimson 1963), and a number of deliberate nuclear explosions (see, for instance, DNA 1958), the effects of which were felt both in outer space and on the Earth.

Article IX also includes an obligation to engage into consultations if an interference with the activities of other State or States is about to occur or if it actually occurs.

When applying the provisions of Article IX, we have the problem of its vagueness. Neither "due regard" nor the other key terms are defined anywhere, and there is a wide margin of interpretation of these terms. We also suffer from the lack of any previous State practice in applying those provisions so far. OST Article IX has rarely been invoked, and no international consultations have ever taken place on the basis of its provisions.

As a result, there is no guidance on how such duties would actually apply to optical pollution and interferences caused by space objects. We will not know for sure until a State or group of States takes the step of invoking Article IX and calls out for consultations related to this matter.

5. The need for new international rules with regard to light pollution of the night sky

Apart from the preceding provisions of the OST, and the potential application of international environmental law to outer space via OST Art. III, at present there are no international rules that apply to the problem of light pollution caused by space objects.

Given that current regulation is clearly insufficient, the potential elaboration of new legal rules addressing specifically this problem was also considered by the WG.

Inspiration for those new rules can be found in the provisions protecting radio astronomy, which have existed for many years and are contained in the Radio Regulations of the International Telecommunication Union (ITU). Protection is achieved by excluding from commercial purposes a number of frequencies and bands to be used exclusively by radio telescopes. National and local laws enable establishing so-called radio-quiet zones around radio observatories[7]. However, no international protection similar to ITU's is found anywhere with regard to optical interference. This lack of specific standards limiting the visual impact of space objects is an important legal void.

This is not necessarily a legal but a space governance problem, in the sense that there are actually some norms already that could apply to the situation, most notably OST Article IX, as pointed out, but for some reason they are not being implemented by States.

In the context of global space governance, let us mention the Statement issued by the Board of Directors of IISL in July 2021, "Consideration of the Interests of the Public and other Stakeholders in the Authorization and Continuing Supervision of Commercial Space Activities." This Statement highlights the importance for national licensing authorities to consider the interests of the global public while granting licenses to private sector applicants. It also declares that, "*space activities should be developed in accordance with the rule of law, for peaceful purposes, and in a manner that is sustainable for the present and future generations.*"[8]

6. Adoption of non-binding standards

Eventually, having international rules, standards or guidelines that are applied uniformly will become necessary. If the international community decides to create such international standards, a place to do so would be the UN COPUOS, which offers the best forum to propose such an uniform approach.

As a first step towards the adoption of a common framework to limit light pollution due to space activities, the WG considered the path of adopting a series of non-binding guidelines in COPUOS. Such rules could take as a model the Space Debris Mitigation Guidelines and the Long-Term Sustainability (LTS) Guidelines adopted in COPUOS in 2007 and 2019, respectively. The aim would be to reflect in those new guidelines a compromise between the satellite operators and the needs of astronomers. The guidelines could then be picked up by national governments as licensing regulations and be gradually applied through their national space laws.

The LTS Guidelines are a useful model for addressing the visual interferences. These guidelines are adopted through a bottom-up approach, offering a platform for all stakeholders to express their views. This is a model that offers flexibility to each country, since they are voluntary norms and, at the same time, certainty to governments and operators.

The LTS Guidelines can fill the gaps of the existing UN Outer Space Treaties and serve as a basis for further development of the standards into binding laws and treaties. If followed by the practice of States, they can become binding customary international rules. The LTS Guidelines have been envisaged as an evolving instrument that should be updated as new challenges are identified. Accordingly, they can serve as a path for addressing the current challenge of establishing a preliminary, non-binding regulation of large satellite constellations with an aim to mitigate their impact on dark and quiet skies.

[7] Such as e.g. South Africa's 'Astronomy Geographic Advantage Act' of 2007: https://www.sarao.ac.za/about/astronomy-geographic-advantage-act/

[8] https://iisl.space/index.php/2023/11/07/iisl-bod-statement-on-the-consideration-of-the-interests-of-the-public-and-other-stakeholders-in-the-authorization-and-continuing-supervision-of-commercial-space-activities/

The actual elaboration of guidelines on optical interference and light pollution caused by space objects could take place in a new, separate expert group. In the discussions during the sessions of COPUOS in 2023, the vast majority of Member States favoured creating a specific group for this purpose.

7. Adoption of international agreements

The IISL WG explored the possibility of adopting some legal protection of the night sky at the international level. In the longer term, the international community might decide to conclude agreements protecting the quality of the night sky as a world heritage. Attempts have been made in that respect, and some soft law documents already exist, notably the Starlight Declaration, available in Starlight Initiative (2007).

We also have the precedents of the United Nations Education, Science and Culture Organization (UNESCO) protecting astronomical heritage sites[9], or the IDA conferring dark-sky status to certain areas around the world[10].

These international initiatives are in the right direction leading to creation of awareness and recognition of the light pollution problem. Although any binding legislative attempt for night sky protection on a global scale is yet to be seen, these initiatives suggest that such attempts are very likely in the near future.

A possibility would be the adoption of an international agreement protecting the dark sky as a world natural and cultural heritage. This would be done under UNESCO, given its traditional role in protecting humankind's natural and cultural heritage.

The emergence of a new fundamental human right of a cultural nature that would protect stargazing and the enjoyment of a pristine night sky was also discussed by the WG. Certain international texts already in force acknowledge the importance of participation of the individual in cultural life and science, as well as the right of Indigenous peoples to perform ceremonies and maintain their traditions, some related to the night sky. The 1966 International Covenant on Economic, Social and Cultural Rights, (UNGA 1966a, in particular Article 15) and the 2007 United Nations Declaration on the Rights of Indigenous Peoples (UNGA (2007b), in particular Articles 11.1, 12.1 & 13.1) contain provisions that could be interpreted in the sense that astronomy (or stargazing generally) is a fundamental cultural right that has to be protected.

8. Existing national laws

Solutions to the current situation could also come from national space law, by introducing licensing rules and legal standards that would be imposed on operators of satellite constellations, taking into account the obligation for the relevant States Parties to the OST to authorise and constantly supervise private space activities.

In this context of national legislation, it is worth noting again the Statement issued by the Board of Directors of IISL in July 2021, which asks national regulators to consider the interests of all stakeholders, including the scientific community and the general public, when authorising activities such as large satcons.

The IISL WG looked for any existing national regulation limiting the optical brightness of space objects, but none was found. The only related national rule is a United States federal law enacted in 2001 prohibiting obtrusive space advertising[11], enacted by U.S. Congress following a suggestion from the IAU in 2000 to COPUOS Member States (COPUOS (2001)). This law does not apply exactly to our case, but it provides a precedent of States listening to astronomers and protecting the value of the global dark night sky.

[9] https://www.iau.org/news/pressreleases/detail/iau1006/
[10] https://darksky.org/what-we-do/international-dark-sky-places/
[11] 51 U.S. Code § 50911. See at: https://www.law.cornell.edu/uscode/text/51/50911

The WG identified a large number of national and local laws and regulations in force preserving the quality of dark skies. The IISL report provides an overview of such laws and regulations, albeit from ground-emitted light pollution only.

Although not directly applicable to our current problem, these laws are evidence of an *opinio iuris* of States in terms of acknowledging the value of dark skies, and they constitute a precedent for a worldwide protection of the night sky. These laws may provide elements relevant to the regulation of the visual impact of satcons.

If no limits are set to the brightness of space objects, particularly at the international level, the progressive deterioration of the global night sky will defeat the purpose of all these national laws and regulations and render them useless.

9. Mitigation measures and collaboration between industry and astronomy

Section 8 of the IISL report contains an analysis of what is actually being done in order to address light pollution of the night sky caused by satcons. Collaboration between the satellite industry and the astronomical community has led to a set of best practices. These include implementing changes to satellite designs to make them less reflective; modifying the orientation of satellites in orbit-raising and orbit-lowering phases in order to minimise the sunlight reflected; and providing positional data of the satellites to astronomers so that they can plan their observations accordingly.

The WG examined to what extent these voluntary measures are working. It is acknowledged that this channel of communication between astronomers and operators must be preserved. However, there is a general perception that the mitigation measures adopted so far have not been sufficient. Even if they had been, the continuity of such measures is dependent on the goodwill of the companies. Self-regulation should not replace the international institutions and instruments governing problematic matters. It can better be conceived as a complementary tool in the joint efforts to preserve the dark skies.

The WG also considered the recent initiatives on space sustainability rating (SSR), as a potential additional measure adopted by governments and companies to promote a responsible use of outer space.[12] The rating of companies based on their respect for sustainability of the orbital environment could also take into account any measures taken with the intention to reduce light pollution caused by satcons and by space objects.

10. Conclusions

Satellite constellations and astronomy are both lawful forms of exploring and using outer space, according to international space law. A coordination in the exercise of these freedoms is necessary, and for the moment it is being achieved through consultations between the space industry and the astronomers and through the adoption of voluntary mitigation measures. However, a proper regulation of the brightness of space objects as seen from the ground will be necessary to fully solve the problem.

Although there is a lack of rules specifically addressing the visual impact of space objects, certain provisions of the 1967 Outer Space Treaty apply to this issue, notably Articles I, III, VI and IX. These could serve as a basis for future international regulations.

A first step towards an international regulation could be the adoption in COPUOS of non-binding guidelines to limit the light pollution caused by space objects. These would then be picked up by national regulators and imposed at the national level. Also the inverse process could happen: Relevant standards might be set up first in the national space law of key countries and then international guidelines developed.

[12] https://spacesustainabilityrating.org/

In the longer term, formal international agreements could be envisioned to protect astronomy and dark and quiet skies, either as a human heritage or as a human right. Such conventions would be an indispensable complement to the many national laws and regulations that are protecting at the local level the quality of the night sky.

References

AAS (American Astronomical Society) 2020, Impact of Satellite Constellation on Optical Astronomy and Recommendations Toward Mitigation, https://aas.org/sites/default/files/2020-08/SATCON1-Report.pdf

Boley, A. & Byers, M. 2023, Who Owns Outer Space?, Cambridge University Press, Ch. 3, https://www.cambridge.org/core/books/who-owns-outer-space/megaconstellations-and-international-law/B7453D99235EB7C62A29C9191087F612

COPUOS (Committee on the Peaceful Uses of Outer Space) 2001, Obtrusive space advertising and astronomical research: Background paper by the International Astronomical Union, UN Document A/AC.105/777, 18 December 2001, https://www.unoosa.org/pdf/reports/ac105/AC105_777E.pdf

COPUOS (Committee on the Peaceful Uses of Outer Space) 2021, Recommendations to Keep Dark and Quiet Skies for Science and Society, https://www.unoosa.org/res/oosadoc/data/documents/2021/aac_105c_12021crp/aac_105c_12021crp_17_0_html/AC105_C1_2021_CRP17E.pdf

COPUOS (Committee on the Peaceful Uses of Outer Space) 2022a, Summary of discussions on dark and quiet skies for science and society, https://www.unoosa.org/oosa/oosadoc/data/documents/2022/aac.105/aac.1051257_0.html

COPUOS (Committee on the Peaceful Uses of Outer Space) 2022b, Report on the United Nations/Spain/International Astronomical Union Conference on Dark and Quiet Skies for Science and Society, https://www.unoosa.org/oosa/oosadoc/data/documents/2022/aac.105/aac.1051255_0.html

DNA (Defense Nuclear Agency) 1958, Operation Argus, United States Atmospheric Nuclear Weapons Tests: Nuclear Test Personnel Review DNA 6039F, prepared by the DNA as Executive Agency for the Department of Defense, https://babel.hathitrust.org/cgi/pt?id=uiug.30112075683737&view=1up&seq=10&skin=2021

IISL (International Institute of Space Law) 2023, Keeping the Night Sky Clear: IISL Working Group on the Light Pollution of the Night Sky from a Space Law Perspective, Final Report, https://iisl.space/index.php/2023/07/05/iisl-publishes-report-of-iisl-light-pollution-wg/

Kocifaj, M., Kundracik, F., Barentine J.C. & Bará, S. 2021, MNRAS, 504, L40

Moro Aguilar, R. 2024, Mega-constellations of Satellites and their Impact on Astronomy: Exploring the Role of Article IX of the Outer Space Treaty, Proc IISL 65

Starlight Initiative 2007, La Palma Declaration in Defence of the Night Sky and the Right to Starlight, https://www.starlight2007.net/index_option_com_content_view_article_id_185_itemid_80_lang_en.html

The Harvard Crimson 1963, May 24th 1963, https://www.thecrimson.com/article/1963/5/24/project-west-ford-pfour-hundred-million/

UNEP (United Nations Environment Program) 1972, Declaration on the Human Environment ("Stockholm Declaration"), Environmental Law: Guidelines and Principles, vol. 1, https://wedocs.unep.org/bitstream/handle/20.500.11822/29567/ELGP1StockD.pdf

UNGA (United Nations General Assembly) 1963, Declaration of Legal Principles Governing the Activities of States in the Exploration and Use of Outer Space, adopted by UNGA Resolution 1962 (XVIII) of 13 December 1963, https://www.unoosa.org/oosa/en/ourwork/spacelaw/principles/legal-principles.html

UNGA (United Nations General Assembly) 1966a, International Covenant on Economic, Social and Cultural Rights, adopted by UNGA on December 16, 1966 by Resolution 2200A (XXI), in force since January 3, 1976, https://www.ohchr.org/en/instruments-mechanisms/instruments/international-covenant-economic-social-and-cultural-rights

UNGA (United Nations General Assembly) 1966b, Treaty on Principles Governing the Activities of States in the Exploration and Use of Outer Space, Including the Moon and Other Celestial Bodies, adopted by UNGA Resolution 2222 (XXI) of 19 December 1966, opened for signature on 27 January 1967, entered into force 10 October 1967, United Nations Treaty Series, vol. 610, p. 205 ('The Outer Space Treaty'), https://www.unoosa.org/oosa/en/ourwork/spacelaw/treaties/introouterspacetreaty.html

UNGA (United Nations General Assembly) 2007a, Resolution 62/200, of December 19, http://www.worldlii.org/int/other/UNGA/2007/248.pdf

UNGA (United Nations General Assembly) 2007b, United Nations Declaration on the Rights of Indigenous Peoples, Resolution 61/295, of December 13, https://www.un.org/development/desa/indigenouspeoples/declaration-on-the-rights-of-indigenous-peoples.html.

Walker, C. et al. 2021, Dark and Quiet Skies for Science and Society: Report and recommendations, Zenodo, `https://doi.org/10.5281/ZENODO.5898785`

Walker, C. & Benvenuti, P. (eds.) 2022, Dark and Quiet Skies II Working Group Reports, Zenodo, `https://doi.org/10.5281/ZENODO.5874725`

Astronomy and Satellite Constellations: Pathways Forward
Proceedings IAU Symposium No. 385, 2026
P. Grimley, J. Barentine, D. Galadí-Enríquez & C. Walker
doi:10.1017/S1743921324000863

Astronomy and satellite constellations: an international lens

Giuliana Rotola[1], Carmen Artigas Moreno[2], Tamara Blagojevich[3], Roohi Dalal[4], Adam Dong[5], Andrew Falle[6], Charlotte Hook[6], Christopher Johnson[7], Jonathan Kaley-Isley[8], Mukulu Kioko[9], Rayan Khan[10], Anne-Sophie Martin[11], Christopher Martin[12], Rafael Moro Aguilar[13], Tiffany Nichols[14], Oleksandra Ostapenko[5], Andrew Simon-Butler[15], Sarah Thiele[4], Marieta Valdivia Lefort[16], Andrew Williams[17], Yana Yakushina[8] and Aaron Boley[5,*]

[1]Sant'Anna School of Advanced Studies, Piazza Martiri della Libertà, 33, 56127 Pisa PI, Italy

[2]Expert on mission UN, Uruguay

[3]OSA Consulting & Space Court Foundation, Serbia

[4]Department of Astrophysical Sciences, Princeton University, 4 Ivy Lane, Princeton, NJ 08544, USA

[5]Department of Physics and Astronomy, University of British Columbia, 6224 Agricultural Rd, Vancouver BC, V6T 1Z1, Canada

[6]Department of Political Science, University of British Columbia, 1866 Main Mall C425, Vancouver, BC, V6T 1Z1, Canada

[7]Secure World Foundation, USA

[8]Space Court Foundation, Netherlands

[9]Square Kilometre Array Observatory, United Kingdom

[10]Cosmic Tribe SMC Pvt Ltd, Pakistan

[11]Institute for International Legal Studies (ISGI-CNR), Rome, Italy

[12]Vela Science, 1 Sunningvale Lane, Christchurch 8053, New Zealand

[13]Florida International University, Miami, Florida, USA

[14]History Department, Princeton University, 129 Dickinson Hall, Princeton, NJ 08544, USA

[15]Interdisciplinary Studies Graduate Program, University of British Columbia, 312-6174 University Blvd, Vancouver BC, V6T 1Z3, Canada

[16]Institute of the Americas, University College London, 51 Gordon Square, London WC1H 0PN

[17]European Southern Observatory, Garching, Germany

[*]Corresponding author: aaron.boley@ubc.ca

Abstract. This article summarizes recent findings of the CPS Work Package 2 Working Group concerning international law and policy relevant to the protection of dark and quiet skies. It identifies existing gaps in the understanding and application of relevant international instruments, extracts lessons learned from previous analyses, and proposes future steps to address orbital light and spectrum pollution. In particular, the article emphasizes the need to use multiple approaches alongside space law, such as policy advocacy and public engagement, to diversify strategies and achieve more effective solutions.

Keywords. Dark and quiet skies; protection of dark skies; orbital light and spectrum pollution; astronomy; international law; international policy

1. Introduction

During the IAU385 Symposium on "Astronomy and Satellite Constellations: Pathways Forward", the IAU Centre for the Protection of the Dark and Quiet Sky, Policy Hub Work Package 2 Working Group (hereafter, WG2) presented a broad review of international and national instruments that could shape efforts to protect dark and quiet skies (D&QS). This article summarizes the WG2 analysis of relevant international law and policies. The work builds on previous studies such as SATCON2 (SATCON2 2021) and Dark and Quiet Skies II (DQSII 2021). The results for the national analysis are presented in a separate article (Yakushina et al., this volume).

Central to the discussion of international protections for D&QS is the complexity inherent in space governance, including the interpretation of key treaties, such as the Outer Space Treaty (OST 1967), and roles of international fora, such as the UN Committee on the Peaceful Uses of Outer Space (COPUOS). Emerging from this complexity is a need to consider alternative avenues, alongside established mechanisms, to make progress on timescales that match the rapid industrialization of outer space.

Primary questions discussed within WG2 are whether astronomical observations should be considered an exercise in the "exploration and use" of outer space under the OST and if not, whether their inherent need to access outer space still provides protections under the OST.

Another question concerns whether there is potential for the International Telecommunication Union (ITU) to aid in mitigating light pollution, in addition to serving its critical role in managing the radio frequency spectrum. Germane to this is the discussion around whether orbital allocations in low Earth orbit (LEO) are possible and whether this too could be done primarily through the ITU.

Environmental principles are also brought into the space context, with a focus on light pollution and the application of the no-harm rule, as well as the precautionary principle in environmental protection. These considerations are essential to understanding and addressing the environmental impacts of space activities.

Liability issues, particularly in the context of satellite constellations, are examined to understand how these activities fit into the liability regime established by the OST and the Liability Convention (1972), especially regarding the concept of "damage".

The work has shown how the evolving nature of space law necessitates the involvement of multiple international institutions and legal instruments. This article delves into identified gaps in international law and policy, lessons learned, and possible avenues for the protection of D&QS.

2. Identified gaps

2.1. *Legal classification of astronomy*

States have yet to make a clear position regarding the legal status of astronomy within the framework of international space law. This is arguably because it simply has not mattered before. However, now that other forms of the exploration and use of outer space (e.g., satellite communications and exploration spacecraft) are poised to cause substantial interference with ground-based astronomical observations, the question of legal status matters for understanding whether there already exists an obligation for satellite launching states to show "due regard" and to avoid "potentially harmful interference" with the activities of states that are conducting astronomical observations under Article IX, OST. Having a widely agreed position on this issue further helps to identify paths toward resolving any corresponding interference, across the electromagnetic spectrum, between activities. Studies show through policy analyses (Rotola and Williams 2021) and a treaty interpretation of the OST (Byers and Boley 2023, ch. 3) that astronomical observations are a form of exploration of outer space under international law. Other WP2 members, while stopping short of this conclusion, do find merit in the interpretation that the legal provisions of the OST, particularly Article IX, apply to protecting astronomical observations.

Nonetheless, a gap remains, which suggests that there is a need for a dedicated effort to clarify the legal status of astronomical observations, as well as the duty of states to show due regard toward astronomical observations. Such an effort could involve the exploration of various legal frameworks, including environmental law, heritage protection, and space law, to identify where and how astronomy best fits. Establishing legal obligations of states toward those which conduct astronomical observations could pave the way for more targeted and effective policies and regulations, both at the national and international levels, to safeguard D&QS for astronomical research and humanity.

2.2. *Environmental principles in space law*

The integration of environmental principles into space law, particularly in relation to light and radio pollution, is an area that remains underdeveloped. While terrestrial environmental law has evolved significantly in recent decades, space law has lagged in incorporating these principles effectively. There are visible steps towards addressing these issues, such as the adoption of the Guidelines for the Long-Term Sustainability of Outer Space Activities (LTS 2019), Planetary Protection Policies (COSPAR 2021), and the principles embedded in Article IX of the OST. However, these measures are simply not enough. Issues such as space debris have received increasing attention, but the impact of human activities in space on the astronomical environment is inadequately covered. The challenge is not only pollution from debris or atmospheric alteration but also orbital light and radio interference, which can significantly hinder astronomical observations.

Addressing this gap requires a concerted effort to develop and integrate environmental principles specifically tailored to the space context. This could involve the creation of new legal instruments or the adaptation of existing ones, taking into account the unique characteristics of space activities and their impact on astronomy. Such principles should aim to ensure sustainable use of space and minimize harmful interference, thereby protecting the interests of the global astronomical community.

2.3. *Role of ITU and resource management*

The International Telecommunication Union (ITU) has a well-established role in managing the radio frequency spectrum, but is facing new and complex challenges in managing spectrum as ever greater allocations are demanded for space assets. Associated with this is the sheer number of satellites being filed, with multiple countries filing for thousands to hundreds of thousands of satellite slots, in many cases with no clear plan or intent to utilize those reservations (Falle et al. 2023). This practice expands the uncertainty associated with the near-future orbital density and the steps most needed to address the looming problem.

Yet, even for the satellites that are plausibly expected to be launched, concerns about orbital overcrowding remain. This exacerbates light and spectrum pollution issues while further raising the risk of space debris and collisions. There is thus a pressing need for the ITU or the international community to develop mechanisms to understand when orbits are becoming too crowded and to potentially limit filings for new satellite constellations based on this information. The ITU is expected to have a role in this process, whether through regulation, its study groups, or both.

In addition to addressing radio spectrum concerns, there has been discussion about whether the ITU could also play a role in orbital light pollution regulation. Light pollution, like radio frequency interference, has the potential to be a substantial threat to astronomical observations and it is part of the electromagnetic spectrum. However, light pollution and its consequences have so far been interpreted as lying outside the ITU's mandate.

Expanding the mandate of the ITU, at least in some way, to include the management of light pollution, as well as addressing the issue of orbital overcrowding in the LEO region, could be a significant step forward. This could involve examining the feasibility and implications of such an expansion, including the technical, legal, and political aspects, and could be part of a broader space sustainability initiative. Moreover, some action could be taken through ITU study groups without, at this time, creating any reorganization or expansion of the ITU's mandate. Overall, the ITU's experience and expertise in managing finite resources in outer space, specifically the radio spectrum for satellites and geosynchronous orbits, could provide valuable insights into how similar principles and practices could be applied to the management of light pollution and LEO. Such a development would not only help to protect the astronomical environment, but would also contribute to the sustainable and responsible use of outer space.

3. Lessons learned

3.1. *Beyond Article IX of the Outer Space Treaty*

Regardless how states view astronomical observations vis-à-vis international law, Article IX of the OST is sufficiently broad that the analysis team found that it may create obligations for states to safeguard astronomical interests. In particular, the article establishes the principle that states must show "due regard" toward the interests of other states in the exploration of outer space. It further prohibits "harmful interference" with the peaceful use and exploration of outer space by states, as well as expresses a requirement for states to seek international consultations in cases of known possible interference. Along with Article I of the OST, these obligations together establish general principles of coordination and cooperation and recognize that there is a diverse set of activities that are used to carry out state interests in outer space.

With this in mind, using Article IX of the OST should not be the only legal avenue pursued. While the provision is significant, its broad language and the interpretative flexibility it offers can lead to varied implementations, which may not always align with the specific needs of protecting astronomical environments.

For this reason, Article IX can be seen as offering a foundation for environmentally motivated solutions, highlighting the importance of protecting the space environment for safe and equitable access to outer space, as well as its peaceful exploration and use. Much more effort is needed to identify actionable paths forward on this point.

Looking beyond these legal interpretations opens up further possibilities. For instance, framing astronomical observations within the context of environmental protection or cultural heritage could offer new perspectives and legal avenues. This approach could lead to the development of specific guidelines or treaties tailored to the preservation of D&QS. It also encourages looking at space law not just in isolation but in conjunction with other areas of international law, thereby fostering a more holistic approach to the governance of space activities. Furthermore, it naturally brings in moral and ethical questions concerning the alteration of the environment, especially when those alterations are driven by a few states and impact all states and peoples.

3.2. *Liability and political strategy*

The concept of liability in space law has the potential to be a strong mechanism for enforcing good norms of behaviour in the exploration and use of outer space, particularly with its provision of "absolute liability" (not fault-based) for damages that occur on the ground, at sea, or in aeroplanes in flight. However, liability of space activities in the context of damage caused by pollution is a complex issue under the Liability Convention, where "damage" means "loss of life, personal injury or other impairment of health; or loss of or damage to property of States or of persons, natural or juridical, or property of international intergovernmental organization". Applying light or spectrum pollution under this definition is limited, if possible at all. For example, whether there is a way to include damage to the sky under the concept of the property of a juridical person remains to be seen but does not appear to be tenable at this time.

One possible way forward, however, is to consider the damage to observational data, which are widely recognized to be a form of property and can be plausibly damaged as a result of another state's activities in outer space. Yet, even with this, a potential limitation is that the Liability Convention does not offer a clear way to incorporate cumulative effects that, while leading to considerable damage, are not attributable to a single event. Moreover, even with a plausible definition of damage, states may not spend the political capital needed to invoke the Liability Convention over something such as data loss.

In summary, while the concept of liability for space-related activities could be interpreted to offer some protection against the impacts of satellite constellations, this is not straightforward. The current legal definitions and frameworks make it a less feasible avenue for addressing these broader concerns.

However, just as political concerns could limit action by some states, others could effectively wield the threat of invoking the Liability Convention as a tool, serving as a form of leverage in international negotiations. It is further a way to drive home the urgency of establishing meaningful regulations to protect D&QS. By highlighting the potential risks and liabilities, states and organizations can be motivated to adopt more responsible practices. This approach can also be instrumental in advocating for the development of new norms and guidelines that address the specific needs of dark and quiet skies, thus indirectly influencing the legal landscape in favour of astronomical conservation.

3.3. *Simultaneous national and international focus*

The analysis of international law and policy highlights the importance of addressing astronomical conservation through both national and international channels. Nationally, countries could implement specific regulations and policies that would directly protect D&QS, tailored to their unique geographical (e.g., dark sky reserves) and socio-political contexts. These national actions could serve as models or catalysts for international initiatives, demonstrating practical methods for mitigating light and radio spectrum pollution.

Internationally, the challenge lies in harmonizing these diverse national approaches and elevating them to a global platform to achieve a common outcome, whether through international standards, agreements, or some other process. This harmonization is particularly important in recognizing and legitimizing astronomical observations as a form of exploration of outer space or otherwise providing protections. Furthermore, it is crucial in providing mechanisms for the protection of the space environment and the assessment of the impact of satellite and other space activities. Such international collaboration could lead to a clearer understanding of the aggregate impacts of satellite activities and pave the way for global solutions for space environmental protection.

4. Future steps

4.1. *Enhanced international dialogue*

Regular and meaningful dialogue among countries is crucial for recognizing and prioritizing the value of astronomy and astronomical observations. This dialogue should be multifaceted, involving not just space agencies and governments, but also satellite operators, international organizations, scientific communities, and civil society. The aim is to foster a global understanding of the importance of astronomy and the threats it faces from light and radio pollution. International conferences, symposia, and working groups could be key platforms for these discussions.

Part of this dialogue involves understanding how to effectively utilize existing international fora, such as COPUOS, while also recognizing their limitations. For instance, COPUOS primarily allows for state representation, which excludes other crucial stakeholders in the astronomy community. Therefore, while leveraging such platforms, there is also a need to create new, more inclusive ones, much like what was achieved during events like the SATCON and D&QS conferences. Establishing new platforms in a more institutionalized way could facilitate a comprehensive and inclusive approach to addressing astronomical conservation.

Part of this dialogue should also focus on developing a shared vision and commitment to protecting D&QS. This can be achieved through the creation of international agreements or treaties specifically dedicated to this cause, or by incorporating these issues into existing international space governance frameworks. Such efforts would also benefit from the inclusion of educational and public awareness campaigns to broaden the understanding and support for astronomical conservation at all levels.

4.2. *Expanding legal interpretations*

To accommodate the unique requirements of astronomy, there is a need to explore broader interpretations and applications of existing law and policy. This involves analyzing existing treaties and agreements under the context of the rapidly expanding access to outer space and all the activities and changes such expansion produces. This could lead to identifying what under existing law already applies to D&QS, and what needs to be adapted or expanded. For instance, an innovative interpretation of the OST's principles in the context of environmental protection and sustainability could provide clarity on legal avenues for safeguarding astronomical interests. A prominent example is generally clarifying concepts such as "due regard".

Such an approach should further involve the exploration of new legal concepts and frameworks that could be specifically designed for astronomical protection. This could include the development of guidelines or protocols for space operations that minimize light and radio pollution under various conditions.

4.3. *Diversifying strategies*

In addition to enhancing international dialogue and expanding legal interpretations, there is an urgent need to diversify the approaches for the protection of the dark and quiet sky. While the use of international space law remains critical, its limitations are evident.

By incorporating environmental protection strategies, the focus can be extended to include not only space activities but also their impact on both the space and terrestrial environments. Heritage conservation can play a role in highlighting the cultural and scientific value of the night sky, thus fostering a broader appreciation and motivation for its protection.

Liability frameworks or the threat of their use can be leveraged to hold entities accountable for activities that cause harmful interference with astronomical observations. The creation of public awareness campaigns is essential to increase the general understanding of the importance of dark and quiet skies, thereby garnering public support for protective measures.

Additionally, defining space environmental metrics such as orbital capacity and seeking space environment protections under bodies such as the ITU could be an important step in managing the increasing number of satellites and reducing the potential for light pollution and radio interference. This approach would require a comprehensive assessment of orbital use and the development of guidelines or regulations to ensure sustainable and responsible activities in space.

By exploring these diverse approaches in parallel to traditional space law discussions, a more robust and effective strategy can be developed for the protection of dark and quiet skies, aiming to create a holistic framework that encompasses various aspects of space activities and their impacts, ensuring the long term sustainability and accessibility of space for astronomical purposes, as well as humanity.

References

Byers, M. & Boley, A. 2023, Who Owns Outer Space? International Law, Astrophysics, and the Sustainable Development of Space, Cambridge University Press.

(COSPAR) Committee on Space Research 2021, "COSPAR Policy on Planetary Protection", COSPAR Bureau, 3 June 2021 https://cosparhq.cnes.fr/assets/uploads/2021/07/PPPolicy_2021_3-June.pdf

(DQSII) Dark and Quiet Skies II Working Group Reports 2021, IAU, UNOOSA, Spain & NSF NOIRLab https://doi.org/10.5281/zenodo.5874725

Falle, A., Wright, E., Boley, A., & Byers, M. 2023, Science, 382, 150

(Liability Convention) United Nations 1972, "Convention on International Liability for Damage Caused by Space Objects", RES 2777 (XXIV), 1971 https://www.unoosa.org/oosa/en/ourwork/spacelaw/treaties/introliability-convention.html

(LTS) United Nations 2019, "Guidelines for the Long-term Sustainability of Outer Space Activities", Report of the Committee on the Peaceful Uses of Outer Space, Annex II, A/74/20 https://www.unoosa.org/oosa/en/oosadoc/data/documents/2019/a/a7420_0.html

(OST) United Nations 1967, "Treaty on Principles Governing the Activities of States in the Exploration and Use of Outer Space, including the Moon and Other Celestial Bodies" (Outer Space Treaty), RES 2222 (XXI), 1966 https://www.unoosa.org/oosa/en/ourwork/spacelaw/treaties/introouterspacetreaty.html

Rotola, G. & Williams, A. 2021, Air and Space Law, 46, 4/5, 545

(SATCON2) SATCON2 Working Group Reports 2021, NSF NOIRLab https://baas.aas.org/report-of-the-satcon2-workshop-1216-july-2021

Astronomy and Satellite Constellations: Pathways Forward
Proceedings IAU Symposium No. 385, 2026
P. Grimley, J. Barentine, D. Galadí-Enríquez & C. Walker
doi:10.1017/S1743921324000838

National approaches to the protection of dark and quiet skies

Yana Yakushina[1], **Marieta Valdivia Lefort**[2], **Chris Martin**[3], **Tiffany Nichols**[4], **Anne-Sophie Martin**[5] **and Aaron Boley**[6]

[1]Faculty of Law and Criminology, Department of European, Public and International Law
University of Ghent, 9000 Gent, Universiteitstraat 4, Belgium
email: ya.yakushina@gmail.com

[2]Institute of the Americas, University College London
51 Gordon Square, London WC1H 0PN, United Kingdom
email: marieta.constanza@gmail.com

[3]Vela Science, 1 Sunningvale Lane Christchurch 8053 New Zealand

[4]LSST-DA Catalyst Fellow, Presidential Postdoctoral Research Associate
Princeton University, Department of Astrophysical Sciences

Department of History, Princeton, NJ, USA
email: tiffany.nichols@princeton.edu

[5]Department of Political Sciences, Sapienza University of Rome
Piazzale Aldo Moro, 5 - 00185 Rome, Italy

[6]Department of Physics and Astronomy, University of British Columbia
6224 Agricultural Rd, Vancouver BC, V6T 1Z1, Canada, email: aaron.boley@ubc.ca

Abstract. In recent years, the protection of dark and quiet skies (D&QS) has become an increasingly urgent concern, driven by the proliferation of artificial light at night (ALAN) and the deployment of mega-constellation satellites in low-earth orbit (LEO). This paper presents preliminary research findings from the IAU CPS Policy Hub National Analysis Team, focusing on the national policy and legal approaches to safeguarding D&QS for astronomical observations worldwide. It synthesises the key interim results with a primary focus on light pollution mitigation, outlining measures to combat light pollution and emerging regulatory trends. The study also identifies obstacles to mitigating light pollution and offers examples of national approaches for D&QS protection, along with future research prospects.

Keywords. Dark and Quiet Skies; Protection of dark skies; Light pollution; Orbital light pollution; Astronomy; National regulations

1. Introduction

During the last few years, the protection of dark and quiet skies (DNQS) has received increasing attention at different political and regulatory levels. This has been prompted by the proliferation of artificial light at night (ALAN) and the number of mega-constellation satellites being placed into low-Earth orbit (LEO). (Perez 2023) These changes have created a complex set of concerns, ranging from impacts on astronomical activities to environmental protection that can no longer be ignored by national governments.

This paper introduces preliminary research results conducted by the IAU Centre for the Protection of the Dark and Quiet Sky from Satellite Constellation Interference (CPS)

Policy Hub National Analysis Team. It identifies policy and legal approaches to the protection of D&QS for astronomical observations in various jurisdictions across the globe. The paper summarises the key intermediate findings of the research focusing on light pollution mitigation, which includes, *inter alia*, key measures to address light pollution and the associated regulatory trends. It also discusses known obstacles to addressing the growth of light pollution. Additionally, the paper provides some examples of national approaches to the protection of D&QS and gives a future research perspective.

Previously conducted research, such as that in the Dark and Quiet Skies II report (DQSII 2021), has shown that several countries have adopted legislation to preserve the sky for astronomical activities. Building upon that work, the current report provides a detailed synthesis of adopted policies and legislation in a broader range of countries. During the first research stage, the main research questions were: (1) whether a given country has an existing framework for the protection of D&QS for astronomy; (2) whether light pollution is recognised as a problem at the national level; and (3) whether there are any relevant regulatory frameworks applicable to address the concerned issue. This approach enables an understanding of existing national action towards light pollution mitigation and the identification of potential regulatory improvements in dark sky protection for astronomical observations.

In addition, this report sheds light on other aspects of the national protection of dark skies for astronomical activities, e.g., recognition of astronomy as a space activity, licensing procedures of large constellations, incorporation of related international law concepts into national legislation, and beyond.

Altogether, the paper brings a broader understanding of legal and policy challenges the astronomical community faces and can experience across different countries, and provides initial recommendations on how to potentially overcome these obstacles to preserve D&QS in the future.

2. Research section

Analysing national regulatory approaches to the protection of D&QS requires an in-depth review of relevant formal written policies and legal documents. In addition to such formal regulatory frameworks, the analysis also considers domestic and local initiatives that may facilitate and promote the adoption of D&QS policies and laws at the national level. This comprehensive approach not only provides insight into the current state of affairs but also enables an assessment of the prospects for enacting innovative legislation aimed at the protection of D&QS.

The national analysis includes three stages. The first stage reviews the overarching D&QS frameworks adopted by different countries. This is done by (1) identifying which countries to include for the analysis and (2) exploring how those countries address the preservation of D&QS.

The selection of countries for analysis was determined by their membership in the United Nations Committee on the Peaceful Uses of Outer Space (UN COPUOS), which currently comprises 102 member states. The research also includes a small number of countries with astronomical facilities that are not part of COPUOS to provide a more comprehensive perspective on D&QS concerns. To investigate national regulatory approaches at the first stage, the work addressed a range of research questions (RQs) including but not limited to: (1) are astronomical activities considered as space activities within the national legislation? (2) are there any specific norms to protect astronomical observations? (3) are there any light pollution/dark sky protection laws? and (4) what other laws can be applied to protect astronomical observations?

In the second phase of the national analysis, yet to be conducted, the research will go beyond general D&QS frameworks by examining legal and policy approaches related to satellites, including authorisation, licensing, registration, and the assessment of their potential impact. To help achieve the goals of the second research stage, the following RQs will be addressed: (1) what are the national regulations for satellite authorisation? (2) what are the national regulations for satellite licensing? (3) are there any specific requirements for satellite mega-constellations? and (4) are there any considerations of potential impacts on D&QS? This stage is especially important for launching states. Finally, in the third phase, the overall conclusions elucidating how and to what extent national states safeguard D&QS recommendations will also be presented to suggest strategies for building stronger and more equitable D&QS and related environmental protections.

Another important aspect of the research is related to the use of terminology, primarily because of the distinct nature of national space legislation. Unlike the international space legal regime such as the Outer Space Treaty of 1967 (OST 1967), which introduces the term "peaceful use and exploration of outer space", individual nations tend to use the term "space activity" to mean activities that engage in the peaceful use and exploration of space. Thus, this research will analyse the protection of D&QS for astronomical observations also by examining how the concept of space activity is defined within national frameworks. Such examination will help to assess the potential for legal recognition of astronomy as a space activity.

3. Intermediate results

These intermediate findings are in response to the primary RQs described in Section 2. Additional questions are also posed and answered, which collectively contribute to an enhanced understanding of national approaches to the protection of D&QS.

3.1. *Are astronomical activities considered an exercise of the exploration and use of outer space within the national legislation?*

As previously noted, the international concept of "peaceful use and exploration of outer space," is often articulated at national levels through the term 'space activity.' In addressing RQ1, the research is focused, in part, on scrutinising the definition of 'space activity,' shedding light on which specific activities are seen as the use and exploration of space within the national context.

Based on the national legislation analysis, with attention to the primary space law and policy of each country with a definition for space activity, there is a lack of clarity regarding the way countries view the nature of astronomical observations.

In several of the examined countries, the definition of space activity does not explicitly designate astronomy or astronomical observations as a space activity per se, although it does not intrinsically exclude astronomical observations from being categorised as such, either (Russian Federation 1993). Consequently, through a legal interpretation of the term 'space activity', there exists potential recognition of astronomical observations as a legitimate type of space activity. Moreover, in certain cases, such as in Chile, astronomy and its advancement are primarily associated with the use of space for terrestrial applications via technologies like satellites (Chile 2020a,b,c). Chile, which does not have a national space policy as such in place, incorporates astronomy into other existing policies such as defence and science. This is particularly done in the context of earth observation or space-based sensing for purposes that include monitoring, border control, communication, defence and military functions, and state security. However, these policies do not explicitly address astronomy or astronomical observations as a scientific discipline, particularly in terms of the exploration of space.

For some nations, such as in the case of Azerbaijan (Azerbaijan 2023), space activity is strictly defined as an activity associated with space objects. Limiting the space domain in such a way may prohibit categorisation of astronomical observation as a space activity.

With the above in mind, it is crucial to note that the discussion regarding the legal recognition of astronomy and/or astronomical observations as a space activity is currently unfolding both at the international and national levels, as evidenced, for instance, by the COPUOS meetings.† Nonetheless, the adoption of new legislation by countries in response to political initiatives flowing from agreements at international fora could take considerable time.

3.2. *Does the legal recognition of astronomical observation and/or astronomy as a form of exploration and use of outer space serve as a prerequisite for protecting D&QS?*

Although the legal recognition of 'astronomy' and/or 'astronomical observation' as a space activity could offer substantial protection for D&QS, obtaining this recognition requires collective efforts from different actors and from various jurisdictions, and it may not be the most efficient or effective way to achieve D&QS protection in the short to medium term. It could also be time-consuming, especially at the international level, depending on the approach. Consequently, in parallel with longer-term efforts or as a separate approach, alternative legal instruments from various fields, such as environmental protection laws or light pollution mitigation regulations, could play a crucial role in reducing light pollution and upholding the protection of D&QS for astronomical observations. Leveraging pre-existing legal frameworks becomes a necessary strategy for addressing this issue, with the examined states showing varying perspectives on its implementation.

3.3. *Are there national norms to protect astronomical observations? Do countries have legislation for light pollution mitigation/dark sky protection? Is light pollution addressed in national legislation?*

Certain countries have implemented legal frameworks and regulations aimed at protecting D&QS for astronomical observations. Notable examples include (1) Spain, which enacted The Law on Protection of the Astronomical Quality of the Observatories of the Institute of Astrophysics of the Canary Islands (Spain 1988) to preserve the sky for astronomical facilities on La Palma and Tenerife; (2) France, which passed a Decree for the protection of exceptional astronomical observing sites (France 2018a); (3) and the city of Saint Petersburg of the Russian Federation, which adopted a local regulation on the protection of D&QS above the Central Astronomical Observatory of the Russian Academy of Sciences in Pulkovo, again for astronomical observations (Soviet Union 1945; Russian Scientific Council 2015). Notably, the majority of these regulations are predominantly found at the local or regional levels, particularly in the vicinity of observatories. A long-standing approach to ensure the preservation of D&QS at least for terrestrial-based light pollution, includes the establishment of protection zones, including dark or quiet areas, within which light or radio emissions are limited. This specific practice cannot be extended to orbital light pollution.

† See for instance, A/AC.105/C.1/2023/CRP.18/Rev.1. United Nations Committee on the Peaceful Uses of Outer Space (COPUOS). Conference Room Paper on the Protection of Dark and Quiet Skies for science and society. 15 February 2023; available online: `https://www.unoosa.org/res/oosadoc/data/documents/2023/aac_105c_12023crp/aac_105c_12023crp_18rev_1_0_html/AC105_C1_2023_CRP18Rev01E.pdf`

Additionally, many of the analysed countries (Bulgaria, Chile, Dominican Republic, Peru, Slovakia, Spain, and South Africa, among others) have initiated national conversations on the necessity to protect D&QS for astronomical observations, following the discussions within COPOUS (UN COPUOS 2023). It is crucial to emphasise that the current status of the protection of D&QS in many states primarily revolves around discussions rather than the existence of concrete, formalised documents.

In this context, there is a growing trend in the adoption of light pollution laws across the globe (e.g., Chile, Croatia, Czech Republic, Italy, Japan, Oman, Spain, USA, etc.; for further information, see Crotia 2019; Chile 2012; Slovenia 2007; Germany 2009; France 2018b; Kazakhstan 2023). These legal measures span various regulatory fields, encompassing urban planning and environmental conservation, rather than being exclusively dedicated to the protection of astronomical observations. Nevertheless, these laws have the ancillary effect of reducing artificial light at night (ALAN) levels and benefiting astronomy. Furthermore, these local and regional efforts do have relevance for orbital light pollution, as they can, in principle, help raise awareness of satellite light pollution by reconnecting people with the night sky, something many have already lost.

3.4. *Is there any country with a regulatory framework for protecting D&QS that could potentially be used as a model for other countries?*

No singular "best approach" or model law has so far been identified during this research, nor is there an expectation to find one, as regulatory approaches vary significantly and are based on their unique national contexts. Since many of the national laws and policies have only recently been enacted, it is premature to assess their implementation or their potential to reduce ALAN. Additional assessments, including accurate measurements of, for example, night sky brightness and constant monitoring of light pollution, will be essential for determining which legislation is effective in this regard.

Although a model framework may not presently exist, there is a discernible trend wherein countries scrutinise the regulations of other countries while formulating their own regulatory frameworks. If a particular country enacts an exceptionally effective law, it would be advantageous for the astronomical community to proactively share this example with other states and stakeholders.

3.5. *Are there any other laws applicable to protect D&QS?*

Among existing national regulatory frameworks, the primary area of legislation for protecting D&QS appears to be environmental law. Within the realm of environmental protection, most countries provide a definition of "pollution," which incorporates "energy" as a pollutant and implies various principles such as the precautionary principle and prevention principle. Certain countries, like Germany†, have amended their general environmental law to include provisions for reducing environmental impacts associated with light pollution. Other examples include France and Kazakhstan‡, which have also taken measures to reduce nighttime brightness. Other regulatory fields that may be helpful in promoting D&QS include space law (e.g., space sustainability, space debris mitigation) and laws on protection of cultural heritage, as well as certain provisions of urban planning and building laws.

† The Federal Nature Conservation Act (Germany); available online: `https://www.gesetze-im-internet.de/bnatschg_2009/`

‡ See, for instance, Decree on the prevention, reduction and limitation of light pollution (France); available online: `https://www.legifrance.gouv.fr/jorf/id/JORFTEXT000037864346`; Environmental Code of the Republic of Kazakhstan; available online: `https://online.zakon.kz/Document/?doc_id=39768520#pos=634;-86.33332824707031`.

3.6. *Is light pollution a form of pollution under the national environmental law?*

Based on the analysis of various national jurisdictions, at this time, classifying light pollution as a form of pollution can occur mainly through one of two means. It may be inferred through an interpretation of national environmental laws that include "energy pollution" as a broader term, encompassing the effects of light pollution (e.g., Slovakia, Poland, Russian Federation, Bulgaria, and China). Some other countries, such as Croatia, Germany, France, Spain, and Slovenia, directly identify light pollution as a form of pollution or identify light/ALAN as a pollutant within their environmental legislation.

3.7. *Do countries have initiatives to protect dark and quiet skies?*

Several types of initiatives in various countries were identified during the research process. Although such initiatives do not constitute legislation, they could contribute to regulatory changes that protect D&QS.

Most of the analysed countries have growing or emerging local and national initiatives designed to promote the preservation of night skies from different perspectives. These initiatives encompass astro-tourism, environmental conservation, and the advancement of astronomical observations. The list of countries with such initiatives includes, but is not limited to, Chile, Ecuador, Brazil, Japan, Poland, Costa Rica, the United Arab Emirates, New Zealand, Colombia, Australia, several African nations, Canada, and the United States, among others.

4. National examples

The above sections summarise the results of the national analysis in aggregate. Here, brief descriptions are given of select national laws and regulations that address light pollution and impacts on astronomy.

In Chile, the *Nueva Norma Lumínica (New Lighting Standard) D.S. N° 1/2022 MMA* (Chile 2023a,b), approved in October 2023, represents a significant step in reducing light pollution nationwide. It extends the regulation established by Decree 43 from 2014 to cover the entire country, introducing new technical requirements and emphasising protection for biodiversity and human health.

The Czech Republic amended the Act of the *Czech National Council on Nature and Landscape Protection in 2017*† to address light pollution in national parks. Additionally, light pollution has been considered in Environmental Impact Assessments since 2020‡ and the Ministry of the Environment updated its public lighting recommendations in 2021§.

South Africa has legislation focused on protecting astronomy from ground-based light pollution and radio interference, though it does not specifically address interference from space objects. (South Africa 2007).

The United Kingdom stands out for its *Clean Neighbourhoods and Environment Act of 2005* (United Kingdom 2005) and the *Society of Light and Lighting (SLL) Code for Lighting 2022*¶, which enable individuals with a "professional interest" in dark skies, including optical astronomers, to take action against "sky glow" interference. However, case resolutions remain uncertain and are determined on a case-by-case basis.

† `https://www.mzp.cz/C1257458002F0DC7/cz/news_20221027-/$FILE/Light_pollution_reduction_measures.pdf`

‡ `https://portal.cenia.cz/eiasea/dokumenty/dokumentSoubor/167/SZ_EIA%20Metodika_final.pdf?lang=cs`

§ `https://www.mzp.cz/C1257458002F0DC7/cz/news_20221027-/$FILE/Light_pollution_reduction_measures.pdf`

¶ The SLL Code for Lighting (London, 2022), `www.cibse.org`, 1.4.4.

The examples above are a small sample of the different ways national legislation is used to address light pollution. However, local jurisdictions often impose stricter rules to protect local astronomical and environmental assets. For example, in Spain, stringent local laws supplement limited national laws (Spain 2007) to protect the Canary Islands' international astronomical observatories (Spain 2005). Similarly, in New Zealand, local ordinances (New Zealand 2022) designate the area around Lake Tekapo as an International Dark Sky Reserve for astro-tourism, benefiting both amateur astronomers and tourism operators while preserving Mt. John's professional observatories.

5. Further research

Further research is needed to provide a more comprehensive analysis of the overarching frameworks available for protecting dark and quiet skies. Particular attention is needed toward addressing satellite interference with radio astronomy (i.e., protection of 'quiet skies'). In addition, an in-depth investigation into national regulations governing satellites, encompassing aspects such as licensing, authorisation, registration, and the assessment of their impact on the outer space environment, will be needed. Further analysis will also include an examination of various initiatives and practices in place for addressing the preservation of dark and quiet skies for multiple purposes. Finally, upon conclusion of this additional research, recommendations will be drafted on actions that can be taken to strengthen protections for dark and quiet skies both within the national context and on a broader, international scale.

6. Contributions

The research is conducted by IAU CPS Policy Hub, WG2, National Analysis team: Yana Yakushina (Conceptualisation, Supervision, Methodology, Writing and Editing), Marieta Valdivia Lefort (Research, Writing, and Editing), Chris Martin (Research, Writing and Editing), Roohi Dalal (Research), Tamara Blagojevich (Research), Tiffany Nichols (Research, Writing, Editing), Andrew Falle (Research), Betty Kioko (Research), Charlotte Hook (Research), Martin Anne-Sophie (Research, Writing and Editing), Lauryn Lee Hallet (Research), Sarah Thiele (Research), and Rayan Khan (Research), Aaron Boley (Conceptualisation, Review and Editing).

References

Azerbaijan, 2023, Law of the Republic of Azerbaijan on space activities of 4 August 2023. `https://president.az/az/articles/view/60774`

Chile, 2012, Decree N43 on Establishing emission standards for the regulation of light pollution. `https://www.bcn.cl/leychile/navegar?idNorma=1050704`

Chile, 2020a, National Defence Policy. `http://163.247.42.118/transparencia/POLDEF/POLDEF2020/POLDEF2020.pdf`

Chile, 2020b, National Science, Technology, Knowledge and Innovation Policy (2020–2022). `https://www.minciencia.gob.cl/politicactci/documentos/Politica-Nacional-CTCi_Chile-2020.pdf`

Chile, 2020c, Action Plan - National Science, Technology, Knowledge and Innovation Policy (2020–2022). `https://www.minciencia.gob.cl/politicactci/documentos/Politica-Nacional-CTCI_Plan_Accion_Chile_2020.pdf`

Chile, 2023a, Decree 43 of the Ministry of the Environment of Chile. `https://www.bcn.cl/leychile/navegar?idNorma=1050704`

Chile, 2023b, New Lighting Standard (2023). `https://www.diariooficial.interior.gob.cl/publicaciones/2023/10/18/43679/01/2391423.pdf`

Crotia, 2019, Act NN 14/2019 on Protection against Light Pollution. `https://narodne-novine.nn.hr/clanci/sluzbeni/2019_02_14_271.html`

Czech Republic, 2022, Light pollution reduction measures in Europe (2022), working paper for the international workshop Light Pollution 2022, during the Czech Presidency of the Council of the European Union Prepared by the Ministry of the Environment of the Czech Republic https://www.mzp.cz/C1257458002F0DC7/cz/news_20221027-/$FILE/Light_pollution_reduction_measures.pdf

Czech Republic. 2023, Methodical Guidelines for preventing and reducing light pollution adopted by the Department of Environmental Impact Assessment and Integrated Prevention of Czech Republic. https://portal.cenia.cz/eiasea/dokumenty/dokumentSoubor/167/SZ_EIA\%20Metodika_final.pdf?lang=cs

(DQSII) Dark and Quiet Skies II Working Group Reports, 2021, IAU, UNOOSA, Spain & NOIRLab. https://doi.org/10.5281/zenodo.5874725

France, 2018a, Decree on the prevention, reduction and limitation of light pollution. https://www.legifrance.gouv.fr/jorf/id/JORFTEXT000037864346

France, 2018b, Decree setting the list and the perimeter of exceptional astronomical observation sites in the application of Article R.583-4 of the Environmental Code, 27 December 2018. https://www.legifrance.gouv.fr/jorf/id/JORFTEXT000037864375

Germany, 2009, The Federal Nature Conservation Act. https://www.gesetze-im-internet.de/bnatschg_2009/

Kazakhstan, 2023, Environmental Code of the Republic of Kazakhstan. https://online.zakon.kz/Document/?doc_id=39768520#pos=634;-86.33332824707031

New Zealand, 2022, Lighting Restrictions for the Mackenzie District to protect the Aoraki Mackenzie International Dark Sky Reserve. https://www.mackenzie.govt.nz/__data/assets/pdf_file/0007/677959/Light.pdf

Perez, A., 2023, Science, 380, 1136

Russian Federation, 1993, On space activities, 20.08.1993 N 5663-1. http://pravo.gov.ru/proxy/ips/?docbody=&nd=102025742

Russian Scientific Council, 2015, Regulations on the astroclimate: Scientific Council of the Russian Academy of Sciences Rules for making decisions on the coordination of economic and construction activities in the protective park zone of the observatory 2015. http://puldb.ru/ac/docs/2015-12-25%20reglament.pdf

Slovenia, 2007, Regulation on limit values of light pollution of the environment 2007. http://www.pisrs.si/Pis.web/pregledPredpisa?id=URED4520

South Africa, 2007, Act on Astronomy Geographic Advantage 21 of 2007. https://www.gov.za/documents/astronomy-geographic-advantage-act

Soviet Union, 1945, The Council of People's Commissioners of the USSR Order dated March 11, 1945 N 4003r. https://www.derev-grad.ru/zakon-stroitelstvo/zakonodatelstvo-o-stroitelstve-sssr4/rasporyazhenie-snk-sssr-ot-11-03-1945--4003r-o-vosstanovl-eni.pdf

Spain, 1988, Law 31/1988 on the Protection of the Astronomical Quality of the Observatories of the Institute of Astrophysics of the Canary Islands of 31 October 1988. https://www.boe.es/eli/es/l/1988/10/31/31

Spain, 2005, Law 31/1988, of October 31, on the Protection of the Astronomical Quality of the Observatories of the Institute of Astrophysics of the Canary Islands. https://www.boe.es/eli/es/l/1988/10/31/31

Spain, 2007, Law 34/2007, of November 15, on Air Quality and Protection of the Atmosphere. https://www.boe.es/eli/es/l/2007/11/15/34/con

United Kingdom, 2005, Clean Neighbourhoods and Environment Act 2005 c. 16, Part 9, Statutory Nuisance, Section 102. https://www.legislation.gov.uk/ukpga/2005/16/section/102

(UN COPUOS) United Nations Committee on the Peaceful Uses of Outer Space, 2023, General exchange of views on dark and quiet skies for science and society, A/AC.105/C.1/2023/CRP.18/Rev.1, COPUOS, STSC, 6-17 February 2022, Item 17. https://www.unoosa.org/res/oosadoc/data/documents/2023/aac_105c_12023crp/aac_105c_12023crp_18rev_1_0_html/AC105_C1_2023_CRP18Rev01E.pdf

(OST) United Nations, 1967, "Treaty on Principles Governing the Activities of States in the Exploration and Use of Outer Space, including the Moon and Other Celestial Bodies" (Outer Space Treaty), RES 2222 (XXI). https://www.unoosa.org/oosa/en/ourwork/spacelaw/treaties/introouterspacetreaty.html

Astronomy and Satellite Constellations: Pathways Forward
Proceedings IAU Symposium No. 385, 2026
P. Grimley, J. Barentine, D. Galadí-Enríquez & C. Walker
doi:10.1017/S1743921324001972

General principles of international environmental and space law applicable to the protection of dark and quiet skies

Jonathan Kaley-Isley[1], Tamara Blagojevich[2] and Christopher D. Johnson[3]

[1]Executive Director of the Space Court Foundation; Visiting Researcher at the International Institute of Air and Space Law at the University of Leiden; contributing member of the International Astronomical Union Centre for Dark Sky Protection (IAU CPS)
email: jonathan.kaley-isley@spacecourtfoundation.org

[2]Associate Director for Legal Research and Special Projects at the Space Court Foundation; Junior Associate and Research Analyst at OSA Consulting LLC; Affiliate member of the IAU CPS as a researcher and writer
email: tamara.blagojevich@spacecourtfoundation.org

[3]Space Law Advisor at the Secure World Foundation; Adjunct Professor of Law at the Georgetown University Law Center; Mr. Johnson has written widely on space law and policy issues and represents the Secure World Foundation at the United Nations Committee on the Peaceful Uses of Outer Space (COPUOS) He is also a contributing member of the International Astronomical Union Centre for Dark Sky Protection (IAU CPS).
email: cjohnson@swfound.org

Abstract. The ambitious deployment of swarms of spacecraft into Low Earth Orbit (LEO) has resulted in the rapid and increasing degradation of the night sky visibility. While commercial rationales exist for this rapid rise of large satellite constellations (so-called "megaconstellations") their negative effects on astronomy, and other uses of the night sky, have implications on the future sustainable and equitable uses of outer space. The special regimes of international space law, as found in the 1967 Outer Space Treaty and subsequent United Nations treaties on space, as well as in national regulation, have not provided the requisite guidance. Likewise, the regime of international environmental law, arguably more robust, fails to provide necessary guidance on the protection of the night sky from anthropogenic space objects. This article surveys the regimes of space law and environmental law for commonalities and points of convergence in an effort to strengthen the protection of the night sky and to offer pathways forward for concerned stakeholders.

Keywords. space law, environmental law, light pollution, outer space treaty, Stockholm declaration, Rio declaration, space debris, astronomy, astronomical observation, sustainability

---

1. Introduction

International environmental law and international space law have a number of similarities, and both contain sets of norms governing state behaviour outside of a state's territorial boundaries. Therefore, well-established rules from one area might be taken as examples to better understand the development and application of the other. The application of overarching norms of international environmental law may prove instrumental for the protection of dark and quiet for science and society.

However, not every norm can be exemplary; only *ius cogens* rules can truly have an *erga omnes* and direct applicability across subject areas. Therefore, norms may be recognised as law where there is a common legal acknowledgment, stemming from appropriate international fora (or multiple national fora) in synchronicity, or in a comparable and harmonised fashion. To source such rules in one special regime of international law and apply them in another, one must look into common best practices and find the official national compilation of the legal acknowledgment, or *opinio iuris*.

With light pollution and skyglow in mind, this article will examine the confluence of space and environmental law in order to assess the development of *opinio iuris* and applicability of well-tested principles and customs of environmental law and their application in the space domain. This article will find that by utilizing the functional and law-in-context methods of comparative legal analysis, the principles of environmental law can be useful in facilitating the progressive development of the space law framework, consistent with the intention of original space treaties.

2. The nexus of space and environmental law

Both environmental law and space law have established norms which enshrine broadly similar values, encouraging their further development within the framework of international law. Such norms have received international recognition and have had sufficient time to be explored in practice.

Article I of the Outer Space Treaty declares that "The exploration and use of outer space, including the moon and other celestial bodies, shall be carried out for the benefit and in the interests of all countries (...) and shall be the province of all mankind. Outer space, including the moon and other celestial bodies, shall be free for exploration and use by all States without discrimination of any kind, on a basis of equality and in accordance with international law (...)". This treaty article indicates that outer space is a domain outside of state territory, but one which all users may access on the basis of equality. In that sense, outer space is akin to the High Seas and the deep seabed. And, as a domain outside of state territory, outer space, along with these other areas, is (in principle) protected by appropriate provisions of the special regime of international environmental law.

A similar spirit is visible in many provisions of environmental law, such as Principle 21 of the 1972 Stockholm Declaration, which provides that "States have, in accordance with the Charter of the United Nations... the responsibility to ensure that activities within their jurisdiction or control do not cause damage to the environment of other States or areas beyond the limits of national jurisdiction".

Similarly, Stockholm Principle 22 reinforces the justification for functional interpretation, to clarify and elaborate regime and norms regarding extra-territorial liability. This principle obligates States to "cooperate to further develop the international law regarding liability and compensation for the victims of pollution and other environmental damage caused by activities within the jurisdiction or control of such States to areas beyond their jurisdiction".

Stockholm Principle 24 further provides that "[transnational] matters concerning the protection and improvement of the environment should be handled in a cooperative spirit by all countries, big and small, on an equal footing. Cooperation through multilateral or bilateral arrangements or other appropriate means is essential to effectively control, prevent, reduce, and eliminate adverse environmental effects resulting from activities conducted in all spheres, in such a way that due account is taken of the sovereignty and interests of all States."

The Stockholm Declaration, by adopting phrases such as "beyond the limits of national jurisdiction", and "activities conducted in all spheres" indicates the necessity to take into consideration environmental principles concerning one country's behavior in all extraterritorial spheres, including outer space. Additionally, the cooperative spirit and purposes of the principles correspond very well with some of the main principles of space law—namely international cooperation and due regard.

Similarly to Stockholm's Principle 21, which provided the rationale for applying the no-harm and due regard rules in outer space, there is growing opinion that Principle 2 of the 1992 Rio Declaration should apply to space debris and related issues. Moreover, the integrative approach supported by the similarity in values and purposes is enshrined in Rio's Principle 25, which states that "[p]eace, development and environmental protection are interdependent and indivisible. " Additionally, adequacy for application of the following Rio principles to space debris-related issues has been assessed by various academics: the no-harm rule, the precautionary principle, and the polluter-pays principle. Bearing in mind that the cause of space debris-related issues can be traced to space traffic congestion, which can contribute to or cause light pollution, it can be concluded that the same applicable principles can in correlation be appropriate for light pollution mitigation. This is further evidenced by the details of the "Space2030" Agenda, established by Committee On the Peaceful Uses of Outer Space (COPUOS).

3. *Opinio iuris*: recognition of in-space light pollution

Since the emergence of space law, subsequent practice in the application of the Outer Space Treaty has led to the adoption of various environmentally oriented soft law instruments. While there is no explicit provision in any of the space treaties addressing in-space pollution, COPUOS has developed a set of Space Debris Mitigation Guidelines. The COPUOS Guidelines are widely based on the Space Debris Mitigation Guidelines promulgated by the Inter-Agency Space Debris Coordination Committee (IADC). These recommendations generally follow the space-specific principles of due regard to the interests of all States, non-contamination, and non-interference. Similarly, the Long-Term Sustainability Guidelines, concluded in 2019, provide norms addressing space safety and sustainability, offering voluntary measures regarding long-lived debris.

Discussions in the Scientific and Technical Subcommittee (STSC) of COPUOS showcase emerging *opinio iuris* related to prevention, precaution, and issues concerning the protection of dark and quiet skies in the sustainable development of outer space. The STSC in its 2023 session considered the "General exchanges of views on dark and quiet skies for science and society" as a single agenda item, in accordance with General Assembly resolution 77/121 on international cooperation in the peaceful uses of outer space. The Subcommittee also had before it a conference room paper on the protection of dark and quiet skies submitted jointly by Chile, Spain, Slovakia, Bulgaria, the International Astronomical Union (IAU), European Southern Observatory (ESO), and the Square Kilometre Array Observatory (SKAO). This activity and attention indicates clear support from regional organisations and States.

Additionally, more than 30 States made explicitly positive statements specifically related to dark and quiet skies. These included the US, UK, India, China, France, Germany, Austria, Spain, Russia, and other representative States. Additionally, some delegations expressed the view that dark skies must be protected as a common cultural and natural heritage. Some delegations expressed that their States' astro-tourism industries relied on dark skies for development. Other delegations expressed that a proposed "expert group" should include interested member States as well as private satellite operators, scientists and academicians to evaluate challenges and propose solutions. Similar trends are visible in some delegations' views that the observers for the International Telecommunication Union (ITU) and IADC should participate in such discussions in order to consider the established guidelines and practices in related fields. Finally, the Subcommittee acknowledged the in-space sustainable development applications, noting that various national and international efforts exist, balancing the provision of satellite services with astronomical observation activities.

At its 2024 session, the STSC adopted the protection of dark and quiet skies as an agenda item through 2027. This is a high level of recognition by geopolitically relevant space-faring nations and engaged stakeholders indicating that the issues related to dark and quiet skies protection are integral to in-space sustainable development and in line with achieving the purposes of existing space law.

4. OST Article IX and the interpretation of environmental principles

When it comes to hard law, the most clearly environmentally oriented provision of the Outer Space Treaty is likely Article IX, often cited as the key provision addressing space debris. Embedded in its language appear to be indications of well-established international environmental law principles such as sustainable development, equity, the prohibition of transboundary harm (the no-harm rule), and principles of prevention and precaution. In surveying Article IX, a number of interlinked concepts arise. Most relevant for the discussion of Dark and Quiet Skies are the principles of cooperation and mutual assistance; due regard to corresponding interests; the prohibition of harmful contamination of the space domain; the obligation of consultations when harmful interference is threatened; and the corresponding right (or privilege) to request consultations when there is a threat of harmful interference.

Article IX provides that States must carry out their activities in the exploration and use of outer space on the basis of cooperation and mutual assistance, and must show due regard for the corresponding interests of other States. They must also avoid harmful contamination of outer space. If a State Party has reason to believe that its activity would cause potentially harmful interference with the activities of other State parties, it must undertake appropriate international consultations before proceeding. In addition, a State that has reason to believe that space activity planned by another State would cause such potentially harmful interference may request appropriate international consultations, the only remedy provided in Article IX. Unfortunately, the term ‘consultations’, simple and unequivocal as it may sound, is subject to a wide variety of interpretations. Most agree that such consultations would involve asking an opinion on, or a common examination of, a specific problem. However, there is no obligation for States to follow up or take action. It is doubtful whether Article IX, considered in isolation, provides sufficiently strong language to serve as the basis for binding legal rules to prevent congestion and pollution of outer space.

Considering that environmental damages are often characterised by multiple causation and that the concept of ‘light’ refers to multiple spectra of electromagnetic radiation which can stem from various sources, a nexus is to be noticed. From a top-down view, space object collisions cause space debris, which clutters space traffic, and further affects night sky visibility. However, even the sole reflectiveness or brightness of a particular array or constellation of space objects can itself be the source of light pollution. In-space-generated light pollution in the form of skyglow is in fact reflectiveness originating from outer space objects.

Article IX of the Outer Space Treaty contains similar sounding yet distinct concepts of harmful contamination and harmful interference. Harmful contamination refers to a need to protect the space domain itself, whereas harmful interference refers to the space activities of other States. While such phrases might be considered as merely aspirational, and are to some extent vague, they still are binding.

The principles embodied in Article IX might be tied to the no-harm rule enshrined in Rio Declaration Principle 14: "States should effectively cooperate to discourage or prevent the relocation and transfer to other States of any activities and substances that cause severe environmental degradation or are found to be harmful to human health".

Rio Declaration Principle 18, known as the precautionary principle, is also of importance when it comes to light pollution. It states that: "In order to protect the environment, the precautionary approach shall be widely applied by States according to their capabilities. Where there are threats of serious or irreversible damage, lack of full scientific certainty shall not be used as a reason for postponing cost-effective measures to prevent environmental degradation."

Article IX of the Outer Space Treaty encourages States to "pursue studies ... and conduct exploration of outer space... so as to avoid harmful contamination and adverse changes in the environment of Earth". As avoidance requires one to take action before an event occurs and scientific certainty is not therefore required to create treaty obligations, these preventive mechanisms fully embrace the precautionary approach. There is a functional correlation between "avoid[ing] harmful contamination" and "discouraging or preventing ... activities and substances that cause severe environmental degradation". This correlation may allow in-space light pollution, as an adverse change to Earth's environment, to fall under the protective umbrella of existing environmental laws and principles.

As there are fundamental disagreements regarding the origins, consequences, and types of light pollution, there is no official internationally accepted, legally binding definition of light pollution. Therefore we may at most consider their common element, which is an acknowledgment of light as energy emitted within specific wavelengths of electromagnetic radiation. As Article IX's treatment of 'harmful contamination' has often been interpreted in legal doctrine as its ordinary meaning being limited to 'biological or radioactive contamination', there is no reason that the evolution of this provision would exclude an interpretation that brings such correlative phenomena in a proximate nexus. Additionally, in the many years since, light pollution has been causally linked to numerous adverse consequences and negative impacts on Earth's environment. For this reason, the possible inability of the previous section of Article IX to bring light pollution under its scope can be remedied by the phrase 'adverse changes in the environment of Earth'.

However, it is important to note that the due regard obligation is not merely aspirational. Rather, the obligation is direct: States "shall conduct activities in outer space"... "with due regard to the corresponding interests...". Helpfully, the other obligations of Article IX inform our understanding of the vague phrase 'due regard', and provide some larger context of what due regard means in the conduct of space activities.

Due regard is not utmost or absolute regard, but just what is considered the 'due' amount, and that amount is left to the States owing this regard to consider and quantify for themselves. Moreover, it seeks to ensure conciliation between rights belonging to different States, especially where those rights are conflicting. Furthermore, due regard is not a requirement for complete equality, but certainly a requirement for some considerations of equity, taking into account the shared nature of the cosmopolitan existence of actors enjoying equal status under the law.

The due regard obligation found in space law exists parallel to a due regard obligation found in other instruments of international law governing the behavior of States outside their territories. For example, Articles 56 and 58 of the UN Convention on the Law of the Sea (UNCLOS) uses this phrase. In the Chagos Marine Protected Area case (*Mauritius v. U.K.*), the Permanent Court of Arbitration Tribunal provided guidance on how to consider due regard, including that its ordinary meaning requires that States have regard for the rights of other States "as is called for by the circumstances and by the nature of those rights", and that the "extent of the regard required... will depend upon the nature of the rights held... their importance... the extent of the anticipated impairment, and nature and importance of the activities contemplated, and the availability of the alternative approaches." They further held that when significant rights are at stake, they are entitled to a "corresponding degree of regard". Also important was a determination that, in assessing the amount of regard owed, "at least some consultation with the rights-holding state" is necessitated "in a majority of cases". Therefore, some consultation, and a "balancing exercise" is warranted. Consequently, while due regard in space law is an under-defined term, some guidance can be found in the terrestrial maritime realm.

The obligation of due regard requires that this regard is given to the 'corresponding interests' of other States in the exploration and use of outer space. It may be initially unclear what 'corresponding' means in this context. Professor Marchisio writes in the Cologne Commentary on Space Law that "The notion of 'corresponding interests' recalls the fact that there are no unilateral interests in outer space and, as a consequence, there cannot be discriminatory treatment. Under this principle, space activities carried out by a given State should be in accordance not only with its own interests but also with the interests and rights of the remaining States Parties to the Outer Space Treaty."

Along these lines, and in the context of Article IX, use of space by one State is a corresponding interest to another State's use of space regardless of the differing nature of such uses—as what unites them is that they are both overlapping interests in a shared domain. In this approach, correspondence does not equal 'identical', nor even 'analogous'. Ensuring that corresponding interests are widely construed will be key in interpreting this due regard obligation in the context of dark and quiet skies, as well as in a variety of other space law contexts. Where is the rationale for this approach? Something may be found in the preamble of the Outer Space Treaty, reflecting the aims and intentions of the treaty, and with the encouragement of international cooperation, and of the respect for scientific investigations found throughout the treaty.

5. Concluding remarks

Space law and environmental law are subsets of international law connected by their shared issues and principles. Both enshrine the requirements of cooperation among States and the balancing of interests to ensure sustainable development.

Perhaps regrettably, the division of the outer space environment has been traditionally regulated with the special regime of space law, while a separate and distinct regime of international environmental law has arisen for terrestrial matters. This artificial division has led to questions of how international environmental law applies to space, whether space is an 'environment' like other environments, and how principles of environmental law might apply to the space domain. These divisions are artificial, but they are nevertheless real.

This article has attempted to extend a bridge between these distinct regimes of international law, by expanding the protections of environmental law to the vague, yet binding, principles in space law. This has been necessary as the issue of protection of the night sky crosses over these distinct domains, as activities in space have effects on the environment on Earth, and as space can also be seen as an environment. The arguments herein are offered as pathways forward for the legal protection of dark and quiet skies.

The preservation and protection of dark and quiet skies is required to achieve balanced and sustainable development of both the Earth and the space domains. Therefore, one must consider best practices in the application of binding norms from each to clarify and inform the imprecise or vague provisions of the other.

References

ICL Draft conclusions on identification of customary international law, 2018, YBILC 2018, Viol II part 2

Treaty on Principles Governing the Activities of States in the Exploration and Use of Outer Space, Including the Moon and Other Celestial Bodies, Jan. 27, 1967, 18 U.S.T. 2410, T.I.A.S. No. 6347, 610 U.N.T.S. 205, Art 1

Rio Declaration on Environment and Development, June 13, 1992, 31 ILM 874 (1992)

Stockholm Declaration on the Human Environment, in Report of the United Nations Conference on the Human Environment, UN Doc. A/CONF. 48/14, at 2 and Corr. 1 (1972), §22

Statute of the International Court of Justice Art. 38, ¶1, 26 June 1945

Space Debris Mitigation Guidelines of the Committee on the Peaceful Uses of Outer Space, Report of the UN Committee on the Peaceful Uses of Outer Space, UN Doc. A/62/20, June 15, 2007, 47–50 [Annex]

Biswanath Gupta, B., & Agasti, T. 2022, J Env Impact Manage Pol, 16

Space2030 Agenda: Space as a Driver of Sustainable Development - Resolution adopted by the General Assembly on 25 October 2021, A/RES/76/3, Seventy-sixth session, United Nations General Assembly, 28 October 2021

Inter-Agency Space Debris Coordination Committee (IADC), Space Debris Mitigation Guidelines, Oct. 15, 2002 (revised in 2007), IADC-02-01

Guidelines for the Long-term Sustainability of Outer Space Activities, A/AC.105/2018/CRP.20, Conference room paper by the Chair of the Working Group on the Long-term Sustainability of Outer Space Activities, Committee on the Peaceful Uses of Outer Space, 27 June 2018

Committee on the Peaceful Uses of Outer Space Scientific and Technical Subcommittee Sixtieth session Vienna, 6–17 February 2023, at XV. General exchange of views on dark and quiet skies for science and society

Resolution adopted by the General Assembly on 12 December 2022, Seventy-seventh session Agenda item 45 International cooperation in the peaceful uses of outer space 15 December 2022

Conference Room Paper On The Protection Of Dark And Quiet Skies For Science And Society, A/AC.105/C.1/2023/CRP.18, 2023

Kocifaj, M., Kundracik, F., and Bilý, O. 2020, MNRAS, 491(4), 5586

Longcore, T., & Rich, C. 2004, Front Ecol Environ, 2(4), 191

UN General Assembly, Convention on the Law of the Sea, 10 December 1982

Chagos Marine Protected Area (*Mauritius v. U.K.*), 31 R.I.A.A. 359, 571 (Perm. Ct. Arb. 2015)

Hobe S., Schmidt-Tedd B. and Schrogl K.-U., Cologne Commentary on Space Law (Berliner Wissenschafts-Verlag 2017)

Astronomy and Satellite Constellations: Pathways Forward
Proceedings IAU Symposium No. 385, 2026
P. Grimley, J. Barentine, D. Galadí-Enríquez & C. Walker
doi:10.1017/S1743921324001297

Update: how proclamations are influencing the grassroots movement for dark & quiet skies

Amy C. Oliver

Center for Astrophysics | Harvard & Smithsonian
email: amy.carissa.oliver@gmail.com

Abstract. Beginning in 2022, light pollution and grassroots advocacy expert Amy C. Oliver FRAS began partnering with DarkSky International (formerly International Dark-Sky Association) to increase traditional and social media coverage of International Dark-Sky Week through the strategic use of proclamations and other ceremonial documents and recognitions throughout their global advocate network. While the year-over-year growth of the proclamations programme and the resultant media attention provide mathematical performance indicators of success, intangible successes have also become evident through the increased confidence of DarkSky advocates and their willingness to engage in both grassroots and policy conversations regarding dark and quiet skies. Previously reported results from ASP2022, Solar Eclipses to Space Telescopes: Communicating Science to Students and the Public, and its forthcoming Proceedings are updated herein, along with additional and new learnings previously unreported.

Keywords. Policy, advocacy, grassroots, proclamation, light pollution, dark skies

---

1. The criticality of community voices in dark and quiet skies preservation

Light pollution is a growing problem, so much so that it is growing at double the rate of world population growth, and the Milky Way is now hidden to more than one third of humanity (Falchi, Cinzano, Durisco et al. 2016). This is causing a big problem for optical astronomy on Earth. At the same time, observational obstructions caused by the rapidly multiplying number of radio frequency interference-emitting devices like mobile phones, televisions, and radios are causing problems for radio and microwave astronomy both on Earth and in space. Astronomers know that to change this, there must first be a widespread change in public policy. However, while the collective voice of astronomy has power it is not always enough to effect that change in favour of dark skies over industry. Enter grassroots advocacy, which relies on the collective impact of communities and regular everyday citizens to effect change, particularly at subnational levels. In other areas of public policy—public health, for example—grassroots advocacy has been regarded for decades as one of the most effective strategies in change-making because it has the power to sway policy even when irrefutable scientific evidence is available. An oft-repeated public health example of this comes with the original failed 2007 Kansas Clean Indoor Air Act, which despite being backed by a report from the U.S. Surgeon General and stacks of research proving the harms of second-hand smoke, was defeated by the testimony of everyday citizens describing the potential negative impacts on their lives and businesses (Hall 2010). That same scenario plays out time and again in communities contemplating dark and quiet skies policies. For grassroots advocates in Arizona, evidence is mounting for the role of the community voice in protecting and managing dark skies.

2. Dark Sky International: including public voices in the change-making process

Founded in 1988, DarkSky International (formerly International Dark-Sky Association) is a non-profit organisation that advocates for responsible outdoor lighting, elevates and recognises communities and entities that engage in responsible lighting practices and provides educational materials and expert resources for individuals, communities, and governments that want to mitigate light pollution and protect dark skies. To date, DarkSky has more than 193,000 supporters, members, and advocates in more than 70 countries who help spread the organisation's mission and make its resources available (DarkSky International 2024). Advocates, members and supporters of DarkSky are empowered to build educational and support networks in their own communities, are strengthened through opportunities to mobilise and act, supported with materials, training, and resources, and help DarkSky keep a pulse on where their communities stand on issues of outdoor lighting and light pollution. These — empower, strengthen, support, and know — are the four cornerstones of any successful grassroots movement (Hall 2010).

2.1. *Leveraging International Dark-Sky Week through a public-driven proclamations project*

In 2022 DarkSky worked with the author, an Arizona-based light pollution and grassroots advocacy expert, to develop a plan to increase awareness of International Dark Sky-Week (IDSW) through the use of proclamations, and leverage that awareness into meaningful conversations about public policy. Also called ceremonial documents, memorandums, or celebratory documents depending on the country or other jurisdiction, proclamations are a favoured, though heavily underutilised, tool of grassroots advocacy projects because the documents and the process of requesting them have no policy implications. Proclamations have no regulatory power and are not tools of enforcement. Proclamations, however, are still official announcements, issued by federal, state, or local governments to recognise events of importance to their communities. In the United States at least, where the issuance of proclamations is a relatively common practice, this can be a time and cost-effective way to engage citizens and leverage awareness. For advocacy groups, these documents serve as educational documents, leverage to generate local and regional media coverage, and conversation starters for matters of public policy. In most cases, proclamations must be requested by a citizen who lives or works in the issuing jurisdiction. For IDSW proclamations, that made DarkSky's wide advocate network central to the project's success. Without them, there would be little access to local, regional, state, and federal level governments outside of Arizona, where DarkSky is headquartered.

2.2. *Moving the needle with increased advocate action*

The 2022 IDSW proclamations project was carried out over just six weeks. Armed with a basic template for writing a proclamation, and instructions for localising verbiage and making a successful request, DarkSky advocates submitted nearly 30 proclamations in the United States and received 23, as previously reported. That effort, in turn, resulted in 16 trackable media mentions with an estimated reach of 22.64 million (Foott & Oliver 2024). At the conclusion of the 2023 programme, advocates had submitted more than 130 requests for proclamations and received 99. Importantly 19 of those recognitions came from jurisdictions outside of the United States where proclamations are not a common practice, and even some that had never issued a document of this kind.

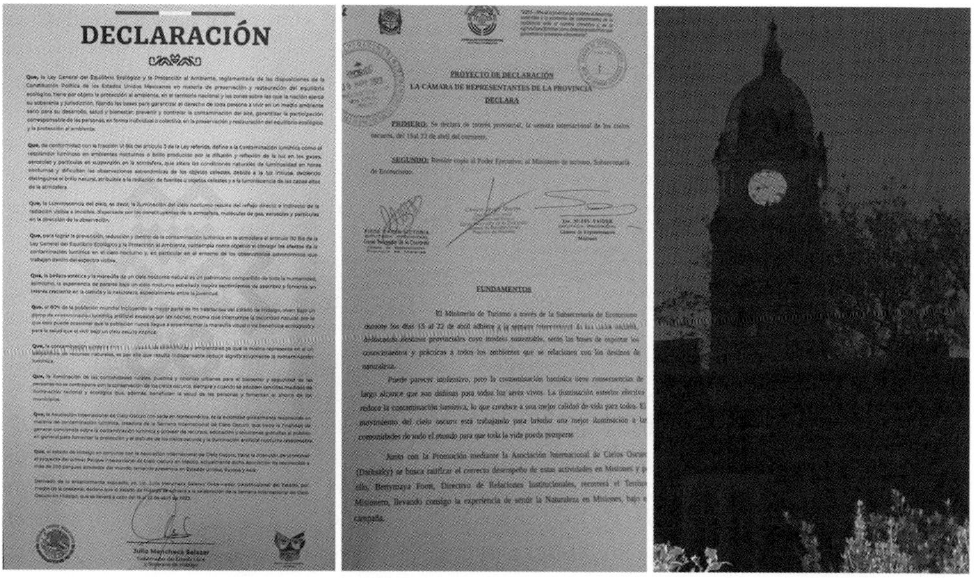

Figure 1. In 2023 the global efforts of DarkSky advocates resulted in the receipt of 99 proclamations, including 19 first-time proclamations from other countries. Pictured at left is a proclamation issued by Julio Menchaca Salazar, Governor of Hidalgo, Mexico. Pictured at centre is a proclamation issued by the ministers of tourism and ecotourism in Argentina. Pictured at right is the red lighting of the town clock in Eastbourne in East Sussex, England.

These included multiple jurisdictions in Italy, the UK, and Canada; Shire Perenjori in Australia; Argentina; and Hidalgo, Mexico. In the UK, advocate efforts also prompted the Eastbourne Borough Council, in East Sussex, England, to unexpectedly light the town clock red from 15 to 22 April 2023 in honour of IDSW. These increased efforts had a marked impact on media coverage before, during, and after the weeklong event, attracting 904 media mentions with an estimated reach of 1.295 billion. In 2024 the advocate network is continuing the project with a new focus on governments that have not yet participated, including the Pitcairn Islands, Brazil, and Metz, France.†

3. Empowering advocates to start conversations in their own communities is what makes change

The number of proclamations requested and received is impressive, and the increased media attention for IDSW resulting from the use of proclamations ensures widespread heightened awareness of light pollution and its mitigations. However, these are not the most important outcomes for the proclamations project. In 2023 the author used Arizona, Washington, and Connecticut proclamations to start conversations with residents about the importance of dark skies not only to astronomy but also to thriving ecology and sustainable environments. This, in turn, led to the formation of Dark SkyNet, a new open-data light pollution monitoring project led by the author, dark skies expert John Barentine, and former astronomer Emilio Falco. The prototype for Dark SkyNet launches in Arizona and Northern Mexico in 2024 and with the support of more than two dozen citizen scientists, will provide evidence for best practices in responsible outdoor lighting policies, and a template for the geographical expansion of the project. Also born of these conversations and launching in 2024, Project GLOWworm, led by the author and the

† 2024 will be Oliver's last report on the programme's outcomes, and the DarkSky team will lead the project in future.

entomologist Joseph Cicero, aims to identify a glowworm sighted at unexpected elevations in the Santa Rita Mountains, and determine whether habitat destruction, including by light pollution, has contributed to the bioluminescent residents' environmental adaptation. After participating in the project in 2023, Southern Arizona advocate Sam Miller was inspired to take on leadership responsibilities in his local chapter, helping to drive education, awareness, and policy measures. Additionally, attention and education surrounding the proclamations project have led to increased interest in participation from individuals otherwise unaffiliated with the dark skies conversation in Connecticut, New York, Vermont, and Texas. These types of growth and interest inform the residents of communities, who as Hall (2010) found in 2007, have an incredible power to make, or deter, change.

Although these examples and experiences have highlighted efforts to create a path forward for dark skies, the same principles can be applied to similar efforts for quiet skies. No matter the topic of discussion, no matter the need, the collective impact of a community's voice has strength and power whereas sometimes scientific evidence does not. Each year of the proclamations project provides growth, support, new learnings, and a better understanding of how to work together to shape the conversation and educate in ways that lead to the adoption of dark and quiet skies best practices.

References

DarkSky International 2024, Who We Are, https://DarkSky.org/who-we-are/

Hall B. 2010, Giving Voice: The Power of Grassroots Advocacy in Shaping Public Policy, Views From the Field, Grant Makers in Health.

Falchi, F., Cinzano, P., Duriscoe, D., et al. 2016, The New World Atlas of Night Sky Brightness, Science Advances

Foott, B. & Oliver, A. 2024 Who We Are, Proclamations as Unexpected Grassroots Policy Products for Dark Skies Preservation, ASP 2022: Solar Eclipses to Space Telescopes: Communicating Science to Students and the Public, accepted.

Astronomy and Satellite Constellations: Pathways Forward
Proceedings IAU Symposium No. 385, 2026
P. Grimley, J. Barentine, D. Galadí-Enríquez & C. Walker
doi:10.1017/S1743921324000449

Merged recommendations for mitigating impact

Richard F. Green
on behalf of the IAU CPS Management Team and Hub Leads

Steward Observatory, University of Arizona, 933 N. Cherry Ave., Tucson, AZ, 85718, USA
IAU CPS
email: rgreen@arizona.edu

Abstract. Expert working groups produced a number of thoughtful technical recommendations to policy makers and industry to mitigate the impact of satellite constellations on astronomical observations. The IAU CPS has undertaken to consolidate those complementary recommendations into a compact set, with advice to industry and policy makers, as well as identification of needs for further definition by the astronomy community.

Keywords. satellite constellations, mitigations

1. Introduction

Growing awareness of the impact of satellite constellations on ground-based astronomy prompted a series of technical workshops. The goal was to assemble technical and policy experts to assess the extent of the impact, to identify further modeling and evaluation required, and to make recommendations to observatories, satellite operators and policy makers. The U.S. National Science Foundation sponsored the workshops SATCON 1 (Walker, Hall, Allen, et al. 2020) and SATCON 2 (Walker & Hall 2021). The UN Office of Outer Space Affairs along with the IAU and the Instituto de Astrofisica de Canarias sponsored the workshops Dark and Quiet Skies for Science and Society (Walker et al. 2021) and Dark and Quiet Skies for Science and Society II (Walker & Benvenuti 2022).

In the reports, the expert working groups listed a number of thoughtful technical recommendations to policy makers and industry to mitigate the impact of satellite constellations on astronomical observations. The IAU CPS has undertaken to consolidate those complementary recommendations into a compact set, with advice to industry and policy makers, as well as identification of needs for further definition by the astronomy community.

2. Reflected sunlight

The specification on the brightness limit of reflected sunlight is based on the assumption that meeting the condition of the telescope with the highest etendue (field of view x effective area) satisfies the limits for most other telescopes. That brightness limit was set by the calibration limit of parallel cross-talk streaks in the Rubin Observatory CCD camera. Within that limit, the streaks can be removed precisely; brighter than that limit, the relationship of the cross-talk to the central streak no longer scales straightforwardly, and custom calibration would be required image by image. The limit is now well known (Tyson, Ivezić, Bradshaw, et al. 2020) to be

$$V = 7^{th} mag + 2.5 * log_{10}(r_{orbit}/550km). \tag{1}$$

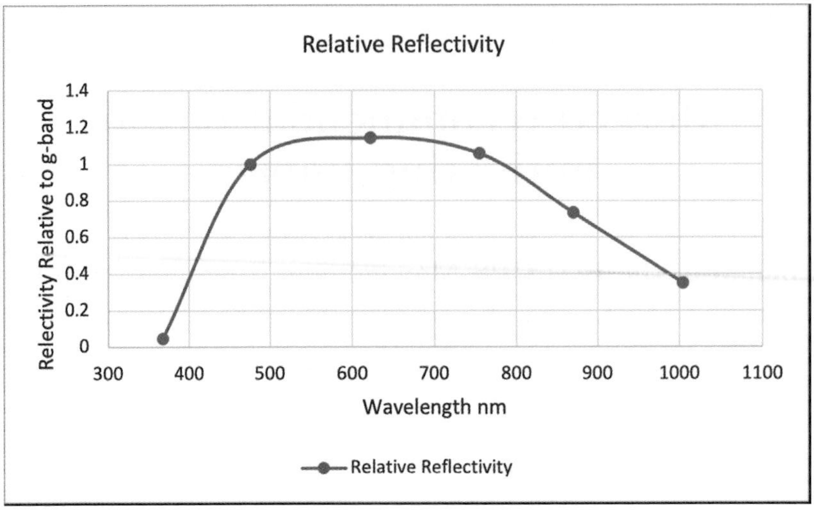

Figure 1. Reflectivity limits in Rubin Observatory filter bands relative to the V (g / 480 nm) band limit of 7th mag in order to calibrate streak cross-talk.

To set engineering-level specifications, it will be necessary to define a tolerance for physical limits as well. Further lab experimentation suggests that the tolerance for the magnitude limit could be as much as 0.25 mag. It is important to note that the Rubin Observatory camera limits in other filter bands are not equivalent to uniform reflectivity across all bands (Figure 1). The same criteria for cross-talk calibration can be applied to the other Rubin Observatory filter bands (Tyson, Ivezić, Bradshaw, et al. 2020).

Although it would be desirable to set firm brightness limits as the basis for regulation, achieving those limits within budgetary constraints has proven very challenging for satellite constellation producers. For example, best efforts to date by SpaceX with Generation 2 minis have just got close to that limit, but regular Generation 2 spacecraft/antennas will be bigger.

3. Flares and glints

Rubin Observatory colleagues are finding through simulations that variability of the reflected sunlight component could be a major difficulty, over and above the DC component. Variable background near a flickering streak or a tumbling piece of space debris could induce false transient detections. The multi-layer insulation (MLI) typically used for thermal management of spacecraft can act like a disco ball in producing low-level glints as a satellite changes viewing aspect. Further simulations are required to respond to the request from industry for a metric as a design goal. Such a metric would specify the limit on variability amplitude as a function of time interval, from low-level flickering to the largest amplitude flares. The ultimate goal would be to include brightness variability as a function of Sun angle and viewing angle as part of predictive modeling data, subject to observational validation.

4. Operational constraints

The consensus goal is to minimise visibility in dark time for the largest aperture professional optical/IR telescopes ($\pm 35^0$ latitude). That motivated the recommendation that operators choose orbital altitudes below ≈ 600 km. It is recognised that such a constraint is not necessarily favorable for observatories at higher absolute latitudes or for programmes that obtain key data in twilight, such as the search for potentially hazardous asteroids. An astronomical constraint is to minimise the number of sunlight streaks by populating the constellation with the minimum number required for reliable bandwidth and latency requirements. This pair of requirements (low altitude and minimal number) make the best conditions for astronomy. It is recognised that, taken together, they are incompatible with industry requirements, where lower altitude requires more numerous satellites to provide continuity of coverage. The mission phases of greatest impact are orbit raise and de-orbit because of significantly lower altitude. The recommendation, as demonstrated in practice by SpaceX, is to minimise visible brightness during those phases by appropriate attitude adjustments to present a lower cross-section to the ground (SpaceX 2022). Operators are also urged to clump their constellations during orbit raise to minimise the transit time of these bright objects through any field of view.

5. Protect radio observatories

The best protection of sensitive equipment such as low-noise amplifiers is for satellite operators to avoid direct illumination of radio telescopes and radio-quiet zones. The U.S. Federal Communications Commission has set a worthwhile example by requiring operators applying for licenses to execute a coordination agreement with the U.S. National Science Foundation that includes a commitment to avoid such down-links. A critical recommendation is support for the development and deployment of technology to minimise side-lobe illumination, both for individual units and in aggregate. Similarly, support is recommended for development and deployment of technology to increase the robustness of low-noise amplifiers and to reduce spurious out-of-band emissions of transmitters. The ultimate goal is to develop the technology and protocols for radio-dynamic zones, with automated avoidance of planned radio telescope observations in direction and frequency.

There are two current major challenges. One is the pressure for direct-to-cell service. The lower frequency of ground-based cellular transmissions compared to currently approved orbital down-link frequencies tends to create beam patterns with more side-lobe emission. To achieve the sensitivity required to receive signals from individual cell phones, the satellite antennas must have significantly larger areas, greatly increasing their apparent brightness. A salient example is the AST SpaceMobile Blue Walker 3 prototype, which appeared as bright as nearly 0th magnitude (Nandakumar, Eggle, Tregloan-Reed et al. 2023). The other challenge is unintended electromagnetic radiation (UEMR). This is low-frequency emission from motors and other unshielded components. Di Vruno, Winkel, Bassa, et al. (2023) report the recent detection of operational StarLink satellites at these low frequencies, which will impact the most remote sites where low-frequency observations are conducted.

6. Impact assessment - information sharing

It will be valuable for satellite producers to predict the total apparent brightness of their units through modeling of reflected sunlight with sun angle and viewing angle. Such modelling would include valid measurements of the bi-directional reflectance distribution function (BRDF) of the surface materials they employ. Ultimately that modelling should extend to prediction of flares and glints. In the radio, it will be important for the producers to provide predictions of effective isotropic radiated power, transmission bandpasses, radiated power as a function of frequency, measured spectral masks, etc.

7. Accurate ephemerides

For observation planning, including avoidance and shuttering when possible, and post-hoc evaluation, astronomers need accurate ephemerides with covariances on an agreed cadence. Desired tolerances are 1-second timing along track, and 1 arcsecond cross-track positional accuracy. The current accuracy of along-track predictions is more than adequate, but cross-track predictions need improvement by an order of magnitude to be useful for narrow-field or multi-fibre instruments. More information is required than is provided in the classic two-line element (TLE punch card) format, but major operators are now providing that additional information; any format acceptable to 18th and 19th Space Data Systems format should be acceptable for astronomers. Achieving higher accuracy should include planned manoeuvres for prediction.

8. Approach

The first phase with the highest probability for near-term success is for the astronomy community to work cooperatively with industry. The IAU CPS is working through its Industry and Technology Hub on identification of and education about problems, jointly advocating for support for technical development for both astronomy and constellation operators. Since the technical ability to meet specific brightness and variability thresholds has yet to be demonstrated, the IAU CPS is recommending the short-term regulatory approach of executing a coordination agreement (or equivalent) with a relevant scientific agency for best efforts as a condition of licensing. As the limits of technical performance become clearly defined, regulations can then be established or modified to include hard limits to be met by pre-launch modeling and post-launch observational validation.

9. Needs

The policy regime requires definition of thresholds for harm. Such thresholds represent the limits beyond which the quality of astronomical data will be significantly compromised. These are urgently needed for the incidence of streaks, for the incidence of flares and glints as a function of amplitude, and for the aggregate impacts on both radio and optical observations. As scientists, we would like to base those levels on well-demonstrated, defensible models. The recommendation is for agency support of the simulation research required to answer those questions, along with astronomy community demand and interest to undertake those projects.

The technical recommendations summary, along with near-term and longer-term implementation approaches, will be available through the CPS website at `https://cps.iau.org/`.

References

Di Vruno, F., Winkel, B., Bassa, C. G., et al. 2023, A&A, 676, A75

Nandakumar, S., Eggle, S., Tregloan-Reed, J., et al. 2023, Nature 623, 938

SpaceX. 2022, Brightness Mitigation Best Practices for Satellite Operators. `https://api.starlink.com/public-files/BrightnessMitigationBestPracticesSatelliteOperators.pdf`

Tyson, J. A., Ivezić, Ž, Bradshaw, A., et al. 2020, AJ 160, 226

Walker, C., Hall, J., Allen, L., et al. 2020 BAAS, 52(2)

Walker, C. & Hall, J. (Eds.) 2021, Report of the SATCON2 Workshop, 12–16 July 2021, BAAS, `https://baas.aas.org/report-of-the-satcon2-workshop-1216-july-2021`

Walker, C. et al. 2021, Dark and Quiet Skies for Science and Society: Report and recommendations, Zenodo, urlhttps://doi.org/10.5281/ZENODO.5898785

Walker, C. and Benvenuti, P. (Eds.) 2022, Dark and Quiet Skies for Science and Society II, `https://doi.org/10.5281/zenodo.5874725`

Astronomy and Satellite Constellations: Pathways Forward
Proceedings IAU Symposium No. 385, 2026
P. Grimley, J. Barentine, D. Galadí-Enríquez & C. Walker
doi:10.1017/S1743921324000644

Astronomy, doughnuts, and carrying capacity

A. Lawrence

Institute for Astronomy, University of Edinburgh
email: al@roe.ac.uk

Abstract. I examine the applicability of ecological concepts in discussing issues related to space environmentalism. Terms such as "ecosystem", "carrying capacity", and "tipping point" are either ambiguous or well defined but not applicable to orbital space and its contents; using such terms uncritically may cause more confusion than enlightenment. On the other hand, it may well be fruitful to adopt the approach of the Planetary Boundaries Framework, defining trackable metrics that capture the damage to the space environment. I argue that the key metric is simply the *number* of Anthropogenic Space Objects, rather than for example their reflectivity, which is currently doubling every 1.7 years; we are heading towards degree scale separation. Overcrowding of the sky is a problem astronomers and satellite operators have in common.

Keywords. Space, Environmentalism, Ecology, Debris, Satellites

1. Introduction

An article in Nature Astronomy in 2022 put forward the scientific and technical case for "Space Environmentalism", a concept that had been developing informally for some time (Lawrence, Rawls, Moriba et al. 2022). I still feel that "Space Environmentalism" is a better term than "Space Sustainability" as, with the latter term, it is not clear what we are trying to sustain, and the term "sustainability" is slowly gathering an aura of greenwashing. An environmental approach captures several key elements: that actions may have damaging consequences elsewhere in the environment; that the true cost of an activity is displaced onto others occupying the same environment; and that damage is incremental and complex. All those aspects mean that we should take a broad environmental approach to our problems rather than a narrow self-enclosed one, and encourage a sense of stewardship of an environment shared by astronomers, by space operators, and by the general public, especially amateur astronomers and indigenous communities. However, Lawrence, Rawls, Moriba et al. (2022) also suggested in passing that space could perhaps be seen as an *ecosystem*, and many discussions since have used concepts carried over from ecology, such as the idea of carrying capacity. In this paper I want to critically examine whether ecological concepts can be sensibly applied to orbital space, and if not, what we should use instead.

2. Applicability of of ecological concepts to orbital space.

Much of what I refer to below comes from standard ecology. A popular ecology textbook is Chapin, Matson & Vitousek (2011), and a very practical one for astronomers new to these topics (like me!) is Lehman, Lohberg & Clark (2021).

2.1. *Is orbital space an ecosystem?*

Ecosystems are normally defined in terms of organisms which interact with each other, and which also interact with their physical environment (nutrients, energy flows, etc). But even if we generalise to more general abiotic complex systems, to usefully think of such a system as an ecosystem we need feedback loops between the components, in which case orbital space and its contents does not look like an ecosystem. Thomson (2023) provides more detail. On the other hand, if we combine the contents of orbital space, the human actors on Earth that launch into space, and the economic and regulatory system they find themselves in, this broader system does look like a human-economic ecosystem. However, it's not clear how this insight helps.

2.2. *How do we define the "carrying capacity" of orbital space?*

Some papers have unpacked the historical development of the term "carrying capacity" and noted how inconsistently and ambiguously it is used (Sayre 2008; Chapman & Byron 2018). It is possible to discern three distinct concepts for which the term "carrying capacity" has been used. (i) The carrying capacity of a train is just the number of people that can fit on before it is full. This isn't a concept that applies to orbital space. (ii) The carrying capacity of a cargo ship is how much you can safely load it with, before it sinks. This seems to be what people sometimes have in mind when talking loosely about carrying capacity in orbital space - we mustn't exceed the carrying capacity or there will be disaster! This "safety level" meaning doesn't obviously apply to orbital space. (iii) In population ecology, the term "carrying capacity" has a very well defined technical meaning. It is an *equilibrium level* which emerges naturally from the behaviour of the Van der Hulst equation used to model the time evolution of a population with density dependent effects. If the population is below this level, it will rise, and if it is above it will fall back. Implicit in this equation are a death rate and a birth rate which depend on some kind of renewable food resource. One can also model competing populations using the Lotka-Volterra equations; each population will find its own equilibrium level, including the possibility of one population wiping out the other. But none of this obviously applies simply to the behaviour of orbiting satellites.

Sturza, Dankberg & Blount (2022) carry out some very interesting work in which they define "carrying capacity" to be the point at which the rate that satellites are consumed by collisions is equal to a specified fraction of the constant rate at which satellites are launched to maintain a constellation. This is related to the "tipping point" concept which we discuss below, and is important and impressive work, but I think that using the term "carrying capacity" in this context is potentially misleading and confusing.

2.3. *Do we expect a tipping point?*

Standard population ecology models assume the availability of environmental resources (such as food) which are renewable. A population or quantity embedded in the environment may saturate, or it may grow indefinitely, depending on the density-dependence of the birth and death rates. If however the system includes a *non-renewable resource*, the population can rise to a maximum and then decline. This was the central insight of the Club of Rome "Limits to Growth" work (Meadows, Meadows, Randers & Behrens 1972; Meadows, Randers & Meadows 2005). In the context of population ecology models, a standard Van der Hulst style carrying capacity could be defined at any one instant, but it would change with epoch, and so could be seen as a "dynamic carrying capacity". This idea is implicit in the qualitative illustrations of Figure 3 of Miraux (2022). There is, however, no obvious non-renewable resource that is being used up in orbital space.

Overall we have to be very cautious about the direct transference of concepts from mathematical ecology. Naively using concepts like "carrying capacity" or "tipping point" is likely to cause more confusion than enlightenment. However, this is not to say that we cannot model the time evolution of the contents of orbital space with simple rate equations embodying things such as launch rate, satellite lifetime, collision probability, and so on. This is exactly what what was done with the JASON debris model (Long 2021; Lawrence, Rawls, Moriba et al. 2022) and in the work of Sturza, Dankberg & Blount (2022).

Figure 6 in Lawrence, Rawls, Moriba et al. (2022) shows an example of a model where too large a launch rate into a particular orbital shell is counter productive in the long term: the gradual accumulation of debris in that shell slowly increases the probability of disabling collisions, and the number of net active satellites starts to go down. However, this process takes several decades, it is not a sudden disaster, and it applies to a specific orbital shell. As we will see below, there are more urgent things to worry about.

3. The planetary boundaries framework and doughnut economics

An interesting suggestion made by both Miraux (2022) and Wilson & Vasile (2023) is that space environmentalism issues could be encapsulated in the Planetary Boundaries Framework (Rockström, Steffen, Noone et al. 2009). In this framework, a variety of metrics are defined that can be used to monitor the health of the planet - for example, ocean acidification, CO_2 concentration, stratospheric ozone depletion, and so on. In the most recent report, Richardson, Steffen, Lucht et al. (2023) argue that the planet is beyond safe boundaries in 6 out of their 9 metrics. A tenth boundary is left for "novel entities". This could perhaps be used for sky brightness (Miraux 2022) or "orbital space depletion" (Wilson & Vasile 2023) but it could also be some metric more closely related to optical or radio interference (see section 5). It would be very fruitful to come to an agreed community definition of what could be used.

As also noted by Wilson & Vasile (2023) the well known "Doughnut Economics" model of Raworth (2017) adds inner boundaries to the planetary boundaries concept. Humanity needs enough activity to produce sufficient food, energy, health, social equity and so on, but not so much that we exceed the safe outer planetary boundaries - we would like to be inside the doughnut, the "safe and just space for humanity". This is a good way to look at the tension between astronomy and space industry that we currently face - we want activity that builds towards goals such as connectivity for all, but without causing too much damage to the sky, or creating so much debris collision risk that space industry becomes unsustainable. But how do we define and agree how much is enough?

One aspect of looking at space environmentalism as an economic issue, and a key issue in Raworth's work, is challenging the assumption that growth is always good. We should be aiming for "thriving" rather than growth *per se*. As physicists, we are aware that a constant rate of growth leads to an exponential. Qualititatively, the population of anthropogenic space objects (ASOs) seems to be accelerating ... does it in fact show exponential growth?

4. Exponential growth of space activity

The left panel of Figure 1 shows the log of the number of currently active satellites versus linear time, using data kindly provided by J. McDowell, extracted from the General Catalog of Artificial Space Objects (McDowell 2023). On such a plot, exponential growth will be of course a straight line. What we see is three distinct historical phases, each showing exponential growth, at different rates. In the original Space Age, the number of active satellites was doubling every 1.2 years. We then have several decades of Normal

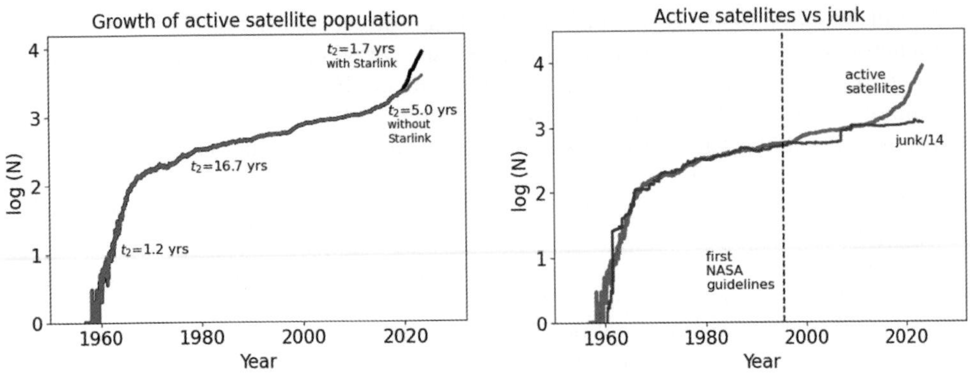

Figure 1. Left: Growth *vs* time of the active satellite population, in log-linear form, using data from McDowell (2023). Approximate doubling times of the three apparent historical segments are marked. Right: active population compared to the "junk" population, divided by 14. See text for details.

Space Activity, dominated by national space agencies and military satellites, but with of course some commercial communications satellites. In this era, space activity was growing much more slowly, doubling every 16.7 years. Finally we reach the New Space era, with activity recently doubling every 1.7 years. This is currently dominated by the Starlink constellation of course, but we can see that the New Space era in fact began before Starlink. With Starlink excluded, we see a doubling time of 5.0 years, starting around 2015. Then the Starlink project begins a second kick upwards from 2019 onwards. Without having to guess the fate of various proposals before the U.S. Federal Communications Commission etc., one could predict the future population empirically by extrapolation.

How has the space junk population grown? Here we define "junk" as the total of tracked debris and various "leftovers" such as rocket bodies, defunct satellites and so on. Once again, the data comes from the General Catalog of Artificial Space Objects (McDowell 2023). In the original Space Age, junk grew even faster than the active population, and then slowed its growth to a rate similar to the active population. The right panel of Figure 1 shows the number of junk objects divided by 14, to (roughly) match the active population. After the very rapid initial rise, the junk curve closely follows the active curve until about 1995. Roughly speaking each active object is associated on average with 14 pieces of junk. (Note that each curve represents the number of objects extant at that time, i.e. does not include objects that had already re-entered.)

After about 1995, the junk curve flattens off and grows much more slowly than the active population. In particular, the recent dramatic growth of active objects is not matched by a similar growth in junk. The year 1995 is when NASA first issued its orbital debris mitigation deadlines which were then adapted and improved over the next few years by the Inter Agency Debris Co-ordination committee (IADC). (See the history summarised at https://orbitaldebris.jsc.nasa.gov/mitigation/.) It seems that *debris mitigation works.*

A large fraction of the junk growth that has been made is due to either the Irridium-Kosmos collision of 2009, or the Chinese ASAT of 2007, which even on this log plot are clearly visible as vertical jumps. On the other hand, the debris caused by the 2021 Russian ASAT is visible only as a temporary blip; at the low altitude concerned, most of the debris made has now re-entered. And of course a major reason why New Space activity has not produced a major burst in net debris is because nearly all this growth

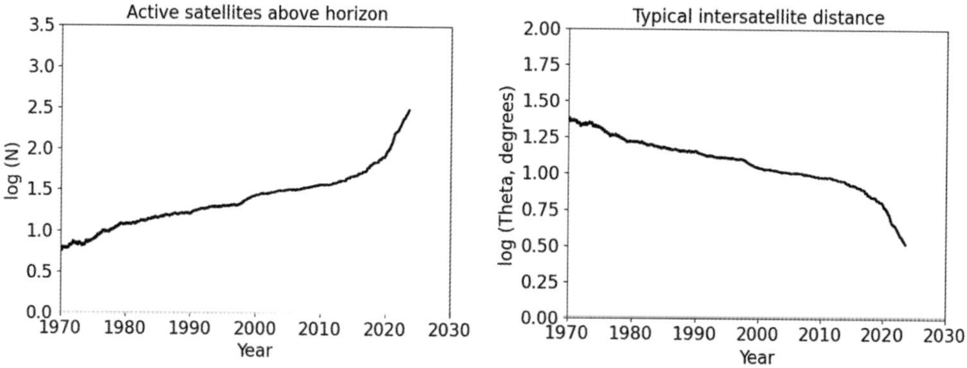

Figure 2. Left: Historical growth of the approximate average number of satellites above the horizon, using a very simplified height and orbit model and data from McDowell (2023). See text for details. Right: the corresponding typical inter-satellite distance on the sky.

has been in low Earth orbit (LEO), and indeed the Starlink project deliberately aims at a *turnover* of population, constantly demising and replacing objects.

Our broad-brush approach here of course masks the fact that the collision risk problem is about specific crowded orbital shells. Also, although some mitigation is achieved partly by "parking" objects in higher orbits after the 25 year lifetime guideline, it is mostly achieved by atmospheric re-entry. The consequence is that junk is now more than ever raining down upon us. We know that burn-up is often incomplete, with large pieces landing on the surface of the Earth, and many more smaller pieces causing an ever increasing risk to air traffic (see Lawrence, Rawls, Moriba et al. 2022). For this reason, and many others, it seems urgent that *Third Party Liability* insurance should become mandatory, just as it is for motor vehicles in most countries.

5. Sky density of active objects

Most mitigation discussions between astronomers and space operators centre on satellite reflectivity, where some progress has been made (at the same time as satellites are getting bigger!). Astronomers also have opinions about orbital height, but this is a bit more subtle, with tradeoffs between hours in sunlight, brightness, and angular speed. Of course from the radio point of view, satellites are visible whenever they are above the horizon. We all know however that the real problem is the *number* of satellites. What is the point of making satellites half as bright if there are fifty times as many of them? Furthermore, large modern telescopes can potentially see all the satellites from here to those in geostationary orbit (GEO) - as long as they are above the horizon - so to a first approximation orbit height doesn't matter. In reality, the fraction of the satellites in a particular orbital shell that are visible above the horizon, and where they are on the sky at a given time of night, depends on the height of the shell, the mixture of orbital inclinations, and the latitude of observation. However we can obtain a simple metric that captures the growth of sky density by ignoring those subtleties. The left panel of Figure 2 shows the growth of this metric, assuming a random distribution of orbit inclinations and placing all satellites at $h = 500$ km. (Using a low height will tend to underestimate the number in the sky). In the 1970s there were typically a handful of satellites in the sky at any one time; today it is over a hundred. (Note that only some of these are visible to the naked eye). The right panel of Figure 2 shows the corresponding decline in the typical distance between two satellites. We are heading towards degree scale separation.

Satellite operators are also affected by the inexorable increase in satellite communications overcrowding, as spelled out by in a recent Viasat White Paper (Viasat 2022)

There are two problems, related to the optical and radio astronomy interference problems. The first problem is the increase in "line of sight collisions", where for example a ground station trying to communicate with a satellite in GEO finds a LEO satellite passing across. For many years, it has been agreed that GEO satellites should be at least one degree apart on the sky, leading to a finite number of highly competitive GEO slots. Of course, LEO satellites are constantly moving, so what we have now is an equivalent statistical problem. This is closely analogous to the problem of satellites every so often streaking across a wide field astronomy optical image. The second issue is the aggregate power problem, from the sidelobes of the many satellites over the sky. This is of course closely analogous to the problem of radio astronomy interference. This problem includes unintentional out-of-band emission, just as radio observatories are now finding. (See elsewhere in these proceedings.)

It seems to me that this problem - sky density communications overcrowding - is going to hit crisis point long before the Kessler syndrome. The effectively useable sky may be captured by a small number of operators, to the detriment of other operators, and astronomers.

6. Conclusions

My main message is: **The problem is the numbers not the brightness.** We should not lose sight of the real problem. But in order to fight that battle, we need a defensible position on this question: **How many is enough?**. Meanwhile, here are some more detailed conclusions:

- We should be very cautious about using ecological concepts uncritically
- It could be useful to see space environmentalism in the planetary boundaries framework
- Satellite population growth shows three distinct historical exponential phases
- The current doubling time is roughly two years
- Space junk growth slowed from 1995 onwards
- Debris mitigation guidelines seem to work
- Material must be descending through the atmosphere at an increasing rate
- We need a new collision liability framework
- The satellite population is approaching degree-scale separation
- Operators have the same interference problems that we do!

References

Chapin, F. C., Matson, P. A., & Vitousek, P.M. 2011, Principles of Terrestrial Ecosystem Ecology, Springer-Verlag, New York.

Chapman, E. J. & Byron, C. J. 2018, Global Ecology and Conservation, **13**, e00365.

Lawrence, A., Rawls, M. L., Moriba, J., et al. 2022, Nat Astro, **6**, 428.

Lehman, C., Loberg, S., & Clark, A. 2019, Quantitative Ecology: a New Unified Approach, University of Minnesota Libraries Publishing. Available online at Libre Texts

Long, G., 2021, The Impact of Large Constellations of Satellites, Document No. JSR-20-2H (JASON Programme Office, Mitre Corporation)

McDowell, J. 2023, General Catalog of Artificial Space Objects, release 1.5.4 (2023 Oct 24). Available at https://planet4589.org/space/gcat/

Meadows, D. H., Meadows, D. L., Randers, J., & Behrens, W. W. 1972, The Limits to Growth, Universe Books, New York.

Meadows, D., Randers, J., & Meadows, D. 2005, The Limits to Growth: the 30 year update, James and James Ltd, New York.

Miraux, L. 2022, Sci. Total Environ. **806**, 150862.

Raworth, K. 2017, Doughnut Economics: seven ways to think like a 21st century economist, Random House Business.

Richardson, K., Steffen, W., Lucht, W., et al. 2023, Sci.Adv., **9**, eadh2458.

Rockström, J., Steffen, W., Noone, K., et al. 2009, Nature, **461**, 472.

Sayre, N. F. 2008, Annals of the Association of American Geographers, **98**, 120.

Sturza, M. A., Dankberg, M. D. & Blount, W. N., 2022, In the online proceedings of the 2022 AMOS conference, available at https://amostech.com/2022-technical-papers/

Thomson, F. 2003, in press. Knowing me Knowing you: a critique of the UK Space strategy, commissioned chapter in Ascending to Outer Space: Foreign Policy Perspectives from New Zealand and Other Nations, ed. Pozza, M., Springer.

Viasat 2022, White Paper Managing mega-constellation risks in LEO, available at https://news.viasat.com/blog/corporate/access-for-all

Wilson, A. R., & Vasile, M. 2023, Journal of Cleaner Production, **423**, 138869.

Astronomy and Satellite Constellations: Pathways Forward
Proceedings IAU Symposium No. 385, 2026
P. Grimley, J. Barentine, D. Galadí-Enríquez & C. Walker
doi:10.1017/S174392132400108X

A solution of a Devil's advocate - put telescopes in space

Lerothodi L. Leeuw

Department of Physics and Astronomy, University of the Western Cape
Modderdam Road, Private Bag X17, Bellville, 7530, South Africa
email: lerothodi@alum.mit.edu

Abstract. We present a solution of a devil's advocate to put telescopes in space, as one possible way to mitigate the impact on astronomy and on the night sky of recent large satellite constellations. The limitations of telescopes in space versus on Earth are noted. However, the solution draws from the history of telescopes that had to be moved from major human settlements to rural areas in the past; and it poses que question if it is possible that the future may lead to similar moves of putting telescopes in space, as satellites become a major human development similar to human development of cities that pushed telescopes out of cities to rural areas in the past.

Keywords. Space telescopes, satellite constellations, night sky, light pollution

1. Introduction

As one possible way to mitigate the impact on astronomy and on the night sky of recent large satellite constellations, we present a solution of a devil's advocate to put telescopes in space. The solution was partially inspired by a conversation on this subject during a live national radio show in South Africa, called SAfm Sunrise, that was a response to various sounds of alarm raised by astronomers on the topic (e.g. Collent 2023; Falchi et al. 2023; Kocifaj et al. 2023; Venkatesan 2023).

2. From doubt to opportunity to caution

Fig. 1 depicts the SAfm Sunrise national radio show podcast and a selected sample of the science topics it has covered as well as some science questions that have been posed on the show. One of the responses is about one possible way to mitigate the impact on astronomy and on the night sky of large satellite constellations. We present a follow up of this response, a solution of a devil's advocate to put telescopes in space. The initial reaction to this was doubt and then a recognition of the opportunity, though with caution. Why consider this at all? If humans have found solutions for building and operating telescopes far from where many humans live daily and sometimes even in hostile locations, like high altitudes and polar latitudes, why not find similar solutions in space, and at large scale? Space has become the next frontier for humans, and why not embrace it and look not only for bigger commercial opportunities like communication satellites but also for growing science activities there like the next planned Moon missions and more? Well, do it with caution and better set best practice guidelines early, but keep open the doors for more! Fig. 2 shows an example of this, a future telescope in space, i.e. the Xuntian, known as the Chinese Survey Space Telescope or the Chinese Space Station Telescope (CSST), proposed to be serviced in space and upgraded over time. One disadvantage of the telescopes in space has been that unlike those on earth, they

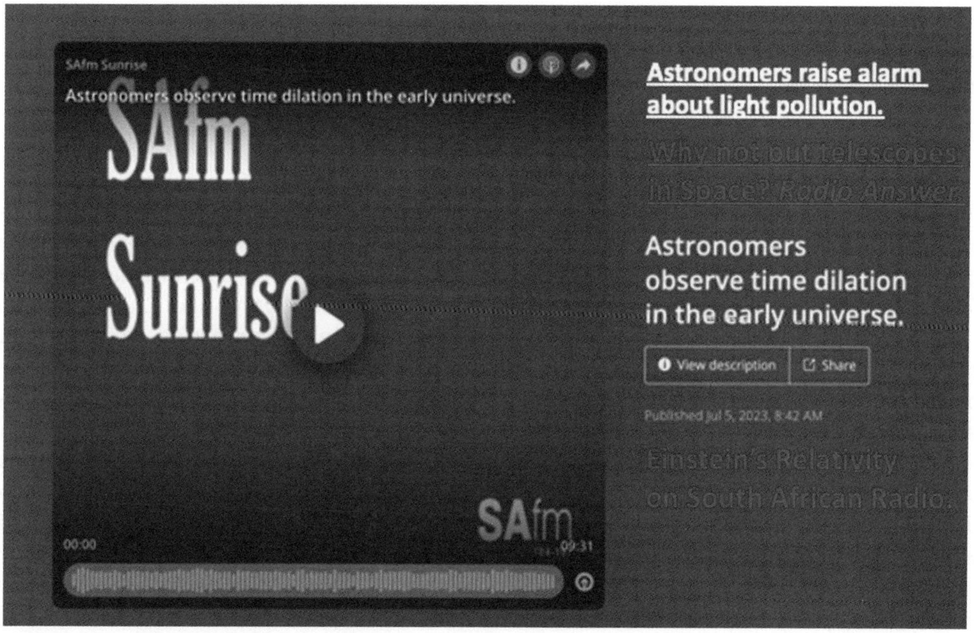

Figure 1. Depiction of the SAfm Sunrise national radio show podcast with a selected sample of the science topics it has covered and insightful science questions posed on the show, including one related to the topic of this paper.

Figure 2. An example of future telescopes in space, i.e. the Xuntian, known as the Chinese Survey Space Telescope or the Chinese Space Station Telescope (CSST), proposed to be serviced in space and upgraded over time (see Xin (2023)).

can't be serviced regularly and upgraded; however, this space telescope is proposed with that option. Will there be more of these examples in the future, including on the Moon and other planets?

3. Science growth globally

As articulated by Leeuw 2024, the global context of science can be described in episodes on logarithmic scale using selected key moments and representative images from literature. Such an example, from modern times to just over 20 000 years ago, is depicted in Fig. 1. Key in the figure is that both scientific instruments and people are shown at different developmental stages of science. The figure shows the telescopes of Galileo Galilei, that he used to observe the moons of Jupiter. These early telescopes are shown in the context of the development of science and instruments used in that development, including observations of Moon phases and other stars in general by eye before Galileo's telescopes. As adopted from Leeuw 2024, the figure lists some individuals and scientific organizations that also played a role in the development of science, to point out that the development is attributed not just to individuals but also organizations that have been pivotal for developing and operating major instruments for doing breakthrough science. The organizations are listed without trying to be exhaustive and are the Royal Astronomical Society (RAS), National Aeronautics Space Administration (NASA), European Space Agency (ESA), European Southern Observatory (ESO) and Square Kilometre Array (SKA). Some of the major instruments of these organizations, including space telescopes, are shown again without trying to be exhaustive, namely the Herschel Space Observatory (Herschel), James Webb Space Telescope (JWST) † and a dish of the MeerKAT Radio Telescope ‡, a pre cursor of the SKA. The instruments show the development of astronomical instruments that include both land and space telescopes and is what prompted the question that inspired the solution of the devil's advocate presented in this paper.

4. Lessons from southern Africa

One of the oldest modern scientific institutions in Southern Africa is the South African Astronomical Observatory § (SAAO), which was founded in 1820 as the Cape Royal Observatory by the British Royal Society for mapping and studying stars of the southern hemisphere. A photograph of the observatory is depicted in the timeline presented in Fig. 3, as one of the scientific developments some 200 years ago, i.e. around the time of Coulomb in the episode that would later include Maxwell, Hertz and Einstein. This development of an observatory at a place distant from its founders, is certainly a good historical example of the development of telescopes in distant locations from their founders as advanced for space in this paper.

Fig. 3 also shows the Southern African Large Telescope (SALT, Buckley et al. 2005) with its 10m class optical mirror built some twenty years ago and the MeerKAT Radio Telescope with its 64 radio antennae now in its 5th year of operation, that we built in South Africa much later and farther away from the Cape Royal Observatory in Cape Town. In this case SALT is some 400 km from Cape Town, because by the time it was built, the city of Cape Town had grown and the resulting light pollution necessitated the building of the telescope 400 km in rural areas with dark skies and good astronomical seeing. MeerKAT is also 700 km outside Cape Town in a radio quiet location away from the many sources of astronomical radio interference or pollution in the city. The headquarters of both observatories are in Cape Town but the actual observatories are far

† NASA Online Pages, www.nasa.gov accessed 10 March 2024
‡ South African Radio astronomy Online Pages, www.sarao.ac.za last accessed 10 March 2024
§ South African Astronomical Observatory Online Pages, www.saao.ac.za last accessed 10 March 2024

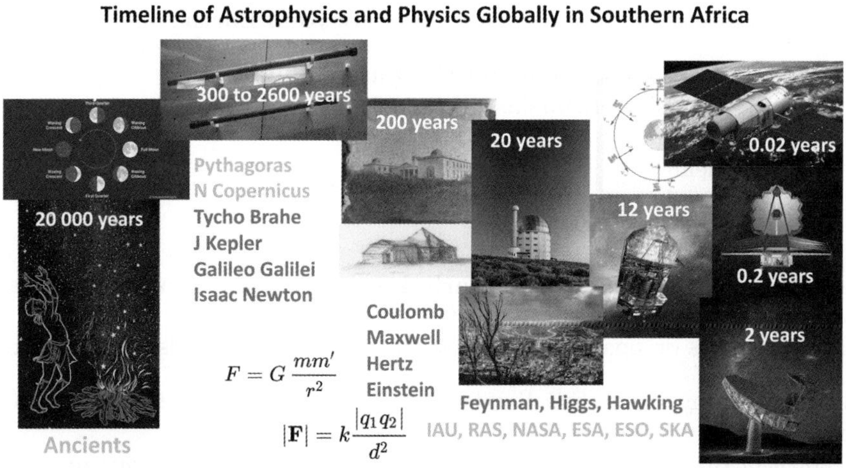

Figure 3. Timeline Depicting the Development of Astrophysics and Physics Globally and in Southern Africa. The timeline is constructed in logarithmic scale from modern times to just over 20 000 years ago using selected key moments and representative images from literature.

from the city. This is indeed another modern example of the solution proposed in this paper of putting telescopes in space away from the light pollution from satellites that interferes with observatories on the ground.

5. Poetic Outlook

The following extract is from poem 'We demand Physics!' by Keith Gottschalk from his book of poems *Cosmonauts do it in Heaven*. The poem is written in response to the Cape Observatory moving from Cape Town to a town in the rural Karoo region about 400 km away from the light pollution in the city of Cape Town. The poem speaks for itself so it's written here without further explanation.

'When the stars fled Cape Town
the astronomers followed
for stars hide in the light
we can only see them
in the brightest dark.'

'We who serve the stars,
left Hartleyvale floodlights,
smog & skyglow to astrologers,
for astrology thrives on opaqueness.'

'Starlight & irony bathe this Sunderland Karoo koppie:
a descendent of British admiralty
now precessed to this desert hill:
tonight, astronomy looks up.'

Below is an adaptation of the poem with the angle presented in this paper of a devil's advocate to put telescopes in space. Again the adapted poem speaks for itself so it's written here without further explanation. In boldface below are the added words to the poem above.

The adapted title is: 'We demand **astro**physics **beyond Earth**', and the poem follows.

'When the stars fled Cape Town **and the Earth's night sky**
the astronomers **and sky enthusiasts** followed
for stars hide in the light
we can only see them
in the brightest dark.'

'We who serve the stars,
left Hartleyvale floodlights **and the Starlink satellite constellation streaks**,
smog & skyglow to astrologers **and youtubers**,
for astrology thrives on opaqueness **and social media on trends and streaks**.'

'Starlight & irony bathe this Sunderland Karoo koppie, **Earth Orbits and the Moon**:
a descendent of British admiralty, **of NASA and Space X Exploration**
now processed to this desert hill, **to space and the Moon**:
tonight, astronomy looks up, **to telescopes beyond Earth**.'

6. Summary

It is noted again that the alternative solution of a devil's advocate to put telescopes in space, as one possible way to mitigate the impact on astronomy and on the night sky of recent large satellite constellations, is presented here deliberately to provoke discussion rather than as a final solution. Some of the limitations of telescopes in space versus on Earth are pointed out and discussed. As noted, the solution draws from the history of telescopes that had to be moved from major human settlements to rural areas in the past, with lessons from Southern Africa; and, asks, if it is possible that the future may lead to similar moves of putting telescopes in space, as satellites become a major human development similar to human development of cities that pushed telescopes out of cities to rural areas in the past. The presentation ends with a poetic outlook that showcases an earlier poem 'We demand Physics!' by Keith Gottschalk and adapts here as 'We demand **astro**physics **beyond Earth**.'

7. Acknowledgement

Lerothodi Leeuw (LL) funding from the South African National Foundation.

Some historical records presented here are critically adopted from the Southern African Astronomical Observatory (SAAO), Square Kilometer Array South Africa (SKA SA), Astronomical Society of Southern Africa (ASSA), and International Astronomical Union (IAU) and Royal Society.

References

Buckley, D. A. H., Charles, P. A., Nordsieck, K. H., & O'Donoghue, D. 2005, *Scientific Requirements for Extremely Large Telescopes*, Proceedings of IAU Symposium No. 232, 1
Collen, J. 2023, "Astronomers sound alarm about light pollution from satellites" (physics.org)
Falchi, F., Bará, S., Cinzano, P., et al. 2023, Nat Astro, 7, 237
Gottschalk, K. 2020, *Cosmonauts do it in Heaven*, Modjaji Books
Kocifaj, M., Kómar, L., Lamphar, H., et al. 2023, Nat Astro, 7, 269
Leeuw, L.L. 2024 in press, "Inclusive astronomy and science in Africa and other non-western regions for the global benefit", EAS Congress 2023 Session: The hitchhiker's guide to astronomy education, public outreach and communication
Venkatesan, A. 2023, Nat Astron, 7, 236
Xin, L. 2023, Sci. Amer., 329

Astronomy and Satellite Constellations: Pathways Forward
Proceedings IAU Symposium No. 385, 2026
P. Grimley, J. Barentine, D. Galadí-Enríquez & C. Walker
doi:10.1017/S1743921324001078

Fraunhofer AVIATION & SPACE and IRIS^2 - a European satellite constellation

Nadya Ben Bekhti-Winkel, David Butgereit, Leon Dehnhardt, Anne Drepper, Thomas Loosen, Maysem Sahbani and Tassja Wagner

Fraunhofer AVIATION & SPACE, Business Unit SPACE, located at Fraunhofer INT, Euskirchen, Germany
email: nadya.ben.bekhti-winkel@int.fraunhofer.de

Abstract. In order to make the European communication infrastructure more resilient and to strengthen the technological sovereignty in space, the European Parliament and the European Council decided in November 2022 to build a new satellite constellation: IRIS^2-Infrastructure for Resilience, Interconnectivity and Security by Satellite. The constellation will be realised through several hundred satellites in multiple orbits and is intended to ensure secure and very fast connectivity for commercial and institutional channels even in places where terrestrial broadband is not feasible or economical. The constellation must also meet the EU's "Green Deal" requirements (avoiding an increase in space debris, protection of other services), use the latest technologies and optimally protect communications against cyber-attacks by applying quantum encryption. The aim of this paper is to provide a deeper insight into the constellation's architecture, regulations and challenges. The importance of working in committees like DG DEFIS and having a large network will also be discussed in order to advance the Dark and Quiet Skies efforts.

Keywords. Fraunhofer Gesellschaft, Fraunhofer AVIATION & SPACE, DG DEFIS, sustainable space economy, National Space Strategy, IRIS^2

1. Introduction

National and international space companies are developing plans for satellite constellations in low earth orbits for a variety of communication purposes. This space-based infrastructure is essential for the functioning of the modern digital society. The expansion of this critical infrastructure is unstoppable. However, it must be recognised that the astronomical and geodesy communities, among others, have expressed concerns about the impact of these constellations on scientific research. Indigenous peoples also fear that the right to a natural night sky will be violated. The various stakeholders from science, business, society and politics must therefore work together in committees to discuss and evaluate the impact of satellite constellations on scientific observations and the natural night sky. Only by openly addressing the challenges that lie ahead will it be possible to raise awareness of the peaceful use of space and develop viable solutions that ensure the harmonious coexistence of the various space players in the future.

2. Fraunhofer-Gesellschaft and Fraunhofer AVIATION & SPACE

The Fraunhofer-Gesellschaft based in Germany is the world's leading applied research organization. Prioritizing key future-relevant technologies and commercializing its findings in business and industry, it plays a major role in the innovation process. Founded in 1949, the Fraunhofer-Gesellschaft currently operates 76 institutes and research units

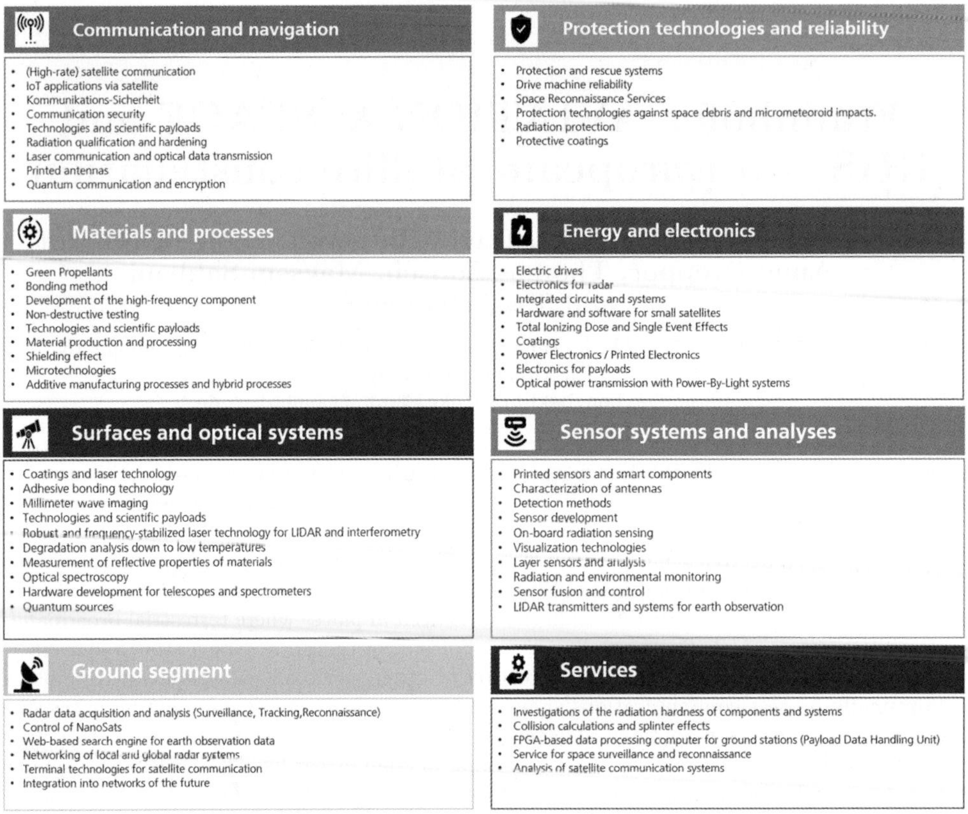

Figure 1. Technology portfolio of Fraunhofer AVIATION & SPACE.

throughout Germany. Over 30 000 employees, predominantly scientists and engineers, work with an annual research budget of 2.9 billion euros. Fraunhofer generates 2.5 billion euros of this from contract research.

At the Fraunhofer-Gesellschaft a very broad scope of research is addressed, ranging from artificial intelligence, bio-economy and quantum technology to industrial aerospace technologies. Interdisciplinary research teams work closely together with partners from industry and governments to solve future challenges and transform novel ideas into innovative technologies. The research projects have a high systematic relevance and strengthen the German and the European economy.

Fraunhofer AVIATION & SPACE is an association of 30 Fraunhofer institutes conducting applied research in the fields of communication and navigation, protection and reliability, energy and electronics, materials and processes, surfaces and optical systems, sensor systems and analysis, ground segments, downstream, and others (see Fig. 1). Fraunhofer AVIATION & SPACE has two separate business units AVIATION and SPACE.

The aim of the Fraunhofer AVIATION & SPACE is to collaborate with industrial and scientific partners in the fields of aeronautics and space. To this end, its member institutes pool their technological expertise to provide a central point of contact for stakeholders in the aerospace domain, such as the German Aerospace Industries Association, the European Space Agency, and the European Commission. In addition, Fraunhofer AVIATION & SPACE provides important contributions to national and European research funding programs, e.g., under the Horizon Europe and the European Space

Agency frameworks and advise space stakeholders in all relevant fields of technology including sustainability aspects.

3. DG DEFIS

The European Union is composed of seven institutions:

- European Parliament
- European Council
- European Council of Ministers
- European Commission
- European Central Bank
- European Court of Justice
- European Court of Auditors

The European Commission consists of of 55 Directorates General, one of which is the DG DEFIS (Directorate-General for Defence Industry & Space). The DG DEFIS was founded in 2020, is assigned to Thierry Breton as Commissioner for the Internal Market and leads the European Commissions's activities in the field of defence industry and space.

In the field of defence industry, DEFIS is responsible for maintaining the competitiveness and innovation of the European defence industry by ensuring the development of a powerful European defence technological and industrial base. In the field of space, DEFIS is responsible for the implementation of the EU Space Programme, which consists of the European Earth Observation Programme (Copernicus), the European Global Navigation Satellite System (Galileo) and the European Geostationary Navigation Overlay Service (EGNOS).

The objectives of DG DEFIS are

- Promoting a strong and innovative space industry,
- Maintaining the EU's autonomous, reliable and cost-effective access to space,
- Promote and further develop European SSA (Space Situational Awareness) activities,
- Implementing the future space programmes for Galileo, EGNOS, Copernicus and GOVSATCOM,
- Exploring ways to make the best use of available assets to achieve the climate goal and improve the crucial link between space, defence and security,
- Implementing and monitoring the European Defence Fund,
- Building an open and competitive European defence equipment market and enforcing EU defence procurement rules, and
- Lead the implementation of the Military Mobility Action Plan.

Early after founding, the DG DEFIS decided to set up expert groups consisting of different organisations with the aim of pooling expertise in the fields of space, defence and aeronautics (especially with EU relevance). The Fraunhofer Gesellschaft was assigned to be part of the expert groups. The expert groups advise DG DEFIS on the development and implementation of measures relevant to the space, defence and aerospace industries, e.g., identifying strategic industrial value chains, critical technologies and supply chain vulnerabilities. Furthermore, they facilitate the cooperation between the Commission and stakeholders on issues related to the implementation of Union legislation, programmes and policies relevant to the space, defence and aeronautics industries, they monitor the industrial and policy developments and promote the exchange of experience and best practices in the space, defence and aeronautics industries.

4. Sustainable space economy – Importance for the protection of astronomy

A paradigm shift has taken place in recent years in aerospace, which used to be characterised by institutions. In addition to state actors, there are numerous private investors who utilise low-Earth orbits to assert their economic interests. Against the backdrop of growing dependence on space infrastructure, however, there is also a rivalry for supremacy in space, in which security policy aspects play a central role. For this reason, 21 guidelines for the long-term sustainability of space activities were developed in 2019 under the auspices of the United Nations Committee on the Peaceful Uses of Outer Space (UN COPUOS; Maclay and Engelhardt 2021; Pelton and Madry 2020; United Nations Office For Outer Space Affairs 2021), analogous to the creation of the Sustainable Development Goals (SDGs). However, these are not binding and merely represent recommendations for action for sustainable space travel. In addition to these guidelines, space law attempts to create an international legal order that, on the one hand, preserves outer space as a "common space" for all states and, on the other hand, aims to take into account the diverse interests of states and the private sector. However, the large number of actors involved with very different interests creates a complex situation in which it is not easy to draw up consensual and binding regulations that are urgently needed for the future sustainable use of near-Earth space.

However, the efforts in recent years to raise the profile of sustainability in the space sector have borne fruit. Space players try to think about sustainability in their missions. The European member states support the announcement of an international "Zero Debris" charter, which aims to minimise emissions and space debris caused by space economy in space itself and to strengthen sustainability. Astronomy also benefits from these efforts. In its ZERO DEBRIS approach, the ESA has recently included point nine "Guarantee dark and quiet skies – Mitigate impacts on ground astronomy".

The German Federal Government also addressed the protection of astronomy in their published national space strategy in September 2023 (Bundesministerium für Wirtschaft und Klimaschutz (BMWK) 2023, Chapter 2.6): "[. . .] Germany attaches particular importance to avoiding space debris and is committed to the preservation of the dark night sky and the protection of radio astronomy".

5. IRIS^2 – a European satellite constellation

The European Union decided in November 2022 to create its own secure communications network for Europe to reach a digital sovereignty, independent of the offerings of e.g. SpaceX/Starlink or One Web: IRIS^2. Fig. 2 shows the aims for a sovereign European satellite constellation. The satellite constellation is to be realized through about 170 satellites, associated ground stations and control centres. This first European multi-orbit (LEO, complemented where appropriate by satellites in MEO and/or GEO) small satellite constellation is designed to ensure secure connectivity even in places where terrestrial broadband coverage is not feasible or economical. The constellation is planned to start offering the full service at the end of 2027.

The satellite network is intended to ensure a secure, very fast and, above all, sovereign exchange for both commercial and institutional communication channels. It will enable high-speed internet, networking of processes (Industry 4.0), autonomous mobility, automatic exchange of information, support for humanitarian and military actions, and other applications for public authorities, businesses and European citizens.

The planned satellite constellation must also meet the EU's "Green Deal" requirements (avoidance of an increase in space debris, and minimisation of disruptive visual

Figure 2. Europe's aims for a sovereign satellite constellation.

brightness), use the latest technologies and optimally protect communications against cyber-attacks through the application of quantum encryption.

IRIS^2 is expected to offer three services:

- Hardgov Services: Services restricted to government-authorised users based on Governmental Infrastructure, which require a high level of security and which are not suitable for Lightgov Services,
- Lightgov Services: Services to government-authorised users based on the Shared Infrastructure and Commercial Infrastructure, and
- Commercial service: Services to commercial users based on the Shared Infrastructure and Commercial Infrastructure.

In order to make the best use of innovative ideas and developments within the European Union, the European Commission launched a call for studies in 2021 aimed at SMEs, start-ups and research institutions. This concept study aimed to exploit new ideas and technologies for an innovative European broadband constellation, to define the requirements regarding data security and interoperability (with systems such as Galileo or Copernicus) and to develop an attractive business plan. At the end of 2021, the international UN:IO (`https://www.unio.global/`) consortium was selected as one of two alliances to conduct the study, each endowed with 1.4 million Euros. The consortium consisted of 20 SME space companies and start-ups (led by Reflex Aerospace, Mynaric and Isar Aerospace) and Fraunhofer AVIATION & SPACE. Fraunhofer AVIATION & SPACE contributed a thorough analysis of the field by investigating scientific publications and patents by employing e.g. machine learning techniques, as well as neutral expert knowledge. Within our work package we also addressed Ethical, Legal and Social Aspects (ELSA) such as Space Debris, Space Law, Equality and Justice in Space, and Dark and Quiet Skies.

In the first half of 2023, it was announced that major space companies such as Airbus Defence and Space, Eutelsat, Hispasat, SES, Thales Alenia Space, Deutsche Telekom, and OHB had applied for the project tender. A final decision with regard to the consortium on the part of the European Union is not yet known.

At this point, two obvious challenges for the final installation of IRIS^2 should be mentioned. Firstly, the EU does not have enough launch capacities and is therefore highly dependent on launch providers such as SpaceX. Secondly, the electromagnetic spectrum is a limited resource. As the increasing number of planned satellite services need more and more available frequencies, the frequency allocation and assignment processes becomes more difficult (as other services must not be disturbed either).

6. Résumé and outlook

The number of satellites in low-Earth orbit is increasing steadily. A peaceful coexistence of all space actors is only possible by openly addressing the challenges that lie ahead. It will then be possible to develop viable solutions that ensure the harmonious and peaceful use of space. Binding regulations would also help to save the important space-based infrastructure, to ensure the sustainable usability of near-Earth orbits, and to protect dark and quiet skies.

It has been shown that the constant discussion of the protection of the Dark and Quiet Skies in various committees and decision-making bodies has been worthwhile. Therefore, it is of utmost importance to continue this committee work in order to further raise awareness for the protection of the natural night sky, even if it is often difficult to decide which of the many committees can best address one's own needs. This is where a large network of people can help. Fraunhofer AVIATION & SPACE can provide support here as a neutral "one-stop shop" for science, industry and government.

References

Bundesministerium für Wirtschaft und Klimaschutz (BMWK) 2023, *Raumfahrtstrategie der Bundesregierung*

Maclay, T., E. W., & Engelhardt, D. 2021, JSSE, 8(2), 150

Pelton, J. N. & Madry, S. 2020, *Handbook of small satellites*, Springer International Publishing

United Nations Office For Outer Space Affairs 2021, "Guidelines for the long-term sustainability of outer space activities of the committee on the peaceful uses of outer space"

Astronomy and Satellite Constellations: Pathways Forward
Proceedings IAU Symposium No. 385, 2026
P. Grimley, J. Barentine, D. Galadí-Enríquez & C. Walker
doi:10.1017/S1743921324001431

National policy with a global impact: satellite constellations and the UK

Robert Massey

Royal Astronomical Society / University of Sussex
email: rmassey@ras.ac.uk

Abstract. The UK is a nation with burgeoning ambitions for its space sector, which sits alongside a strong astronomical community embedded in major international science projects on the ground and in space. As the primary representative body for UK astronomers, the Royal Astronomical Society (RAS) has the challenge of navigating these sometimes competing priorities, working to persuade the British government to give the science of astronomy the protection it requires. This paper summarises where we are, recent policy developments, and thoughts on our future plans.

Keywords. Satellite constellations, UK science policy, UK space policy

---

1. Introduction

As long ago as 1961, just four years into the space age, NASA proposed Project West Ford, the deployment of 480 million copper needles in orbit to serve as dipole antennae for global communications (described in detail by Lisa Ruth Rand earlier in this symposium). The Royal Astronomical Society (RAS) joined protests by astronomers around the world, with our then Council noting: "*Such a step would begin to shut off one of the few wavelengths available to radio astronomy to further our knowledge of distant regions of space. At the present stage of development of radio astronomy it is quite impossible to say that a particular wavelength could be dispensed with, and it would be scientifically a grave disaster if this project were permanent.*" (Bondi 1962; Lovell & Ryle 1962)

Today the RAS continues its keen interest in the impact of satellites and space debris on our science. The UK is a country with a strong astronomy research base, including membership of ESA, ESO and SKAO, and astronomers here use facilities all over the world. Citations for UK astronomy are ranked between second and fourth in the world.† The strength of the field undoubtedly makes a significant contribution to the aspiration for our nation to become a "science superpower" (His Majesty's Government 2023).

Alongside this, the UK Government sees the space sector as a high value sector (worth around £17.5 billion and employing nearly 50,000 people) with significant potential for growth. Government ministers back it with enthusiasm at numerous events, and promote associated work on space launch sites and space-based services; see, e.g., UK Space Agency (2023).

† See e.g. Scimago Journal & Country Rank (Astronomy and Astrophysics, Space and Planetary Science, citations, 2024) `https://www.scimagojr.com/countryrank.php?category=3103&year=2022&order=ci&ord=desc&area=3100` and `https://www.scimagojr.com/countryrank.php?area=1900&year=2022&order=ci&ord=desc&category=1912`.

Figure 1. The first meeting of the RAS Satellite Constellations Working Group, in January 2020. Image: Lucinda Offer / RAS

The industrial development of space and astronomical research are both significant recipients of public funding. The challenge for the RAS is then to offer balanced policy advice on satellite constellations, without being perceived as ignoring the economic benefits they bring, or as an opponent of the commercial use of space more generally.

2. Satellite constellations and the UK

At the start of the first Covid lockdown in the UK in March 2020, the satellite operator OneWeb filed for Chapter 11 protection, as the company was essentially bankrupt. To our surprise, the UK Government stepped in with a £400m bailout package, partly the result of pressure from Dominic Cummings, the then Chief Adviser to Prime Minister Boris Johnson. With additional funding from the Indian conglomerate Bharti Global, the satellite company stayed in business, and in 2023 merged with Eutelsat. The details of the initial procurement process remain controversial (Jones 2020), but at present the UK Government retains its golden share in OneWeb, as does France.

In our view, this gives the UK (and French) Governments a special responsibility for space sustainability, with public investment in ground- and space-based astronomy facilities, and at the same time in a satellite constellation system that interferes with their operations.

3. The RAS response

The RAS is one of the largest astronomy membership bodies in the world, with more than 4,000 Fellows in the UK and in around 70 other nations. Our role includes political advocacy for astronomy and geophysics, and we regularly meet with lawmakers in Westminster as well as civil servants in different government departments.

Like much of the global astronomy community, we were unprepared for the first Starlink launches in the summer of 2019. It would not be unfair to describe our initial reaction as a mixture of outrage and ignorance, as we grappled with this new issue (for example the author began engagement with SpaceX after complaining about the launches on Twitter), while we pondered a more considered response.

In the second half of that year the Society established ad hoc working groups, initially covering the whole area, and later supported by two subgroups for optical and radio astronomy (respectively chaired by Prof. Martin Barstow of Space Park Leicester and Prof. Simon Garrington of Jodrell Bank Observatory). These include a number of Fellows, representatives of observing facilities, from ESO and the European Space Agency, and from the UK Space Agency (UKSA) and Science and Technology Facilities Council (STFC), both of whom fund UK astronomy and space science research. In addition staff from the major satellite operators are usually in attendance.

Early actions included input into the international process led by the IAU, including the SATCON meetings, persuading STFC to fund two projects to assess the impact of satellite constellations (we hope to report on the radio project in the near future, the optical assessment was in line with other work), discussions with UKSA, and seeding parliamentary questions to ministers in the House of Lords.†

More recently we wrote to the Federal Communications Commission, arguing for a pause in the deployment of the second generation Starlink constellation (RAS 2022).

4. UK government and political responses

The most serious phase of the Covid pandemic in the spring of 2020 completely overshadowed our work on satellite constellations, as universities and research organisations transformed their operations, and the focus of national life rightly centred on managing the global crisis.

The four years since the first Starlink launch have in addition been an unusually turbulent period in British politics, covering four prime ministers, with four MPs holding the science minister role, and to further complicate matters the relevant government departments were restructured in 2023. We were also disappointed that the most relevant scrutiny body in the House of Commons, the then Science and Technology Committee, asked for evidence on the impact of satellite constellations in an inquiry in 2021, but chose not to explore this further (RAS 2021; House of Commons 2022).

Despite these setbacks, the UK astronomy community has seen some successes. OfCom, whose remit includes regulation of the radio spectrum, now makes a point of referencing radio astronomy in its consultations, and is proposing to protect radio observatory sites in the UK from interference. A caveat is that this is far only for the bands of the spectrum internationally allocated for radio astronomy (Ofcom 2022).

In 2021 the UK delegation to the UN Committee on the Peaceful Uses of Outer Space (UN COPUOS) supported the IAU-led position on dark and quiet skies, (UK Space Agency 2022) and in June 2022, during his first spell as science minister, George Freeman MP (His Majesty's Government 2022) acknowledged the need to protect astronomy in his speech at the Summit for Space Sustainability.

At least on the face of it, subsequent policy development reflects this position. In January 2023 UKSA worked with the RAS to run a one day conference on dark and quiet skies. and at the G7 Science and Technology Ministers' Meeting in Sendai that May, the UK endorsed its communique, including the reference to continued discussion on satellite constellations and astronomy at UN COPUOS and the ITU (G7 2023).

July 2023 saw the creation of the Astra Carta,‡ developed by the Sustainable Markets Initiative set up by King Charles III. Alongside this initiative, which is intended to be for

† See various in Hansard, e.g., `https://questions-statements.parliament.uk/written-questions/detail/2021-04-21/hl15160`, `https://questions-statements.parliament.uk/written-questions/detail/2020-07-06/hl6487` and `https://questions-statements.parliament.uk/written-questions/detail/2021-05-12/hl156`.

‡ `https://www.sustainable-markets.org/astra-carta/astra-carta-seal/`

private sector partners, the Earth Space Sustainability Initiative (ESSI)§ was launched at the same time, and is now working on sustainability standards for the space sector, with the support of UKSA. The RAS signed the ESSI statement of principles, which explicitly include protection of the night sky, and in that context consideration of the rights of indigenous peoples, were in part developed by the IAU CPS, and we expect a number of our Fellows to serve on its working groups.

As a postscript, we continue to use different political levers to draw attention to the challenge satellite constellations present for astronomy, including in evidence to the new Commons Science, Innovation and Technology Committee, (RAS 2023) and in December 2023 in a debate on the issue in the House of Lords, in response to an oral question asked by a Labour peer (House of Lords 2023).

5. Conclusion

Awareness of the impact of satellite constellations on astronomy now features in UKSA, STFC and OfCom policy, principally within the area of space sustainability and the RAS can take some credit for this, after lobbying through multiple channels over the last four years.

The next challenge for us is to persuade the UK Government to do more than recognise the problem, for example in the licencing regime for launches from British soil, or by operators operating under our national framework. Satellite operators typically prefer voluntary guidance and best practice approaches enforcement, and the free market instincts of the Government concur with this.

In 2022 the RAS joined the IAU CPS as a contributing member, and we give in kind support to its work, with concomitant benefits for ourselves in shaping our dialogue with other stakeholders. The CPS provides much needed global coordination in the astronomy community, and we will continue to draw on its expertise for our public policy response, alongside input from our own working groups.

We should also remain cognisant of the difficulty in making our case to the wider public and (most) elected politicians, not least in a time of major armed conflicts and the climate emergency. In the UK the vast majority of the population lives under light polluted skies, so an apparently small further constraint on their view of the stars, let alone the challenges satellite constellations present to the science of astronomy, may not be significant concerns. Arguably our approach here should focus more on the rights of access to a natural land- or skyscape, and emphasise that private corporations and governments are not the only stakeholders in the development of space.

Irrespective of this, the RAS remains committed to securing the best outcomes for astronomy, including in meeting the challenge of large satellite constellations, and our community can rely on our support in the years to come.

Acknowledgements

The author acknowledges the support of colleagues and interested Fellows of the Royal Astronomical Society, particularly the volunteers in our working groups, members of the CPS Policy Hub who provide support for our work, and the editors of this Symposium for their assistance in preparing the LaTeX files.

References

Bondi, H. 1962, West Ford Project, Introductory Note by the Secretary, QJRAS, 3, 99.

G7 2023, G7 Science and Technology Ministers' Communique, PDF, Group of Seven, https://www8.cao.go.jp/cstp/kokusaiteki/g7_2023/230513_g7_communique.pdf.

§ https://www.essi.org/

His Majesty's Government 2022, Government announces package of new measures to drive space sustainability, Press release, UK Department for Science, Innovation and Technology, https://www.gov.uk/government/news/government-announces-package-of-new-measures-to-drive-space-sustainability.

His Majesty's Government 2023, UK Science and Technology Framework, Policy paper, UK Department for Science, Innovation and Technology, Prime Minister's Office, https://www.gov.uk/government/publications/uk-science-and-technology-framework.

House of Commons 2022, UK space strategy and UK satellite infrastructure, Second Report of Session 2022–23, House of Commons Science and Technology Committee, https://committees.parliament.uk/publications/31490/documents/176763/default/.

House of Lords 2023, Near-earth Satellites, Hansard, 834: debated on Monday 13 November 2023, UK House of Lords, https://hansard.parliament.uk/lords/2023-11-13/debates/48555809-07AB-467F-B043-0252A7F88062/Debate.

Jones, D. 2020, Chair comments on use of ministerial direction for purchase of stake in satellite firm, Technical report, UK Parliament Business, Energy and Industrial Strategy Committee.

Lovell, A. C. B. & Ryle, M. 1962, Interference to Astronomy from Belts of Orbiting Dipoles (Needles), QJRAS, 3, 100.

Ofcom 2022, More spectrum for satellite connectivity: Extending access in the Ku band (14.25-14.5 GHz), Technical report, UK Office of Communications, https://www.ofcom.org.uk/consultations-and-statements/category-1/extending-access-in-ku-band.

RAS 2021, Commons Science and Technology Committee: Inquiry into UK Space Strategy and UK Satellite Infrastructure, PDF, Royal Astronomical Society, https://ras.ac.uk/sites/default/files/2021-11/Space%20Strategy%20and%20Satellite%20Infrastructure%20RAS%20final.pdf.

RAS 2022, Letter to Marlene Dortsch, Federal Communications Commission, PDF, Royal Astronomical Society, https://ras.ac.uk/sites/default/files/2024-01/Dortsch%20FCC.pdf.

RAS 2023, Commons Science, Innovation and Technology Committee: Inquiry into UK Astronomy: RAS response, PDF, Royal Astronomical Society, https://ras.ac.uk/sites/default/files/2023-12/Astronomy%20SIT%20Committee%202023.docx.pdf.

UK Space Agency 2022, UK Statement on Dark & Quiet Skies at the 59th session of the Scientific and Technical Subcommittee of COPUOS, Technical report, https://www.gov.uk/government/news/uk-statement-on-dark-quiet-skies-at-the-59th-session-of-the-scientific-and-technical-subcommittee-of-copuos.

UK Space Agency 2023, First dedicated government fund to build space infrastructure launched with £50 million, Press release, https://www.gov.uk/government/news/first-dedicated-government-fund-to-build-space-infrastructure-launched-with-50-million.

Astronomy and Satellite Constellations: Pathways Forward
Proceedings IAU Symposium No. 385, 2026
P. Grimley, J. Barentine, D. Galadí-Enríquez & C. Walker
doi:10.1017/S1743921324001200

National security interests in relation to dark and quiet skies

Roohi Dalal

Department of Astrophysical Sciences, Princeton University
Princeton, NJ 08544, USA
email: rdalal@princeton.edu

Abstract. The U.S. military has recently begun deploying large constellations of small satellites in low Earth orbit (LEO) for communications and missile warning. This paper describes what such systems may entail, and discusses the impacts that they may have on astronomical and environmental interests. While the deployment of such constellations is being motivated by national security interests, the growth of the satellite population in LEO can also pose threats to national security. This includes not only concerns related to orbital congestion and orbital debris, but also issues that are closely related to the preservation of dark and quiet skies. This paper will highlight the overlapping threats posed by the growth of satellite constellations to both astronomical and national security interests, and describe a potential path forward for dark and quiet skies advocacy that leverages and collaborates with national security interests to establish norms against light and spectrum pollution.

Keywords. Dark and Quiet Skies, National Security

1. Introduction

The U.S. military has recently acknowledged the importance of satellite systems in low Earth orbit (LEO), and has signaled a shift towards deploying large constellations of small satellites in LEO for surveillance, reconnaissance, communications, and missile defense, rather than relying on a less resilient configuration of a small number of large satellites in geostationary orbit (GEO). Given the quickly evolving space security landscape, it will be essential to engage with military users of space in an effort to preserve dark and quiet skies (DQS). These users are both developing satellite constellations in support of national security, and also potentially suffering impacts from light pollution and radio interference from LEO satellites. This paper will focus on the relationship between DQS advocacy and U.S. national security interests through the lens of U.S. Department of Defense (DoD) activities. In particular, I will discuss the activities and plans of the US Space Force (USSF), founded in December 2019, and one of its direct-reporting units, the Space Development Agency (SDA). The SDA is tasked with deploying disruptive space technology, emulating the smartphone industry's approach to upgrades by putting a new set of minimum-viable capabilities on orbit every two years (Machi 2021). As a result of this quick turnaround, DQS advocates should pay close attention to SDA plans, as they will contribute significantly and rapidly to the evolution of the night sky. At the same time, engagement with the SDA offers an opportunity for DQS advocates to develop norms to limit light and spectrum pollution from satellites on a similarly rapid timescale, through the inclusion of mitigation requirements in SDA contracts with satellite manufacturers.

Table 1. Overview of existing U.S. DoD satellite systems in geostationary (GEO) and highly elliptical orbits (HEO). The U.S. currently relies on small numbers of large, expensive satellites in these high-altitude orbits for communications, intelligence gathering and missile warning and tracking.

System Name	Number of Satellites	Purpose
Defense Support Program (DSP) (U.S. Space Force 2020a)	23	Early warning (IR sensor to detect heat from missile and booster plumes)
Space Based Infrared System (SBIRS) (Missile Defense Project 2021; U.S. Space Force 2023)	6 in GEO, 4 in Highly Elliptical Orbit (HEO)	Early warning, theatre and intelligence missions (using short- and mid-IR sensors)
Nuclear Detonation Detection System (U.S. Department of Defense Inspector General 2018)	Not publicly known (hosted on GPS, DSP, and other classified satellites)	Detect, locate, and report nuclear detonation
Military Strategic and Tactical Relay System (MILSTAR) (U.S. Space Force 2020b)	6 (5 operational)	Communications
Advanced Extremely High Frequency System (AEHF) (U.S. Space Force 2020c)	6	Communications
Next-Generation Overhead Persistent Infrared (Next-Gen OPIR) (Space Systems Command 2021; Wolfe 2023)	2 in GEO, 2 in HEO (expected 2026)	Missile warning, tracking, defense

In this paper, I will first discuss the proliferated LEO architecture that is currently being developed by the SDA, as well as other ongoing efforts to better protect national security interests through space activities, highlighting their potential impacts on DQS. I will then suggest how light and radio pollution from satellites might have negative impacts on national security interests, making the case that the DoD should be invested in mitigation measures. Finally, I will present a possible path forward, in which SDA contracts with a number of satellite manufacturers could be leveraged to promote mitigation measures in support of DQS preservation.

2. Why should DQS advocates care about DoD plans?

For many years, the U.S. DoD has relied primarily on systems in geostationary orbit (GEO) for its missile warning/tracking and communications capabilities. An overview of existing GEO systems is shown in Table 1. These systems primarily consist of small numbers of large, expensive satellites, as a result of which the entire system is particularly vulnerable to anti-satellite (ASAT) attacks. For this reason, the SDA announced in September 2022 that the Next-Gen OPIR satellites will be the Space Force's "final missile warning and tracking satellites deployed to that layer [geosynchronous orbit]", and that "the path that the Space Force is marching towards is that ... we'll do away with the GEOs and the big, exquisite, expensive satellites" (Erwin 2022).

Instead, the Space Force is developing a proliferated LEO constellation, named the Proliferated Warfighter Space Architecture (PWSA), with the goal of fielding a more resilient satellite system (Space Development Agency 2023b). As this constellation would consist of a larger number of small satellites ($\sim$1000 satellites by 2026) compared to existing GEO infrastructure, the hope of the Space Force is that "the bullet now costs more than the satellite" and that debilitating attacks on the system would be prohibitively expensive (Machi 2021). Moreover, a large constellation also enables ground-based assets to quickly receive data, which will be crucial in crisis and war-fighting scenarios.

The PWSA will consist of two main layers, the Transport Layer of communications satellites with an optical crosslink, and the Tracking Layer for missile detection and early warning satellites (Machi 2021). These satellites are being procured from multiple satellite manufacturers, including SpaceX, York Space Systems, Northrop Grumman, Lockheed Martin, L3Harris, Raytheon, and Ball Aerospace, with an interoperability requirement (Hadley 2023; Ghouri 2022). Currently, the SDA is launching a demonstration constellation, referred to as Tranche 0, with 23 of the 28 planned satellites on orbit as of October 2023. Tranche 0 will consist of 20 Transport Layer satellites, with 10 each from Lockheed Martin and York Space Systems, and 8 Tracking Layer Satellites, with 4 each from SpaceX and L3Harris (Space Development Agency 2023a). The first operational layer of the PWSA will be Tranche 1, which the SDA plans to field in late 2024. Tranche 1 will include 126 Transport Layer satellites, 35 Tracking Layer satellites, and 12 tactical demonstration satellites, for the demonstration and experimentation of tactical satellite communication and Integrated Broadcast Service capabilites from LEO (Space Development Agency 2023a). This will be followed by the fielding of at least 3 additional Tranche constellations through 2030 (Space Development Agency 2022).

It is not clear whether the brightness of the launched Tranche 0 satellites has been measured. Moreover, the DoD has less incentive to make these satellites difficult to track, given the increased resilience of the system, so it is also not clear whether the DoD has invested in brightness reduction measures for PWSA satellites. As the Tranche 0 satellites are in two orbital planes at an altitude of approximately 1000 km (Space Development Agency 2023a), their brightness is unlikely to exceed that of lower altitude constellations, but they will continue to reflect sunlight for longer portions of the night. With plans for additional Tranche constellations underway, it will be important for DQS advocates to express their concerns about light pollution to the DoD, and present potential mitigation measures, just as we have been doing with commercial satellite operators.

In addition to a proliferated LEO constellation, the SDA has also been developing rapid launch capabilities as part of its Tactically Responsive Space effort. The aim behind this effort is to counter potential adversary offensives by enabling rapid replacement of satellites in LEO. This possibility was successfully demonstrated on September 14, 2023, when Firefly Aerospace launched USSF's VICTUS NOX space domain awareness mission within 27 hours of receiving orders (Erwin 2023). This included all final launch preparations, including trajectory software updates, payload encapsulations, transport to the launch pad, fueling and launch. The potential for a more frequent launch schedule enabled by this rapid turnaround capability will contribute to increased light pollution and sky glow, in addition to a number of other environmental concerns. Moreover, the desire for a rapid turnaround has the potential to stifle regulatory efforts which could slow down the launch approval process. When engaging with the DoD, DQS advocates should be aware of the national security interests that are informing the development of these capabilities, and make the case that national security interests would also benefit from the preservation of DQS.

3. Why should the DoD care about DQS?

In order to engage the DoD on issues related to DQS, it will be important to highlight the relevance of DQS to national security. The proliferation of LEO satellites is broadly a concern for national security because of the overcrowding of orbits and potential for increased debris generation. However, here I focus specifically on four national security issues closely related to light and spectrum pollution from these constellations.

Radio Interference: Interference-free military radio communications are critical for national security and USSF operations (Spire Global Inc. 2022). While much of the space security dialogue has focused on intentional radio interference in the form of jamming satellite communications, USSF operations are also affected by unintentional forms of interference, including unintended electromagnetic radiation (Di Vruno, Winkel,

Bassa, et al. 2023). This problem is exacerbated by the overcrowding of radio frequencies, which can lead to the degradation or denial of security-oriented space capabilities (Roberts Beischl). National security could be bolstered by stricter regulation on spectrum allocation and unintended electromagnetic leakage.

Cislunar Space Situational Awareness (SSA): The National Cislunar Science and Technology Strategy released by the National Science and Technology Council in 2022 suggests that "The United States should advance Cislunar SSA by initially identifying an architecture that optimizes terrestrial, Lunar-based, and space-based sensors, then leveraging existing ground-based sensors and developing new ground-based sensors as needed" (Cislunar Technology Strategy Interagency Working Group 2022). Such ground-based optical sensors would likely need to have a wide field-of-view to cover the large volume of cislunar space, and therefore be similarly impacted by satellite streaks as wide-field astronomical surveys. As these sensors are looking for relatively faint objects and require multiple observations for orbit determination, they will be sensitive to data loss as a result of satellite streaks. Cislunar SSA is becoming an increasingly important and challenging aspect of space security, and in order to have a viable low-cost ground-based solution, it will be necessary for the agency in charge to address the issue of satellite light pollution.

Planetary Defence: As highlighted in the Dark and Quiet Skies II report (Walker & Benvenuti 2022), ground-based telescopes play a crucial role in planetary defense to detect and characterize Near-Earth Objects (NEOs), and if needed, manage or mitigate potential Earth impacts. Defending the planet against a possible NEO impact is certainly both a global and national security interest. Light pollution from satellite constellations, and the resulting loss of data in astronomical images, could cause delays in the identification of NEOs, risking a significant threat to national and global security.

Interference With Theatre and Intelligence Missions: Many of the GEO satellites listed in Table 1 have mid- and short-IR sensors that provide technical intelligence for theater missions and battlespace awareness (Missile Defense Project 2021). A potential avenue of exploration is whether these satellites have the same problems looking in towards the Earth and through LEO that ground-based observatories have looking out. It is possible that the thermal emission from dense shells of LEO and very low Earth orbit (vLEO) satellites will interfere with these missions. However the sensitivity of the IR sensors on these satellites is not publicly available, and will be necessary to estimate whether such interference is significant.

4. A potential collaborative path forward

The DoD is currently thinking deeply about how it will interact with the commercial space sector, with the USSF Chief of Space Operations announcing that their Draft Commercial Space Strategy has to move from "aspirational to actionable" (Foust 2023). In particular, the DoD is interested in reducing its reliance on purely commercial space capabilities, and is instead developing contracts with private industry for space systems, like the PWSA. This is therefore an opportune moment to engage with the DoD on using these contracts to further DQS preservation, by leveraging the national security interests inherent in mitigating light and spectrum pollution.

The DoD is procuring satellites from a number of different operators and manufacturers, with the addition of particular requirements in these contracts. For example, the SpaceX Starshield contract awarded on September 1, 2023, includes additional requirements such as higher levels of cybersecurity than Starlink (Tingley 2023). Moreover, all satellites that are part of the PWSA must be interoperable, even though they are manufactured by at least seven different companies (Space Development Agency 2023c). DQS advocates can advocate for light and spectrum pollution mitigation measures to be a part of these contracts, which would encourage the development of such measures and best practices across many different manufacturers, and hopefully eventually pave the way for mitigation to become a norm in the industry. If this cannot be an explicit requirement of the contract, then the DoD could instead consider having mitigation measures be one aspect of the bidding process, where manufacturers that have implemented mitigation techniques get a boost in the process of awarding contracts. Such steps would not only further the preservation of dark and quiet skies for astronomy and cultural heritage reasons, but would also be in the interest of national security.

Competing interests

The author declares none.

References

Center for Strategic and International Studies 2021, Space-based Infrared System (SBIRS), https://missilethreat.csis.org/defsys/sbirs/

Cislunar Technology Strategy Interagency Working Group 2022, National Cislunar Science & Technology Strategy, National Science & Technology Council, https://www.whitehouse.gov/wp-content/uploads/2022/11/11-2022-NSTC-National-Cislunar-ST-Strategy.pdf

Di Vruno, F., Winkel, B., Bassa, C. G., et al. 2023, A&A, 676, A75

Erwin, S. 2022, DoD to end procurements of geosynchronous missile-warning satellites, Spacenews, https://spacenews.com/dod-to-end-procurements-of-geosynchronous-missile-warning-satellites/

Erwin, S. 2023, Firefly launches Space Force 'Victus Nox' mission, Spacenews, https://spacenews.com/firefly-launches-space-force-victus-nox-mission/

Foust, J. 2023, Space Force revising commercial space strategy to make it "actionable", https://spacenews.com/space-force-revising-commercial-space-strategy-to-make-it-actionable/

Ghouri, F. 2022, SDA awards contracts to Ball Aerospace and York Space Systems, Seradata, https://www.seradata.com/sda-awards-contracts-to-ball-aerospace-and-york-space-systems/

Hadley, G. 2023, SDA Awards $1.5 Billion for 72 New 'Transport' Satellites to Lock-heed and Northrop, Air and Space Forces Magazine, https://www.airandspaceforces.com/sda-72-new-satellites-2026/

Machi, V. 2021, US Military Places a Bet on LEO for Space Security, Space Development Agency, https://www.sda.mil/us-military-places-a-bet-on-leo-for-space-security/

Missile Defense Project 2021, Space-based Infrared System (SBIRS), Missile Threat, Center for Strategic and International Studies 2021, https://missilethreat.csis.org/defsys/sbirs/

Roberts, M., Beischl, C., & Mosteshar, S. 2019, Small Satellite Constellations: National Security Implications, in Pelton, J. (ed.) Handbook of Small Satellites (Springer, Cham), 1, https://doi.org/10.1007/978-3-030-20707-6_52-1

Space Development Agency 2022, About Us 2020-2021, https://www.sda.mil/wp-content/uploads/2022/04/SDA-About-Us-2020-to-2021_march-2022-1.pdf

Space Development Agency 2023a, Space Development Agency Successfully Launches Tranche 0 Satellites, https://www.sda.mil/space-development-agency-successfully-launches-tranche-0-satellites/

Space Development Agency 2023b, SDA Layered Network of Military Satellites Now Known as Proliferated Warfighter Space Architecture, https://www.sda.mil/sda-layered-network-of-military-satellites-now-known-as-proliferated-warfighter-space-architecture/

Space Development Agency 2023c, SDA Issues Solicitation for Tranche 2 Tracking Layer Space Vehicles, https://www.sda.mil/sda-issues-solicitation-for-tranche-2-tracking-layer-space-vehicles/

Space Systems Command 2021, United States Space Force Next Gen OPIR GEO Program completes Block 0 GEO Space Vehicle Critical Design Review, https://www.ssc.spaceforce.mil/Newsroom/Article-Display/Article/2744261/united-states-space-force-next-gen-opir-geo-program-completes-block-0-geo-space

Spire Global Inc. 2022, Identifying Radio Frequency (RF) and GPS Interferences for Military Applications with Satellite Data from Spire Global, https://spire.com/press-release/identifying-radio-frequency-rf-and-gps-interferences-for-military-applications-with-satellite-data-from-spire-global/

Tingley, B. 2023, SpaceX wins $70 million Space Force contract for Starshield military satellites, Space.com, https://www.space.com/spacex-starshield-space-force-contract

U.S. Department of Defense Inspector General 2018, Evaluation of the Space-Based Segment of the U.S. Nuclear Detonation Detection System (Report No. DODIG-2018- 160)

U.S. Space Force 2020a, Defense Support Program Satellites, https://www.spaceforce.mil/About-Us/Fact-Sheets/Article/2197774/defense-support-program-satellites/

U.S. Space Force 2020b, Milstar Satellite Communications System, https://www.spaceforce.mil/About-Us/Fact-Sheets/Article/2197755/milstar-satellite-communications-system/

U.S. Space Force 2020c, Advanced Extremely High Frequency System, https://www.spaceforce.mil/About-Us/Fact-Sheets/Fact-Sheet-Display/Article/2197713/advanced-extremely-high-frequency-system/

U.S. Space Force 2023, Space Based Infrared System, https://www.spaceforce.mil/About-Us/Fact-Sheets/Article/2197746/space-based-infrared-system/

Walker, C., & Benvenuti, P. (Eds.) 2022, Dark and Quiet Skies II Working Group Reports, Zenodo, https://doi.org/10.5281/ZENODO.5874725

Wolfe, F. 2023, Northrop Grumman Receives $235 Million Contract for Next Gen OPIR Polar Satellites, Via Satellite, https://www.satellitetoday.com/government-military/2023/10/06/northrop-grumman-receives-235-million-contract-for-next-gen-opir-polar-satellites/million-contract-for-next-gen-opir-polar-satellites/

Astronomy and Satellite Constellations: Pathways Forward
Proceedings IAU Symposium No. 385, 2026
P. Grimley, J. Barentine, D. Galadí-Enríquez & C. Walker
doi:10.1017/S1743921324001947

Forecasting numbers of post-mission satellites of major constellations, and mitigation methods

Kenneth MacLeod[1,2,3]

[1]IAU Centre for the Protection of the Dark and Quiet Sky from Satellite Constellation Interference

[2]DarkSky International

[3]American Institute of Aeronautics and Astronautics

email: kenmacleod@earthlink.net

Abstract. Surprisingly large numbers of post-mission satellites will be decaying towards re-entry at any time, after the mega-constellations are highly populated and one mission lifetime has expired. This study shows calculations that there will be many thousands at one time. And these post-mission satellites will have substantial additional issues for astronomers. Mitigation methods to be considered include a larger satellite in each orbital ring, to service or de-orbit Internet satellites that have failed or depleted their propellant, as well as joint research projects to develop and mature subsystems to deploy drag balloons, drag tethers, or other methods to de-orbit more rapidly. Proposals include the potential for a joint research facility at an appropriate existing space research centre.

Keywords. Post-mission satellites, expired satellites, mitigation, de-orbiting

---

1. Introduction

This study aims to answer several questions related to the number of non-active satellites that will be present in orbit once the currently planned mega-constellations are in place: How many active satellites would be in orbit when all the projected Internet constellations are fully populated? How many post-mission Internet satellites would be decaying towards re-entry, at any given time? What additional problems will these post-mission populations pose for the expected dark and quiet skies impacts? And what is the best path forward for mitigation?

2. How many Internet satellites in total?

Current plans foresee a total of approximately 100 000 Internet satellites:

- 44 019 from SpaceX (USA)
- 13 620 from Astra (USA)
- 12 992 from SatNet (PR China)
- 7 606 from WorldVu (UK)
- 5 921 from Boeing (USA)
- 15 818 from the next top 15 constellation requests (many nations)

These figures are based on Oltrogge (2022) (see slide 14 therein). Although other estimates have shown much higher potential numbers, Daniel Oltrogge of COMSPOC Corporation considered them unlikely, and it is better to be reasonably conservative.

3. Intake rate and decay time

The product of the intake rate to the decay spiral and the decay time to re-entry yields the average number of satellites in any decay spiral. For our calculation, we divide the post-mission satellites into two categories: expired satellites on the one hand, and failed plus damaged satellites on the other hand. This approach leads to clearer and more accurate calculations.

The intake rate for expired satellites is equal to the number on orbit divided by the lifetime. The intake rate for failed plus damaged satellites is the number on orbit times the failure plus damage rate.

4. Simplified calculation of the population of decaying post-mission Internet satellites

We aim to compute the number of satellites that are in orbit after the end of life and after having experienced failure or damage. The number results from this expression:

$$D = \frac{N}{L} T_L + N F T_F. \tag{4.1}$$

In Eq. 4.1, D is the average number of decaying post-mission Internet satellites at any time. N counts the total number of active satellites in Internet constellations at completion. L represents the average life of satellites in units of time, for instance in years. T_L means the average time, in the same units, to decay to re-entry after end of life. F measures the average rate of failure plus damage in units of inverse time, for instance, years^{-1}. Finally, T_F indicates the average time (e.g., in years) to decay to re-entry after failure on orbit.

We may now introduce some reasonable values for the parameters.

As indicated in Sect. 2, we set $N = 100\ 000$, approximately. We estimate $L = 5$ years, a limitation that arises from the limited mass of propellant for orbit-keeping and station-keeping. A good approximate average for T_L would be 0.5 years, although this value ranges from 4 weeks to several years. If we estimate a 2% failure plus damage rate on orbit per year, we get $F = 0.02$ year^{-1}. Finally, T_F ranges from 1 to more (or much more) than 5 years to re-entry after failure, and we may accept 3 years as a conservative average.

Entering these values into Eq. 4.1, we get:

$$D = \frac{100\,000}{5} 0.5 + 100\,000 \cdot 0.02 \cdot 3 = 10\,000 + 6\,000. \tag{4.2}$$

So, we get an estimation of some 16 000 decaying post-mission Internet satellites, 10 000 of them after having completed their lifetime and 6 000 more having experienced failure or damage. This estimation applies after the constellations are fully populated and after one lifetime cycle has been completed. This is certainly only a small fraction of the total 100 000 active constellation satellites, but these numbers are quite important due to the additional observation problems that these objects imply.

5. Post-mission satellites cause additional problems for astronomy observations

The orbital parameters of post-mission satellites change continuously, and not uniformly, as the objects' orbits decay. These satellites may lose attitude control, which leads to increasing brightness due to loss of reflective positioning. They may even tumble, causing frequent bright flashes that are difficult to predict or remove.

Post-mission satellites will cause numerous re-entry fireballs, very bright in optical and IR, and reflective in radio. The computations shown indicate that there may be as many as 60 re-entries per day.

Some of these objects may transmit powerful radio signals to ALMA, Meerkat, SKA, etc., if muting or navigation functions fail while the transmitter is still working.

Their decreasing altitude leads to increasing brightness, due to the $1/r^2$ relation, but it also makes these satellites spend a larger fraction of their orbit in the Earth's shadow.

Post-mission satellites could potentially collide with each other, yielding rapid increases in space debris.

6. Unparalleled next-generation observatories at risk of reduced science output

Deprecating the next generation of observatories would effectively waste a portion of the funding and technical efforts under way, and deter future funders and developers.

There are seven new billion-dollar/euro-class Earth-surface astronomy facilities coming to fruition:

- Vera Rubin Telescope for optical all-sky survey and transient events
- Giant Magellan Telescope for deep optical observations
- Thirty Meter Telescope for deep optical observations
- Extremely Large Telescope for deep optical observations
- ALMA microwave array
- Meerkat radio array
- Square Kilometer Array radio array

All these facilities were designed before the existence of any of the mega-constellations of Internet satellites, but now they must deal with millions of glints, streaks, occultations, and radio-frequency (RF) bursts. Post-mission satellites will require substantial additional mitigation arrangements.

7. Existing and potential mitigations – first generation

Dimming has been the first obvious mitigation action for optical astronomy, whether using dark coatings, visors, dielectric mirrors, or attitude control for angling away from sunlight.

Even dimmed satellites cause occultations, which will be worse if the satellites grow much larger. Occultations will become more of a problem as observations include more temporally dynamic targets, rather than merely time exposures.

Some mitigation actions may not be effective on post-mission satellites with limited attitude control.

Reducing the mandatory time span for de-orbiting would help. For instance, the US Federal Communications Commission (FCC) de-orbiting time limit has recently been reduced from 25 years to 5 years after end of mission for satellites that will transmit in the RF bands while above the US territory. Best practices documents for satellite launch and operation have been developed and published by the National Aeronautics and Space Administration (NASA), American Institute of Aeronautics and Astronautics (AIAA), United Nations Office for Outer Space Affairs (UNOOSA), European Space Agency (ESA), and others.

Bilateral agreements offer another way to explore. The recent agreement of the National Science Foundation (NSF) with SpaceX Starlink is one example that will be followed soon by similar agreements with OneWeb and Kuiper.

Reducing constellation populations would reduce their impact, considering brightness as a factor. However, the reduction of the population is limited by the $1/r^2$ factor if one wants provide service to small mobile terrestrial terminals.

All these first-generation mitigation measures are necessary and welcome, but probably not sufficient.

8. Potential mitigations – next generation

It is worth considering additional mitigation measures. These may include new systems for rapid de-orbiting of expired satellites. Starlink has proposed to de-orbit in 4 weeks, which would be a big step forward.

Installing redundant systems, and armour, would reduce failure and damage rates.

It would be also useful using a watchdog timer, and backup control to trigger a de-orbit burn upon failure or damage. A watchdog timer could be used, too, to deploy a drag tether or a drag balloon upon failure or damage.

We may consider deploying a service/cleanup satellite in each ring. The deployment of numerous satellites in each discrete ring offers the opportunity to employ a single orbiter to refuel or de-orbit many satellites in the constellation. Such a cleanup satellite could drift around a ring to reach any troubled satellite. It would require a very low delta-V propulsion system, for incremental changes in altitude.

9. Accelerating a path forward

All next-generation mitigation methods described in the previous section require development and testing, then systems engineering and trade studies. Expertise and funding may not be adequate at any one satellite operator. Instead, a joint development and testing organization could be more effective, something similar to Cable Labs, WiFi Alliance, or IEEE 802 for terrestrial communications. The organization could be operated as ad- junct to an existing organization. We may mention several examples from the United States of America: Aerospace Corporation, Jet Propulsion Laboratory (JPL), Argonne National Laboratory, Southwest Research Institute (SwRI), American Institute of Aeronautics and Astronautics (AIAA).

10. Conclusions

From our study we draw the following main conclusions:

- The total number of Internet satellites being planned for the large constellations is approximately 100 000.
- The total number of post-mission Internet satellites decaying downward towards re-entry after end of life or failure on orbit would be about 16 000 at any time.
- Such large numbers of post-mission satellites will cause additional problems for astronomical observations.
- Several promising mitigation methods may be effective.
- Systems engineering and trade studies are needed to define optimum mitigation methods.
- A joint lab for testing and demonstrations could be helpful to this end.

References

Oltrogge, D. 2022, "Backdrop and Overview of Collision Avoidance Analysis and Decision Making", Low Earth Orbit Kinetic Space Safety Workshop, Lausanne, Switzerland, 4-5 May 2022, `https://kineticspacesafety.com/wp-content/uploads/2022/05/KSSWSlides Oltrogge.pdf`

Astronomy and Satellite Constellations: Pathways Forward
Proceedings IAU Symposium No. 385, 2026
P. Grimley, J. Barentine, D. Galadí-Enríquez & C. Walker
doi:10.1017/S1743921324001091

Optical observations of PMS stars from the Rozhen Observatory in Bulgaria

Evgeni H. Semkov

Institute of Astronomy and National Astronomical Observatory, Bulgarian Academy of Sciences, Sofia, Bulgaria
email: esemkov@astro.bas.bg

Abstract. For several decades we have been performing photometric monitoring of some of the star formation regions using the telescopes in Rozhen Observatory. We consider that the study of photometric variability of pre-main sequence (PMS) stars is of great importance in understanding stellar evolution. A number of young stellar objects show brightness variability with large amplitudes that can be registered with small and medium-sized telescopes. But in recent years, bright traces of satellites have increasingly appeared in our CCD images. So far they are not much of a problem since most of our objects and the standard stars around them are point sources of small apparent size. But in the future, the possible increase in the number and brightness of satellite tracks may affect the quality of the photometric data we receive.

Keywords. telescopes, stars: pre–main-sequence, stars: variables: T Tauri, FU Orionis, UX Orionis

1. Introduction

National Astronomical Observatory (NAO) Rozhen is situated in the Rhodope Mountains at 1750 m altitude and is a leading astronomical centre in the southeast Europe. The observatory is equipped with four telescopes for optical observations: a 2-m RCC telescope, a 50/70 cm Schmidt telescope, a 60-cm Cassegrain telescope, and, from the summer 2023, a new robotic 1.5-m RC telescope (Fig. 1). All telescopes are equipped with new CCD cameras and filter wheels with wide-bandpass filters for photometry in the Johnson-Cousins system. For spectral observations, the 2-m RCC telescope is equipped with two high-dispersion spectrographs: Coudé and Echelle. In the near future, we plan to equip the new 1.5-m RC telescope with a low-dispersion spectrograph for observations of faint objects. Since the Rhodope Mountains are one of the few places in Europe preserved from light pollution, we plan to develop optical observations in the future as well.

2. Results and discussion

A significant part of our scientific program is devoted to the study of variable stars of the type of FU Orionis (FUors) (Audard, Ábrahám, Dunham et al. 2014) and UX Orionis (UXors) (Natta, Grinin, Mannings & Ungerechts 1997). These two types of young variable objects show changes in brightness with large amplitudes and play an important role in the study of stellar evolution. The outbursts of FUor-type are caused by a substantial increase in the rate of accretion from the circumstellar disk onto the stellar surface (Hartmann & Kenyon 1996). Young solar-mass stars probably undergo several tens of episodes of such enhanced accretion, whereby up to 50% of the protostellar mass can be added due to the FUor phenomenon. Outbursts usually last for several decades or a

Figure 1. The new robotic 1.5-m telescope at NAO Rozhen.

century, and the rise in brightness of the star is much shorter than the time for the star to fall in brightness.

The explanation of variability of UXors is variable extinction from orbiting circumstellar clumps or clouds of dust or from an edge-on circumstellar disk (Grinin, Kiselev, Minikulov et al. 1991). During eclipse, at low brightness levels, a reversal of the direction of change of the color indices, characteristic of these stars, is observed. The results of our observations of PMS stars strongly suggest that UXor type of variability is a widespread phenomenon among young stellar objects of different masses and spectral types. In many cases we observe different types of variability exhibited by the same object for which long series of optical photometric data are required for study.

One of the most significant results of our observations from Rozhen Observatory is the discovery of a new FUor-type star, V2493 Cyg (Semkov, Peneva, Munari et al. 2010, Semkov, Peneva, Munari et al. 2012). This object turned out to be very important for the study of the FUor phenomenon, since its maximum brightness has lasted for thirteen years (Semkov, Ibryamov & Peneva 2021). Another object of this type V733 Cep (Reipurth, Aspin, Beck et al. 2007), which is well studied by us and shows a light curve that is almost symmetrical in contrast to the rest of FUor stars (Peneva, Semkov, Munari & Birkle 2010). Our most interesting result from studying UXor-type stars is that this phenomenon is common among low-mass T Tauri stars such as V1184 Tau (Semkov, Tsvetkov, Borisova et al. 2008), GM Cep (Mutafov, Semkov, Peneva & Ibryamov 2022a) and V1180 Cas (Mutafov, Semkov, Peneva & Ibryamov 2022b). But there is also a young stellar object V582 Aur that shows in its photometric behaviour a combination of a FUor outburst and a UXor type of eclipses (Semkov, Peneva, Munari et al. 2013).

In recent years we have been registering more and more bright traces of satellites on our optical photometric images (Fig. 2). Our experience shows that frames obtained in the red (R) and yellow (V) filters are mainly affected. So far, these traces do not significantly affect the accuracy of the photometry, unless they pass in close proximity to the studied object. But the expected increase in these phenomena could cause problems in the near future. In order to avoid the unpleasant influence of the constellations of satellites, it is necessary: (1) to obtain more than one image in each filter and (2) to calibrate a larger

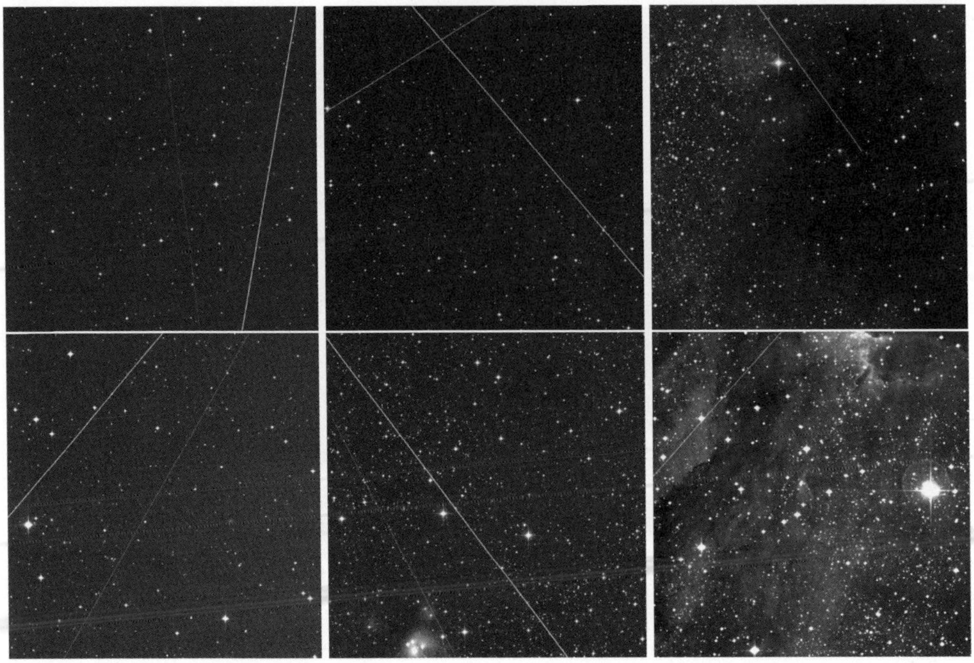

Figure 2. Examples of images affected by satellite constellations. All images were obtained with a 50/70 cm Schmidt telescope at NAO Rozhen.

number of comparison stars around the objects we observe. In this way, we will receive reliable photometric data.

Acknowledgment

The research infrastructure this research was done with is funded by the Ministry of Education and Science of Bulgaria (support for the Bulgarian National Roadmap for Research Infrastructure). This work was partly supported by the Bulgarian Scientific Research Fund of the Ministry of Education and Science under the grant KP-06-N78/5 (2023).

References

Audard, M., Ábrahám, P., Dunham, M. M., et al. 2014, in Protostars and Planets VI, ed. H. Beuther et al. (Tucson, AZ: Univ. Arizona Press), p. 387
Grinin, V. P., Kiselev, N. N., Minikulov, N. Kh., et al. 1991, Ap&SS, 186, 283
Hartmann, L., & Kenyon, S. J. 1996, ARAA, 34, 207
Mutafov, A., Semkov, E., Peneva, S., & Ibryamov, S. 2022a, Bulg Astron J, 36, 3
Mutafov, A., Semkov, E., Peneva, S., & Ibryamov, S. 2022b, RAA, 22, 125014
Natta, A., Grinin, V. P., Mannings, V., & Ungerechts, H. 1997, ApJ, 491, 885
Peneva, S. P., Semkov, E. H., Munari, U., & Birkle, K. 2010, A&A, 515, A24
Reipurth, B., Aspin, C., Beck, T., et al. 2007, AJ, 133, 1000
Semkov, E. H., Tsvetkov, M. K., Borisova, A. P., et al. 2008, A&A, 483, 537
Semkov, E. H., Peneva, S.P., Munari, U., et al. 2010, A&A, 523, L3
Semkov, E. H., Peneva, S. P., Munari, U., et al. 2012, A&A, 542, A43
Semkov, E. H., Peneva, S. P., Munari, U., et al. 2013, A&A, 556, A60
Semkov, E., Ibryamov, S., & Peneva, S. 2021, Symmetry, 13, 2433

Astronomy and Satellite Constellations: Pathways Forward
Proceedings IAU Symposium No. 385, 2026
P. Grimley, J. Barentine, D. Galadí-Enríquez & C. Walker
doi:10.1017/S1743921324000942

Spectral evolution of Iridium satellites recorded by AMOS spectral camera system

Katarína Sabolová, Danica Žilková, Jiří Šilha, Pavol Matlovič and Juraj Tóth

Comenius University in Bratislava
email: sabolova76@uniba.sk

Abstract. Spectroscopy of artificial space objects is a method utilised to retrieve the reflectance spectra of satellites, providing valuable information about the surface properties and material composition of objects. However, the reflectance spectrum of an object tends to slightly change over time, in a way that implies the surface shifting to redder colour hues and darkening overall. The causes of this reddening effect are up to this day unknown, although the leading explanation is the deterioration of surface materials due to exposure to the harsh environment of space. Large satellite constellations are ideal for the analysis of this effect. The studied data were obtained by AMOS - All-Sky Meteor Orbit System - developed by the Comenius University in Bratislava, Slovakia. Specular glints of Iridium satellites recorded over the course of six years are analysed to provide further insight into the explanation of the reddening effect.

Keywords. Space debris, Reflectance spectroscopy

1. Introduction

Spectroscopy of artificial space objects is a suitable method for obtaining information about surface properties and material composition of objects. This information can be used for distinguishing unknown man-made space objects from natural ones, as well as deducing the object's purpose while it was still functional. Spectra of bright specular glints of artificial objects usually fit into one of three distinguished categories: convex spectra, spectra with increasing reflectivity towards longer wavelengths, or straight spectra. As seen in Figure 1, each of these categories represents a characteristic material type. Convex spectra can be associated with solar cell material; spectra with increasing reflectivity towards longer wavelengths with golden or copper-coloured foils used for thermal insulation; and straight spectra with mirror-like, highly reflective metal foils.

2. AMOS satellite spectra

The All-Sky Meteor Orbit System (AMOS), together with its spectral cameras, is a fully-automated worldwide system developed for detection, orbit determination and spectroscopic observation of meteors, (Tóth, Šilha, Matlovič et al. 2019). Specular glints of artificial objects are recorded as a by-product of AMOS observation. Raw spectra are cleared of background noise, solar analogue spectrum, atmospheric extinction, and camera sensitivity, to retrieve the reflectance spectrum of the object. The process is further described in (Žilkova, Šilha, Matlovič & Tóth 2023). Based on the camera sensor's sensitivity curve maximum the wavelength range was constrained to 500-650 nm, and only bright spectra with signal-to-noise ratio greater than 3.5 were selected to minimise errors.

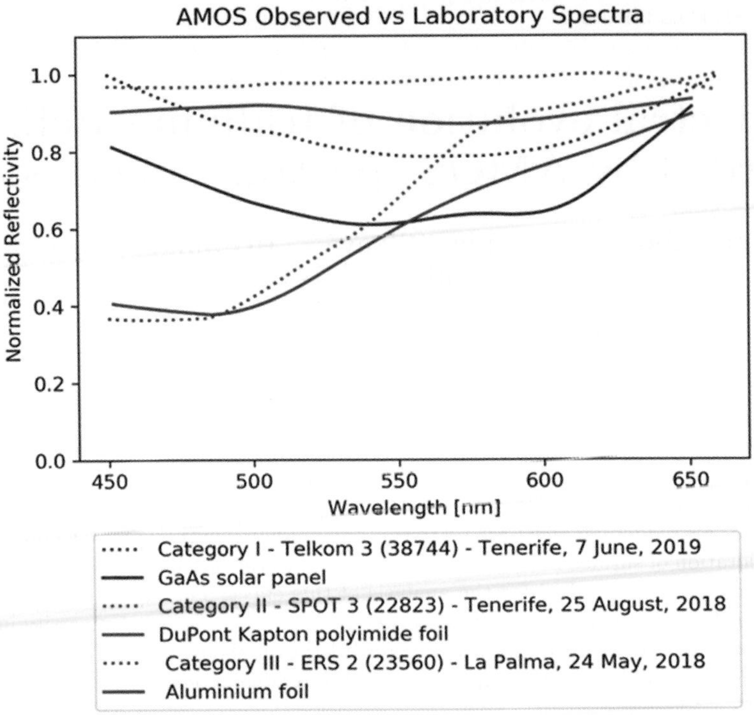

Figure 1. Laboratory measurements of different aerospace materials (solid lines) compared to real observations (dotted lines), (Žilkova, Šilha, Matlovič & Tóth 2023)

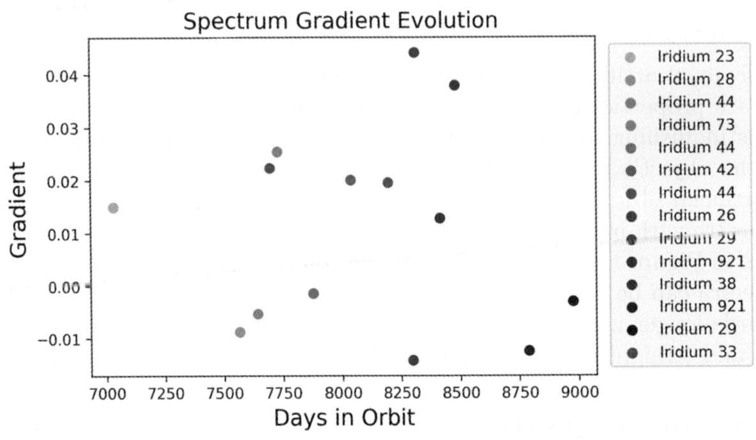

Figure 2. Slope of each reflectance spectrum in relation to the amount of days spent in orbit. The red point indicates the Iridium 33 satellite after collision.

3. Iridium satellites and space weathering

Repetitive spectral measurements of the same object show an unexplained change in the shape of the object's spectrum. There are many possible factors causing this effect, such as the incidence angle of the incoming light and the amount of time the object had spent in space environment. Long exposure to space environment, known as space weathering, leads to overall darkening and a shift to redder colour hues in the object's spectrum (Abercromby, Guyote, Okada & Barker 2005). This is referred to as the reddening effect, as the object's surface appears to become redder in colour. Ideal candidates for the analysis of this effect are satellites of the same design launched in large quantities. In this case, Iridium constellation satellites were selected. These defunct telecommunication satellites are enveloped in three large aluminium antennas oriented towards the Earth's surface, which create bright, abundant glints in the night sky, called Iridium flares, (Maley & Pizzicaroli 2003).

Reflectance spectra of the Iridium satellite constellation collected by AMOS world-wide system over the span of six years are analysed. The spectra are represented by the slopes of their spectral functions in Figure 2. The greater the slope of the function, the more shifted the spectra are towards red. The spectrum of Iridium 33 – marked with red in Figure 2 – was recorded after the collision with a non-functional satellite. It is expected that the surface of this satellite has been deformed during the collision, therefore the shape of its spectrum cannot be taken into account in the analysis (Kelso 2009). The overall slope of the satellites' spectra seems to be constant, or slightly increasing, after already spending 20 years in orbit. As the space weathering effect is most prominent soon after the object's launch into space and decreases over time, different – more recent – artificial satellites would be more suitable candidates for further analysis of the reddening effect.

References

Abercromby, K., Guyote, M., Okada, J., & Barker, E. 2005, 2005 AMOS Technical Conference

Kelso, T. S. 2009, Adv Astron Sci, 135, 1099

Maley, P. D. & Pizzicaroli, J. C. 2003, Acta Astron, 52, 629

Tóth, J., Šilha, J., Matlovič, P., et al. 2019, Neo and Debris Detection Conference, 1

Žilková, D., Šilha, J., Matlovič, P. & Tóth, J., 2023, Adv Space Res, 71, 3249

Astronomy and Satellite Constellations: Pathways Forward
Proceedings IAU Symposium No. 385, 2026
P. Grimley, J. Barentine, D. Galadí-Enríquez & C. Walker
doi:10.1017/S1743921324000589

Amateur astronomers' perspectives on bright satellites in the age of megaconstellations: affected observations and willingness to act

Vladimir Pashkovsky and Stanislav Karpikov

StealthTransit Laboratories, StealthTransit, Germany
email: communications@stealthtransit.com

Abstract. This paper presents an overview of the attitudes and awareness within the amateur astronomical community regarding modern satellite megaconstellation projects. A series of interviews and polls to assess the perspectives of this large community was conducted, aiming to uncover their concerns and issues. Additionally, the potential to involve respondents in community-driven projects focused on satellite data acquisition and processing was investigated. The objective is to enhance data quality for astronomical imagery protection and improve existing tools for scientific research, including satellite tracking, trajectory prediction, and the detection of space debris and orbital objects.

Keywords. Starlink, Satellites, Observations, Tracking, Astrophotography

---

1. Introduction

The issue of Low Earth Orbit (LEO) satellite impact on optical astronomy is now widely addressed in studies and scientific publications within the global community of professional astronomers. In-depth analyses of the mechanisms and consequences of such impacts have been undertaken (see Williams, Hainaut, Otarola et al. (2021)). A significant concern is the interference caused by bright satellite trails appearing in captured images. This issue is expected to intensify in the coming decades due to the increasing number of satellites orbiting Earth, which may reach 80,000 by 2030 (Williams, Hainaut, Otarola et al. (2021)). Notable constellations contributing to this surge include Starlink, OneWeb/Telesat, Amazon Kuiper, IRIS^2, AST Mobile/BlueWalker, China's G60, Roscosmos, and others. Starlink alone is responsible for an annual increase of 1,500-2,000 satellites.

While Starlink is a well-known contributor to this issue, numerous smaller projects, garnering less attention, also pose challenges due to their creators' limited interest in mitigating light reflection from the objects. The impact of modern megaconstellations extends beyond professional scientists to include amateurs engaged in astrophotography.

Various mitigation methods have been proposed (e.g., by Tyson, Ivezić, Bradshaw et al. (2020)), ranging from post-processing models to remove trails after image capture to implementing strict regulations and certification for satellite operators to minimize object reflection (Walker, Hall, Allen et al. (2020); Lalbakhsh, Pitcairn, Mandal et al. (2022)). Unfortunately, most measures fall short of solving the problem entirely (Walker, Hall, Allen et al. (2020)), prompting ongoing discussions and active development of new methods. One of such methods involves predicting the positions of concerning satellites and halting image capture during their passes. A straightforward solution is the use of an active shutter in front of the scientific camera or the telescope, which operates without interfering with the scientific camera, effectively eliminating the trails (e.g., Hainaut & Williams (2020)).

The success of such methods hinges on the accuracy and completeness of satellite tracking data. Current databases like NORAD, however, do not provide precise tracking for some satellites and space objects within required limits as shown by Walker, Hall, Allen et al. (2020). Leveraging the amateur astronomy community to collect and aggregate data emerges as a potential solution. Millions of amateur telescopes, utilized for tens of millions of observation hours monthly, can generate extensive and accurate data, including statistics and measurements.

2. Goals

In this study, our primary objective was to estimate the extent of the impact of satellites on the amateur community, addressing the following aspects:

• Assessing the awareness of the community about the highlighted problem and understanding their attitudes.

• Conducting a geographical analysis to identify the distribution of amateur astronomers, as well as their types of equipment and engagement with the satellite interference issue.

• Evaluating the community's stance on various options for solving the problem, ranging from "halting satellite launches" to "continuing launches despite potential negative consequences."

Geographical assessment is crucial as the development of a community-based tracking system requires coverage over a significant portion of the Earth's surface with detectors.

Our secondary goal is to analyze the potential for astronomical observations made using amateur telescopes to contribute to the global scientific process of space exploration.

3. Method

The study was conducted between September 30, 2021, and February 1, 2022, and comprised the following elements:

• Initial Phase: the study was initiated by analyzing AstroBin (www.astrobin.com), a community of advanced astrophotographers dedicated to sharing experiences related to observation and image processing. This platform was selected as the primary venue for conducting our survey.

• Data Collection: 350 questionnaires were distributed to respondents, providing them with a scale of 11 positions (ranging from 0% to 100%).

• In-Depth Interviews: 13 in-depth interviews were conducted with selected respondents. During these interviews, participants were asked to articulate their attitudes, express specific concerns, and suggest potential ways for their active involvement in addressing the satellite interference problem.

4. Results

The interviews highlighted that astrophotographers closely monitor developments in optics, photonics, and image processing technologies, showing a high level of technical literacy. The survey unveiled a challenging choice, perhaps even a moral dilemma, for this audience. While the satellites cause a negative impact on astronomy, they also contribute significantly to global Internet access, innovative space- and terrestrial-based technologies, and the advancement of the Internet of Things. But over 80% of respondents in the AstroBin community expressed positions such as "Stop Starlink" or "Limit Starlink."

Geographically, respondents from 41 countries participated in the survey, demonstrating the diversity of perspectives. The statistical reliability of conclusions is high, with a measurement error not exceeding $\pm 7.5\%$ at a 95% confidence level.

5. Discussion

5.1. *Community Attitude*

Our research indicates that astrophotographers confront the issue of Low Earth Orbit (LEO) satellite impact on images of the night sky. While it may not be an acute problem for most at present, 21% of astrophotographers reported damage to at least half of their images. The awareness of the problematic influence of LEO satellites on astronomy and the anticipation of increasing satellite numbers in the future are evident among the community.

Notably, European astrophotographers are more inclined to call for a complete halt to the Starlink project compared to their counterparts in the United States. Respondents with more than 6 years of experience in astrophotography are also more likely to support a complete shutdown, while those with less than 5 years of experience lean towards advocating a reduction in satellite number and brightness. A correlation exists between the proportion of images affected by satellites and astrophotographer attitudes towards Starlink; those with more than 20% of their images spoiled tend to oppose the Starlink project.

5.2. *Proposals and Future Projects*

A potential avenue for improving the community's attitude towards satellites involves engaging them in projects for data collection, with potential rewards from companies and government organizations utilizing the data.

We propose an approach that facilitates the extraction of data from these images to enhance Space Situational Awareness (SSA), benefiting the space industry, collision warning systems, and the protection of professional scientific observations. This approach includes:

- Collecting a massive amount of data from amateur telescopes in real-time automatically.
- Extracting relevant trajectory data of all captured satellites using specialized algorithms.
- Comparing this data with other sources, including Two-Line Element Sets (TLEs) from NORAD, to determine the degree of deviation of the real orbit from the orbit described in TLEs.
- Compiling a database of deviations that exceed a certain threshold, signaling critical and potentially concerning situations.
- Making this data publicly available for satellite operators to take action to prevent collisions and for astronomers to safeguard their observations.

A prototype system that implements the above approach for a single telescope has been developed and tested by the authors (Pashkovsky, Karpikov & Balashov (2021)). This system includes a device for detecting and processing trails, along with a local database.

In addition to creating an open database of critical deviations, this approach enables the formation of an alternative public repository of real trajectories compared to NORAD. This repository allows recording and retrieving information of bright objects passing through the field of view of any telescope participating in the project at any given moment in the past.

Supplementary material

To view supplementary material for this article, please visit https://doi.org/10.1017/S1743921324000589.

References

Hainaut, O., & Williams, A. 2020, A&A, 636, A121

Lalbakhsh, A., Pitcairn, A., Mandal, K. et al. 2022, IEEE Access, 10, 24383

Pashkovsky, V., Karpikov, S., & Balashov, A. 2021, Andor Camera Supports First System to Mitigate Impact of LEO Satellites, https://andor.oxinst.com/learning/view/article/andor-camera-supports-first-system-to-mitigate-impact-of-leo-satellites.

Tyson, J. A., Ivezić, Ž., Bradshaw, A. et al., 2020, AJ, 160, 226

Walker, C., Hall, J., Allen, L. et al. 2020, BAAS, 52, 0206

Williams, A., Hainaut, O., Otarola, A., et al. 2021 A Report to ESO Council on the Impact of Satellite Constellations

Astronomy and Satellite Constellations: Pathways Forward
Proceedings IAU Symposium No. 385, 2026
P. Grimley, J. Barentine, D. Galadí-Enríquez & C. Walker
doi:10.1017/S1743921324000681

Assessing the interference of mega-constellations on ground-based astronomy using MASCARA

Peter Breslin, Remko Stuik, Ignas Snellen and Koen Kuijken

Leiden University, Leiden Observatory, P.O. Box 9513, 2300 RA Leiden, The Netherlands
email: pbreslin@tcd.ie

Abstract. As the number of large Earth-orbiting satellite networks (i.e., mega-constellations or mega-sats) increases, so does the threat posed to astronomical research. Bright satellite trails in exposures of the night sky can affect the quality of observations and hinder science objectives. Using the MASCARA station to detect satellites from the SpaceX Starlink network as a representative case, this study quantifies the present and future impact of mega-sats on ground-based optical astronomy. We find that further design revisions are required to mitigate the brightness concern of pre Gen2 Starlinks and that hundreds of satellites could be visible in all-sky observations at peak observing times if additional measures are not taken.

Keywords. satellite interference, starlink, optical astronomy, mega-constellations

---

1. Introduction

The onset of cheaper and smaller spacecraft has driven the development of large constellations of Earth-orbiting satellites (mega-sats) in recent years, with major commercial businesses like Amazon and SpaceX leveraging them to provide high-speed internet services. As large population sizes are required to achieve network connectivity on a global scale, the risk of ground-based observations becoming contaminated with satellite reflections is rising (Venkatesan, Lowenthal, Prem & Vidaurri 2020). For optical images taken over long exposures and/or with a wide field of view (FoV), bright streaks or 'trails' may be detected as the satellite moves. A qualitative assessment of these trails could provide key insights into the magnitude of their interference and yield results that could inspire future satellite design strategies with increased consideration for the astronomical community.

The Starlink network was used as a representative sample for this study. SpaceX have attempted to mitigate brightness concerns through several Starlink iterations, such as the DarkSat and VisorSat. Although these changes reduced reflectivity, neither has been sufficient in darkening the satellites to an acceptable level for astronomical operations (Lalbakhsh, Pitcairn, Mandal et al. 2022).

Observations were collected with an exposure and cadence of 6.4 seconds on 24th October 2022 using the Multi-site All-Sky CAmeRA (MASCARA; Talens, Spronck, Lesage et al. 2017) station, a five-camera wide-field system with a combined FoV providing coverage of almost the entire sky down to 20° from the horizon. The contrast between trail and background was improved by superimposing and subtracting two consecutive images such that only pixels that changed between both images remained. Automated routines were implemented to recognize trails, including procedures for assessing satellite

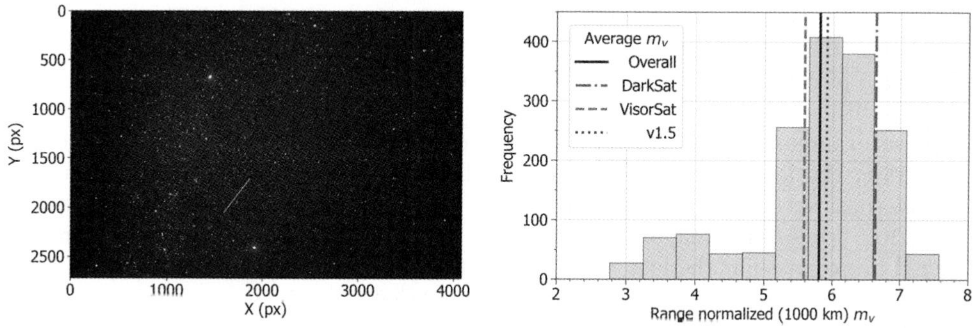

Figure 1. A typical image taken over an exposure of 2×6.4 seconds by one of the MASCARA cameras (FoV of 53×74 degrees) that is polluted with a Starlink trail (left), and the distribution of detected Starlink trail magnitudes scaled to a common distance of 1000 km (right). The line styles indicate the average magnitudes for the different types of Starlinks.

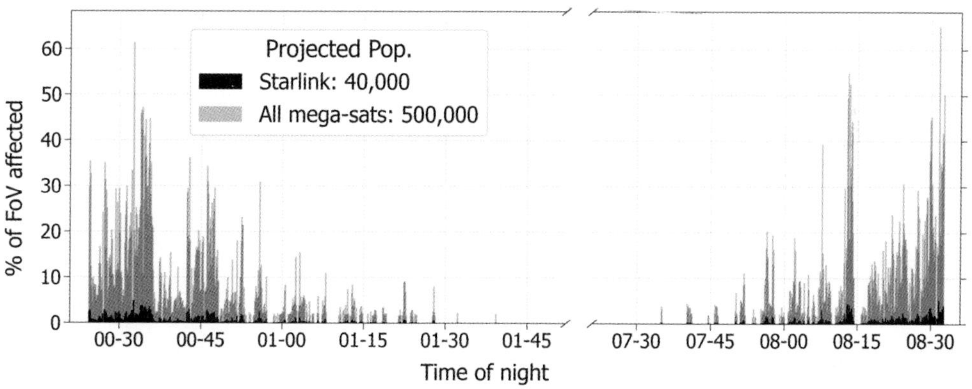

Figure 2. The percentage of the total MASCARA FoV affected by trails for projected population sizes. The conditions of the observing night are used as reference. Black bars approximate the expected Starlink population size (40 000) while grey approximates the expected combined total of all mega-sats (500 000).

visibility, suppressing noise, line detection, and trail extraction. Visual brightness magnitudes (m_v) were then calculated for each trail by comparing a satellite's measured flux to that of reference stars. These were normalised to a common range of 1000km.

2. Results and conclusions

1605 trails were measured and their magnitudes computed for 175 of the 3173 Gen1 Starlinks launched at the time. Their distribution is given in the right-hand panel of Figure 1, where the average over all observations was found to be 5.8 ± 0.10 with a standard deviation of 0.95. Values of 5.6 ± 0.10 and 6.6 ± 0.09 were determined for that of VisorSat and Darksat, respectively. These are all brighter than the goal set by the SATCON1 workshop, which advocates for satellites to be fainter than $m_v = 7$ (Walker, Hall, Allen, et al. 2020). Furthermore, an average magnitude of 5.9 ± 0.09 was found for the v1.5 design, the latest version deployed at the time of observation. We therefore conclude that the mitigation strategies adopted by SpaceX up until this point appear not to have been sufficient in reducing the reflectivity of their satellites.

Images taken when the Sun was at an elevation between 18 and 34 degrees below the horizon were found to be most affected by Starlink trails. As these observations fall outside of twilight hours, it can be concluded that the presence of these trails may interfere with science objectives. The percentage of the MASCARA FoV obscured by trails was examined to better quantify the extent of this interference. While this reached just 0.18%, the level of contamination significantly increases when future population sizes are considered, as depicted in Figure 2. This value reaches 5.2% for the full Starlink population, while a maximum exceeding 65% is found for the combined size across all planned mega-sats. For the latter projection, this may lead to the entire area outside of the Earth's shadow becoming unusable for observations.

To ensure the preservation of uninterrupted astronomical research, timely mitigation strategies should be developed before science operations like large ground-based surveys (e.g., Rubin/LSST) are disrupted.

References

Venkatesan, A., Lowenthal, J., Prem, P., & Vidaurri, M. 2020, Nat Astro, 4, 1043

Lalbakhsh, A., Pitcairn, A., Mandal, K., Alibakhshikenari, M., et al. 2022, IEEE Access, 10, 24383

Talens, G. J. J., Spronck, J. F. P., Lesage, A.-L., et al. 2017, A&A, 601, A11

Walker, C., Hall, J., Allen, L., et al. 2020, BAAS, 52

Astronomy and Satellite Constellations: Pathways Forward
Proceedings IAU Symposium No. 385, 2026
P. Grimley, J. Barentine, D. Galadí-Enríquez & C. Walker
doi:10.1017/S1743921324000656

Experimental procedures for satellite modelling

Eduardo Unda-Sanzana[1], Juan Pablo Colque[1], Nicolás Mac-Lean[1], Christian Adam[1], Jeremy Tregloan-Reed[2], Penélope Longa-Peña[1], Mario Soto[2], Ángel Otarola[3] and Marcos Díaz[4]

[1]Astronomy Center (CITEVA), Universidad de Antofagasta, Chile
email: eduardo.unda@uantof.cl

[2]Instituto de Astronomía y Ciencias Planetarias (INCT), Universidad de Atacama, Chile

[3]European Southern Observatory (ESO)

[4]Departamento de Ingeniería Eléctrica y Laboratorio de Exploración Espacial y Planetaria SPEL, Universidad de Chile, Chile

Abstract. The Astronomy Center (CITEVA) of the Universidad de Antofagasta manages the Ckoirama Observatory and the Astroengineering Laboratory of the Atacama Desert. The former is undergoing an expansion thanks to funding approved by the Chilean Science Agency (ANID) in 2022, with support from other Chilean universities, the Chilean Air Force, and the IAU CPS, so that a new satellite tracking station will begin operations in 2024, specifically designed to collect reflected sunlight from satellites. The latter, on the other hand, is developing an experimental setup designed to test space-grade materials in order to contribute data to produce bidirectional reflectance distribution function models of satellites.

Keywords. satellites, observatories, BRDF

1. The Ckoirama Observatory

Ckoirama (from the Kunza word meaning "Twilight") is the first, and so far only, Chilean-state owned astronomical observatory situated under the skies of the Atacama Desert. It is owned and operated by the Astronomy Center (CITEVA) of the Universidad de Antofagasta on the grounds of an old saltpeter camp called Yungay, located just 70 km southeast of Antofagasta (Chile). Since its opening in 2015, it has supported a range of research, training and outreach activities thanks to its 0.6-m Planewave Telescope, equipped with a FLI Proline 16801 CCD camera and a set of g'r'i' Sloan filters (Char, Unda-Sanzana, Colque, et al. 2016). Since the onset of the astronomical community concerns about the effect of satellite constellations, Ckoirama has participated in observational campaigns that have already led to some key results: the first observations and magnitude measurement of Starlink's Darksat (Tregloan-Reed, Otarola, Ortiz, et al. 2020), the optical-to-NIR magnitude measurements of Darksat (Tregloan-Reed, Otarola, Unda-Sanzana, et al. 2021), and more recently, the high optical brightness of the BlueWalker 3 satellite (Nandakumar, Eggl, Tregloan-Reed, et al. 2023). These results were all obtained using a mount capable only of sidereal tracking, so the observing mode was limited to letting satellites cross the field of view (FoV) and then photometrically analysing the streaks. A project was submitted to the Chilean Science Agency (ANID) to expand the capabilities of Ckoirama with a satellite-tracking station.

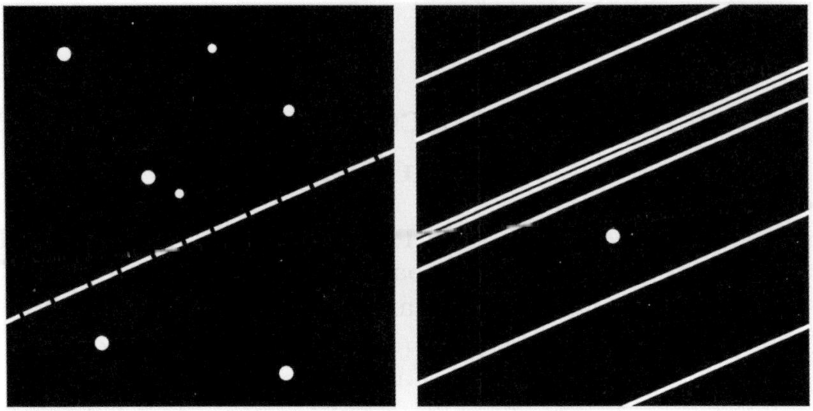

Figure 1. Observing modes of the satellite tracking station.

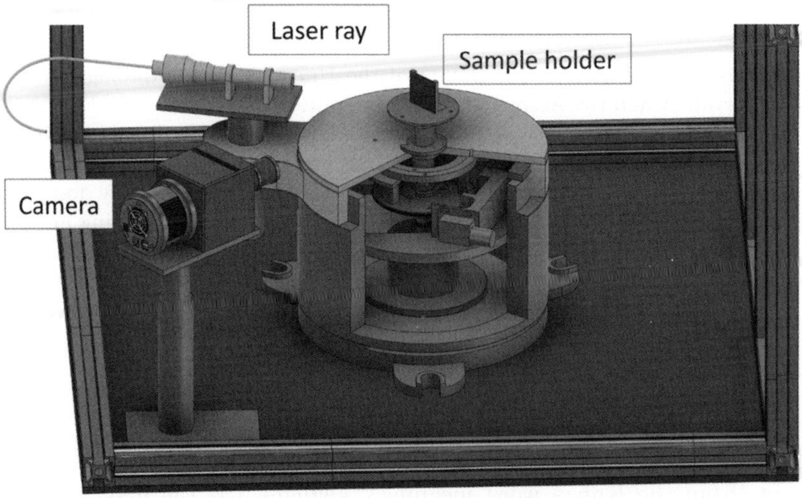

Figure 2. Main components of the reflectivity measuring tool.

The project was supported by other Chilean universities, the Chilean Air Force, and the IAU CPS, and it was selected for funding in December 2022.

2. A new satellite tracking station

The components of the new satellite-tracking station are being manufactured, with delivery expected to Antofagasta by the end of 2023. The new telescope, a 0.5-m Planewave, will sit on a fast, direct-drive mount, designed to accurately tracking fast-moving satellites. Two observing modes will be enabled by the new setup. In the first mode (Figure 1, left) the stars will remain fixed while the satellite crosses the larger than 1 by 1 degree FoV in a series of images (one segment of the satellite's path will be observed in a single image). For each image accurate timings and positions of the satellite will be measured, allowing comparison with the timing and position predicted by the two-line elements. In the second mode (Figure 1, right) the stars will appear as trails while the satellite remains in a fixed position, tracked at high angular velocity.

Some aspects of the operation of the new station will be supported by the Chilean Low Earth Orbit satellite (CLEOsat) group, which will provide software, expertise, and consultation services for the development of satellite tracking and scheduling software for the ground station to use in automatic mode, along with software, expertise, and consulting services to aid in the development of a data pipeline. Full operation is foreseen in 2024.

3. A new tool to characterise space-grade materials

The analysis of satellite data is typically used for bidirectional reflectance distribution function (BRDF) analyses, with the expectation of being able to accurately predict the brightness of a satellite as observed from any angle. This, in turn, is expected to enable observatories to develop avoidance tools that will help them prioritise observations while reducing satellite interference. However, a common unknown that complicates this effort is our lack of detailed information on the designs and material composition of satellites because of commercial confidentiality, so a degree of educated guesswork is involved in this work. Our group is developing a new tool to test the reflectivity properties of samples of space-grade materials. A central device rotates in increments of 1 degree while holding a sample of a suspected space-grade material. A laser fixed to a rotating ring centred on the holder of the material will also move in 1-degree steps while the laser beam strikes a precise point on the sample. Finally, a fixed camera mounted on top of an optical bench measures the reflected light while the ring and device rotate. This setup will be enclosed in a dark, opaque box. The device will thus allow our team to measure the reflectivity properties of any sample from various angles, thereby allowing us to have a full reflectivity profile of each suspected material to be present on the surface of a satellite of interest. This will allow the predictions of theoretical models to be contrasted against experimental data by assuming that the properties of different materials may be combined by weighing the surface fraction they cover. As shown by Fankhauser, Tyson, & Askari (2023), the BRDF data produced by a tool like this can be used to build brightness models that fit well the observations of satellites. The tool is in the stage of prototyping and it is expected to produce first results by the end of 2023.

This work is supported by ANID / Fondo 2022 QUIMAL / 220005

References

Char, F., Unda-Sanzana, E., Colque, J., et al. 2016, Boletín de la Asociación Argentina de Astronomía, 58, 200

Fankhauser, F., Tyson, J. A., & Askari, J. 2023, AJ, 166, 59

Nandakumar, S., Eggl, S., Tregloan-Reed, J., et al. 2023, Nature, 623, 938

Tregloan-Reed, J., Otarola, A., Ortiz, E., et al. 2020, A&A, 637, L1

Tregloan-Reed, J., Otarola, A., Unda-Sanzana, E., et al. 2021, A&A, 647, A54

Astronomy and Satellite Constellations: Pathways Forward
Proceedings IAU Symposium No. 385, 2026
P. Grimley, J. Barentine, D. Galadí-Enríquez & C. Walker
doi:10.1017/S1743921324000619

Classification of orbital anomalies using machine learning models

H. Idelbacha[1], O. Beltramo-Martín[2], M. Kaab[1,3], A. Boskri[1] and Z. Benkhaldoun[1]

[1]Oukaimeden Observatory, High Energy Physics and Astrophysics Laboratory, FSSM, Cadi Ayyad University, Marrakech, Morocco
email: hamid.id.96@gmail.com

[2]SpaceAble, Paris, France

[3]National School of Applied Sciences of Beni Mellal, Sultan Moulay Sliman University, Beni Mellal, Morocco

Abstract. Orbital anomalies, such as unexpected changes in a satellite's orbit, can significantly impact satellite operations and threaten human and financial resources. Detecting these anomalies is a crucial challenge in data analysis within the spatial domain. During satellite operations, a large amount of data is generated, These data allow us to assess the satellite's state at a specific moment. This paper introduces a study aimed at developing a diagnostic algorithm using machine learning techniques to classify orbital anomalies. The proposed algorithm will be able to classify these anomalies based on their origins using various supervised learning models. The performance of the algorithms used was measured by comparing their accuracy and using the confusion matrix over five distinct classes of orbital anomalies.

Keywords. Artificial satellites, SSA awareness, Machine Learning, Anomaly Classification

---

1. Introduction

As the number of satellites in orbit, including satellite constellations, continues to grow, there is an increasing demand to improve trajectory models and enhance operational risk management in order to achieve a deeper understanding of the causes of orbital anomalies. This study employs machine learning models for the classification of anomalies within the orbits of artificial satellites, using their descriptions provided through Two-Line Elements(TLEs). This study will conclude by conducting a comparative analysis of the performance of these different models to analyse the advantages and limitations of the selected models. We assumed that the data provided to the classifier are true anomalies (which we have referred to as 'true positives') This work is a collaborative effort between the Oukaïmeden Observatory† and the company SpaceAble‡.

2. Experiments & methodology

SpaceAble has provided a learning database containing Two-Line Elements (TLEs) of satellites categorized as abnormal. These anomalies fall into five distinct classes: **"Deliberate"**, **"Accident"**, **"Battery"**, **"Deliberate Collision"**, **"Propulsion"**. We have chosen to use three different machine learning models for classification:

† The Oukaïmeden observatory located 50km south of Marrakech is operated by the University Cadi Ayyad and was inaugurated in 2007. https://ouca.uca.ma/an/

‡ SpaceAble is a French startup specialising in Space Situational Awareness services, founded in 2018 by Julien Cantegreil. https://spaceable.org/

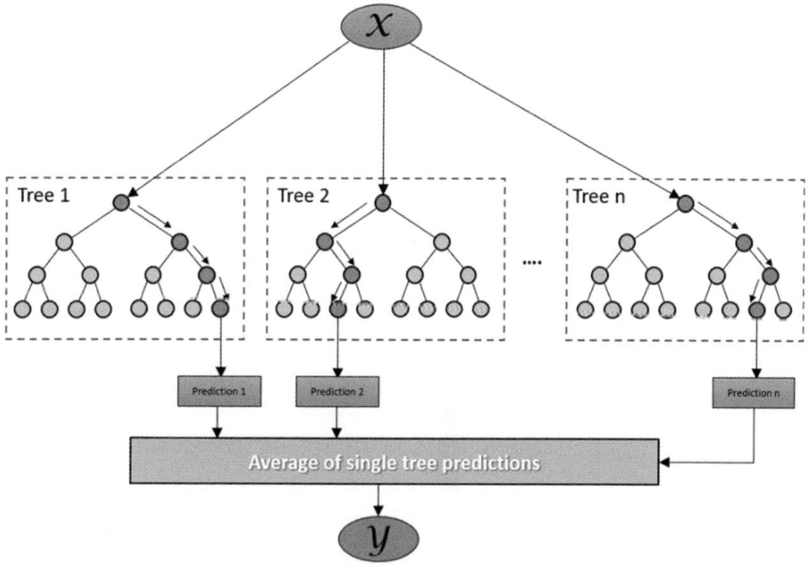

Figure 1. Random Forest Structure.
Milad, Adwan, Majeed, et al. (2021)

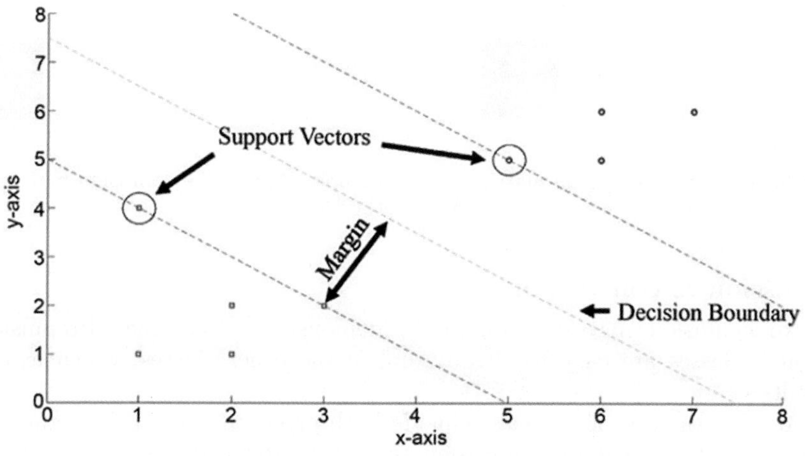

Figure 2. Support Vector Machine Example.
Forghani, Savadjiev, Chatterjee et al. (2019)

Decision Tree(DT), **The random forest(RF)** introduced by Ho (1995) and further developed by Breiman (2001) (see Figure 1) and **Support Vector Machine (SVM; see Figure 2)** a classic machine learning method that aims to find a decision boundary maximising the margin between two classes, providing robustness against noisy data by optimising between margin size and training error Lam, Ling, & Nguyen (2012).

3. Preliminary results

We used classical metrics such as the confusion matrix(Figure 3), precision, recall, and F1 score(see Table 1 and Table 2), as our preliminary results to evaluate our models.

Table 1. The results with the Random Forest method.

	AccidentalCollision	Battery	Deliberate	DeliberateCollision	Propulsion	Average
Precision	1.00	0.00	1.00	0.00	0.69	0.54
Recall	0.50	0.00	1.00	0.00	1.00	0.50
f1-score	0.67	0.00	1.00	0.00	0.82	0.50

Table 2. The results with the SVM method.

	AccidentalCollision	Battery	Deliberate	DeliberateCollision	Propulsion	Average
Precision	0.50	0.00	0.60	0.00	0.75	0.37
Recall	1.00	0.00	1.00	0.00	0.67	0.53
f1-score	0.67	0.00	0.75	0.00	0.71	0.42

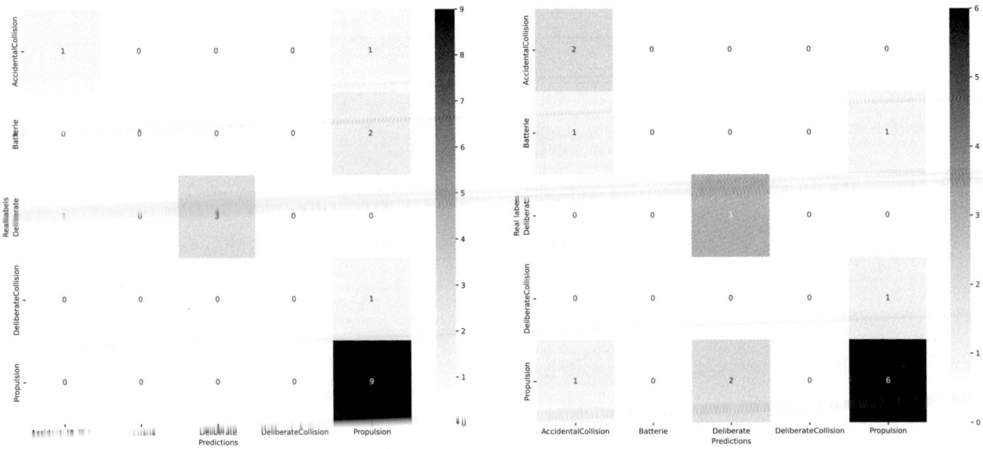

Figure 3. Left: The confusion matrix for RF method. Right: The confusion matrix for the SVM method.

4. Discussion & conclusion

The two confusion matrices (Figure 3) demonstrate that the "Propulsion" and "Deliberate" classes are easy to classify, unlike the other classes, which exhibit very poor results.

These poor performances lead to a decline in both precision and recall averages, though they remain relatively balanced (precision and recall around 0.5)(see Tables 1 and 2). The average F1 score is generally poor.

We are still working on orbital anomaly classification using other models such as neural networks to optimise all models and compare their performance.

The classification of anomalies is possible.

References

Breiman, L. 2001, Machine Learning, 45, 5

Forghani, R., Savadjiev, P., Chatterjee, A., et al. 2019, Comp Struct Biotech J, 17, 995

Ho, T. K. 1995, in Proceedings of 3rd international conference on document analysis and recognition, (IEEE), 278

Lam, H. K., Ling, S. H., & Nguyen, H. T. (eds.) 2012, Computational intelligence and its applications: evolutionary computation, fuzzy logic, neural network, and support vector machine techniques, (London: Imperial College Press)

Milad, A. A., Adwan, I., Majeed, S. A., et al. 2021, IEEE Access, 9, 158041

Astronomy and Satellite Constellations: Pathways Forward
Proceedings IAU Symposium No. 385, 2026
P. Grimley, J. Barentine, D. Galadí-Enríquez & C. Walker
doi:10.1017/S1743921324001303

Solar radiation pressure effect on Telkom-1 using spherical model

N. Pratiwi, D. Herdiwijaya, T. Hidayat and M. I. Ikhsan

Department of Astronomy, Institut Teknologi Bandung, Bandung, Indonesia & Department of Atmospheric and Planetary Science, Institut Teknologi Sumatera, Lampung, Indonesia
email: nindhita.pratiwi@sap.itera.ac.id

Abstract. Telkom-1 is a geosynchronous communications satellite owned by telecommunications company, PT Telkom Indonesia. This paper will discuss solar radiation pressure as one of the most significant perturbations in geostationary orbits. Solar radiation pressure models come in various types, each differing in complexity and accuracy. The simplest solar radiation pressure model is a spherical model, often used for educational or preliminary calculations. Based on our modelling, the results indicate that the semimajor axis values from the spherical model closely approximate the actual data. However, the eccentricity calculated from the spherical model is nearly ten times larger than that derived from the actual data. Additionally, the amplitude of the mean anomaly from the spherical model increases over time, and the spherical model of inclination also demonstrates an increasing trend with time.

Keywords. Solar Radiation Pressure, Spherical Model, Telkom-1, Satellite

1. Introduction

There are two types of perturbations: gravitational and non-gravitational. This research specifically focuses on the effects of solar radiation pressure on orbital satellites, as it is the most significant of non-gravitational perturbations (Baron 2013). The influence of solar radiation pressure on satellite orbital parameters manifests as long-term changes in eccentricity, inclination, right ascension of the ascending node, and argument of perigee (Alessi, Schettino, Rossi & Valsecchi 2018). This study will analyse the impact of solar radiation pressure on Telkom-1, an Indonesian geostationary communications satellite operated by PT Telekomunikasi Indonesia. In solar radiation pressure (SRP) modelling for satellites, the spherical model is a simplification used to approximate the effects of SRP on a satellite's motion. In this model, uniform reflectivity (albedo) is often assumed for the entire satellite surface, meaning that it reflects all incoming solar radiation consistently, regardless of the angle of incidence or location on the sphere (Darugna 2017).

2. Data and method

The inputs for this modelling include real orbital parameter data, solar radio flux data (F10.7 index), geomagnetic activity data (Dst, Ap, and Kp indices), and the geocentric vectors of the satellite and the Sun. Meanwhile, the F10.7, Dst, Ap, and Kp indices were obtained from https://omniweb.gsfc.nasa.gov, covering the period from January to December 2004. The real orbital parameter data were obtained from PT Telkom upon request and include information such as the semimajor axis, inclination, eccentricity, argument of perigee, right ascension of the ascending node, and mean anomaly. In this study, real orbital parameter data are required as an input to calculate the solar radiation

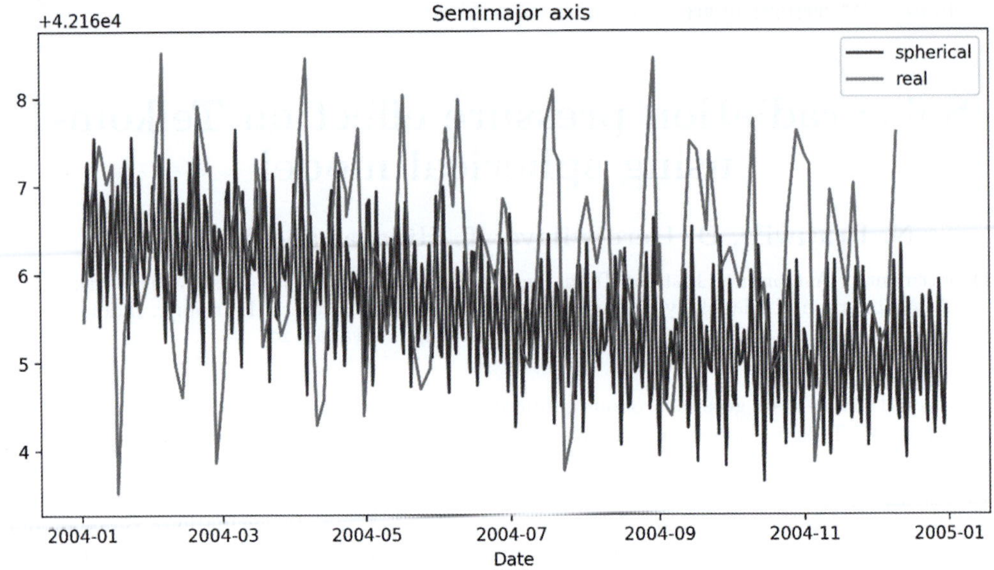

Figure 1. Comparison of the semimajor axis between real data (red) and spherical model (blue).

pressure. We determined the orbital parameters affected by solar radiation pressure using Python and the Freeflyer. This research validates the modelling results with real satellite data and subsequently analyzes the effect of solar radiation on the orbital parameters of Telkom-1. The modelling results are generated daily, while the real data are not available on a daily basis.

3. Results and analysis

The interval in the real orbital parameter data is not uniform, ranging between two and three days. However, we simulated the satellite using SRP at daily intervals in our modelling. Because of the varying intervals, the actual orbital parameter data may lack some level of detail when compared to the spherical model.

Figure 1 shows that the semi-major axis values from the real data and the spherical model are quite similar. The primary difference lies in the trend. The semi major axis values from the spherical model exhibit a decreasing trend, whereas the variations in the real semi-major axis data do not display a clear increasing or decreasing trend. Nonetheless, it is worth noting that the result from the spherical model closely approximates the real data. Moving to Figure 2, we can see that, according to the spherical model, the inclination increases with time. However, real inclination data show some oscillations, suggesting that the real satellite maintains a more stable inclination value. This variability in real inclination can be attributed to the influence of solar radiation pressure, which causes periodic changes in inclination.

Figure 3 compares the eccentricity values derived from actual data and those from the spherical SRP model. The primary distinction between the actual data and the model lies in the eccentricity values. Specifically, the eccentricity calculated from the spherical model is nearly ten times larger than that derived from the actual data. However, upon closer examination, especially in the middle region of the plot, it becomes evident that the eccentricity from the real data and the model exhibit nearly identical periodic behavior. The peaks in both datasets are almost similar in terms of periodicity. Figure 4 compares the real mean and the mean anomalies calculated using the spherical model. The most

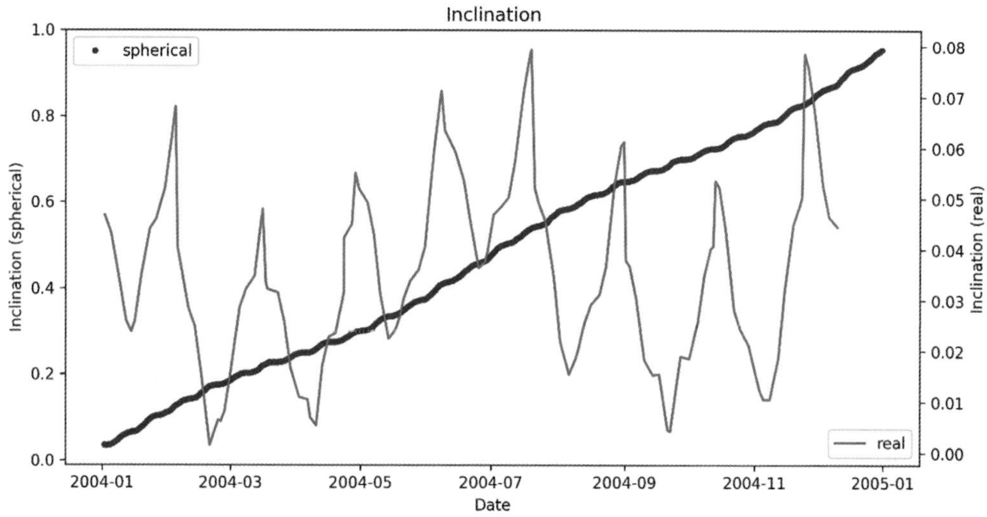

Figure 2. Comparison of the inclination between real data (red) and spherical model (blue).

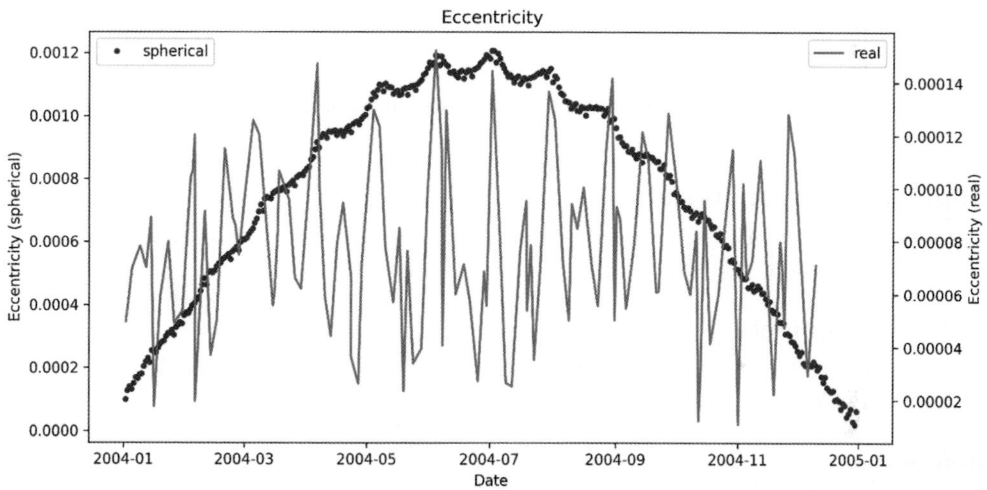

Figure 3. Comparison of the eccentricity between real data (red) and spherical model (blue).

noticeable difference is that the real mean anomaly oscillates between 0 and 360 degrees, whereas the mean anomaly from the spherical model oscillates within a narrower range. However, it is worth noting that the amplitude of the mean anomaly from the spherical model increases over time.

4. Conclusion

We validated our modelling results with actual satellite data and subsequently analysed the impact of solar radiation on the orbital parameters of Telkom-1. Solar radiation pressure influences satellite orbital parameters, leading to changes in semi-major axis, inclination, eccentricity, and mean anomaly. According to our modelling, the results reveal that the semi-major axis values from the spherical model closely approximate the real data. However, it's important to note that the eccentricity calculated from the spherical model is nearly ten times larger than that derived from the real data. Furthermore, the amplitude of the mean anomaly from the spherical model exhibits an increasing trend

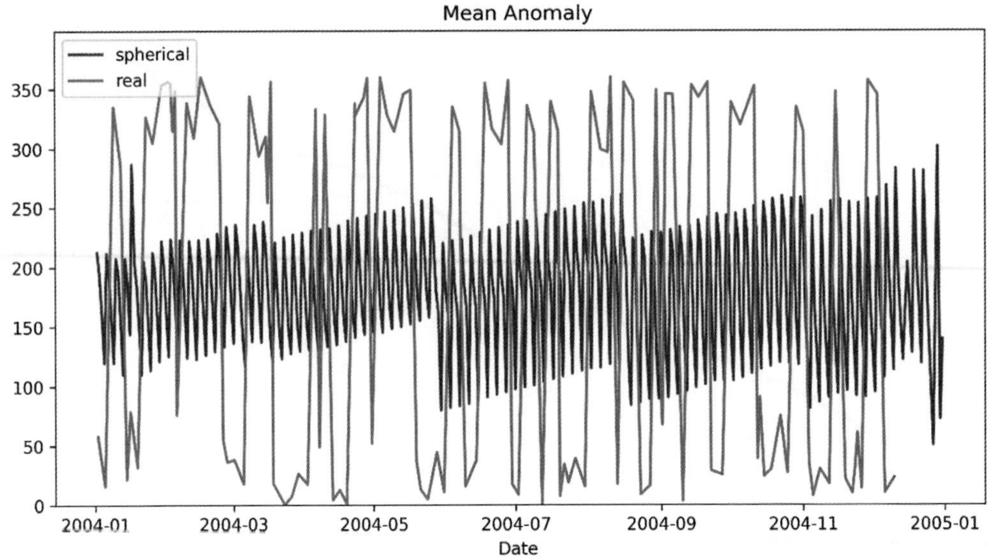

Figure 4. Comparison of the mean anomaly between actual data (red) and spherical model (blue).

over time, and the spherical model of inclination also demonstrates a similar upward trend with time.

Supplementary material

To view supplementary material for this article, please visit https://doi.org/10.1017/S1743921324001303.

Acknowledgments

This research is founded by the 'Dana Bantuan Penelitian Program Beasiswa Tugas Belajar ITERA'.

References

Alessi, E. M., Schettino, G., Rossi, A., & Valsecchi, G. B. 2018, MNRAS, 473, 2407
Baron, A. S. 2013, Baghdad Science Journal, 10(4), 1253
Darugna, F. 2017, Solar radiation pressure modeling for the QZS-1 satellite (Master Thesis), Universita Degli Studi di Padova, 13–22

Astronomy and Satellite Constellations: Pathways Forward
Proceedings IAU Symposium No. 385, 2026
P. Grimley, J. Barentine, D. Galadí-Enríquez & C. Walker
doi:10.1017/S1743921324000930

A catalogue of radio astronomy stations characteristics for sharing and compatibility studies

Boris Sorokin[1,2], **Federico Di Vruno**[1,2] and **Mike W. Peel**[1,3]

[1]IAU Centre for the Protection of the Dark and Quiet Sky from Satellite Constellation Interference, France
email: sathub@cps.iau.org

[2]SKA Observatory, Jodrell Bank, Macclesfield SK11 9FT, United Kingdom

[3]Imperial College London, Blackett Lab, Prince Consort Road, London SW7 2AZ, UK

Abstract. Historically, information about radio astronomy observatories to be used in sharing and compatibility studies was provided by the relevant national administrations responsible for their radio astronomy observatories. The IAU CPS provides a streamlined platform for radio astronomy observatories and satellite operators to collaborate on sharing and compatibility studies prior to enrolling on the national or international regulatory framework, reducing the burden on national administrations. This leads to the requirement to have an alternative solution to have information about radio astronomy observatories. We discussed currently used alternative data collection as well as shared the information on the creation of a CPS-based database to serve as a source of radio astronomy observatories characteristics for sharing and compatibility studies to be carried out within IAU CPS.

Keywords. ITU-R, MIFR, radio astronomy stations, sharing and compatibility studies, database

1. Introduction

Sharing and compatibility studies between different systems are a crucial part of determining the conditions of coexistence between various spectrum users. To enable such studies, parameters of both victim and interfering systems have to be defined. For radio astronomy observatories these parameters could be grouped into three categories each possibly including multiple instances of lower-level ones:

- general (station) parameters:

country, name, type (single dish/array/mixed), coordinates and altitude, min/max frequencies and contact information

- specific antenna parameters:

specific coordinates and altitude, receiver height above the ground, minimum operating elevation angle and min/max frequencies

- specific frequency band parameters:

min/max frequencies, effective antenna area at centre frequency, support of different operation types (continuum, spectroscopy and VLBI), equivalent noise temperature and cryo-cooling capabilities

2. Information access via the ITU

The Radiocommunication Sector (ITU-R) of the International Telecommunication Union (ITU) maintains and hosts the Master International Frequency Register (usually

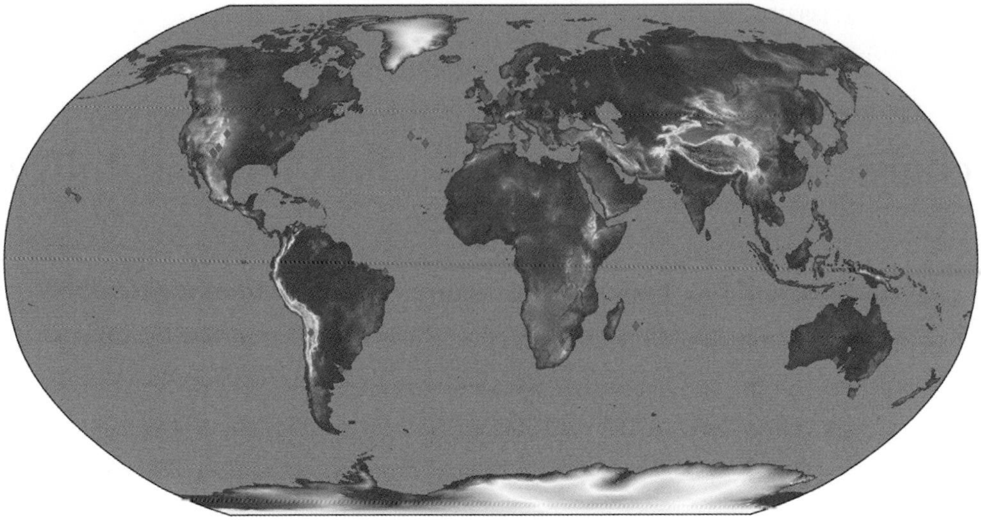

Figure 1. Radio astronomy stations registered in the MIFR.

referred to as MIFR). This database contains frequency assignments based on filings submitted by national administrations in accordance with the Radio Regulations (RR).

2.1. *SNS/SNL interfaces*

Space Network System Online (also known as SNS) and Space Network List (SNL) are common interfaces to retrieve information from the MIFR. These systems require either an annual subscription or ITU-R membership to access.

2.2. *ITU Space Explorer*

The ITU has recently released a new tool, providng a much friendlier user interface compared to the SNS query system. Space Explorer provides users with information on radio astronomy stations' locations as well as their minimum and maximum operating frequencies. More information on this tool can be found on the relevant ITU web page†.

This data is essential for determining the need for sharing and compatibility studies between specific pairs of satellite systems and radio astronomy stations. However, there is not enough information to carry out studies even with default (ITU-R Recommendation RA.1631 for RAS antenna pattern and ITU-R Recommendation RA.769 for RAS protection criteria) parameters. Thus, while this tool provides the ability to determine the affected radio astronomy observatories, it still requires falling back to SNS/SNL to gather the required information.

3. CPS database

The IAU CPS is also developing a database to be used as a source of information for satellite operators who collaborate with the CPS for the protection of radio astronomy stations. The IAU CPS is working together with the ITU on this matter, so the core of this database would be based on the MIFR data snapshots with the potential to be updated according to newer snapshots.

However, not all radio astronomy stations were filed and registered in the MIFR. For example, the South Pole Telescope is not present there, as can be seen in Figure 1.

† `https://www.itu.int/en/ITU-R/space/ITUSpaceExplorer/Pages/default.aspx`

Due to this, the IAU CPS database, unlike MIFR, would also contain additional information provided by the radio astronomy community to keep information up-to-date and also to take into account stations missing from the MIFR.

The database is currently in development, and we expect that the first version will be publicly available in the next year.

Initially, the information to populate the IAU CPS radio astronomy station characteristics database would be derived from MIFR and Space Explorer. All interested radio astronomers are invited to submit any inconsistency they find to ras.database@cps.iau.org.

Astronomy and Satellite Constellations: Pathways Forward
Proceedings IAU Symposium No. 385, 2026
P. Grimley, J. Barentine, D. Galadí-Enríquez & C. Walker
doi:10.1017/S1743921324001194

Challenges to masking satellite streaks in astronomical images

Adam Snyder, Daniel Polin and Tony Tyson

Department of Physics & Astronomy, University of California/Davis, Davis, California, United States
email: aksnyder@ucdavis.edu

Abstract. Bright satellite streaks present challenges for astronomy. No streak detection is prefect and the profile of the streak flux varies along the streak. Using Rubin Observatory's LSST as an example, we undertake full simulations and outline the challenges and the impact on science.

Keywords. Satellite streaks, LSST, streak masking, bogus alerts

1. Introduction

The proliferation of low Earth orbit (LEO) satellites motivated a study of the impact of satellite trails in astronomical images. For this study we use the parameters of Rubin Observatory's Legacy Survey of Space and Time (LSST). Consisting of 2000 fifteen second exposures per night for ten years covering the entire visible sky from northern Chile, LSST can uniquely probe the faint universe as well as discover new types of transient objects which explode or move. LEO satellite interference with LSST science programs can be split into two broad regimes: (A) the impact on faint astronomical object detection and measurement, and (B) bogus astronomical transient alerts. We discuss the issues related to the satellite trail detection and the challenge of trail masking; systematic effects that can occur if trails are not properly identified and/or masked; and bogus transient object alerts.

2. Satellite streak modeling and masking

In astronomical images with many stars and galaxies, satellite streak detection is not 100% efficient. Both initial streak detection and the determination of streak mask width are complicated by the variability in satellite surface brightness profile shapes and variation of the satellite apparent brightness. To study the satellite trail identification and masking implemented by the Rubin Observatory Data Management software stack, it was first necessary to simulate LSST Camera images with realistic trails.

The shape of the satellite trail surface brightness profiles depend on the apparent satellite brightness, the projected size and shape of the reflective elements, the telescope primary mirror size, satellite orbital height, and the observed angle from zenith. First the 2D stationary surface brightness profile is determined assuming the satellite is represented as a disk of uniform brightness that is convolved with the defocus function of a point source and a Von Kármán atmospheric seeing function. Then, the 1D surface brightness profile of the trail created as a result of the satellite's motion across the focal plane is calculated by integrating the 2D stationary surface brightness profile along the direction of the satellite's travel. The code for generating these profiles is implemented as a Python package called LEOSim that uses the open source software Galsim. to create and convolve

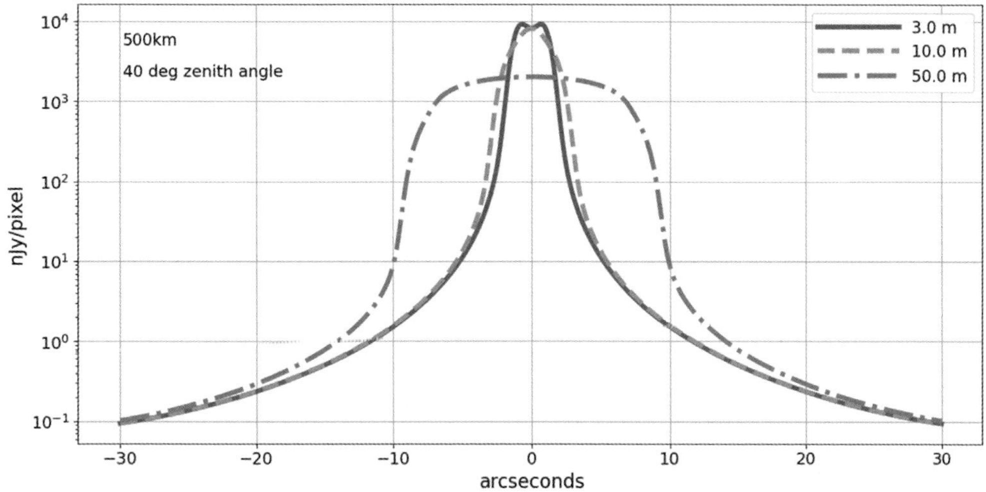

Figure 1. Simulated surface brightness cross-sections of satellite streaks caused by three 5th AB magnitude g-band LEOsats orbiting at a height of 500 km and observed at 40 deg zenith angle by the Rubin Observatory. The extended wings of the cross-sections at greater distances from the streak centers are a result of the Von Kármán atmospheric seeing profile.

the satellite shape, defocus kernel, and atmospheric PSF. Figure 1 shows three examples of 1D surface brightness profiles for different size LEOsats.

Because of this variation in the surface brightness profile shapes, we are therefore interested in "edge" cases where the systematic errors exceed the statistical error (noise), either by failures in either the streak detection or an incompletely masked streak.

3. Satellite trail systematic errors

Hundreds of exposures must be combined in order to detect the faint galaxies required for cosmological science, which are typically 200 million times fainter than LEO satellite trails, in a process known as co-addition. To inform the requirements for satellite trail masking, we studied the conditions under which masked streak residuals in simulated LSST coadd images would result in undesirable bogus galaxy detections. This is due to faint light spilling over the mask and to masking failures. The systematic effects of these bogus detections include contamination of galaxy catalogs and blending of bogus and real galaxies.

An additional consideration is the negative impacts of satellite trails on the study of transient astronomical events. These transient events are detected in single or pairs of 15 second LSSTCAM exposures and the resulting alerts are issued within 1 minute of readout. LEO satellite trails that are partially detected or incompletely masked, as well as satellite momentary reflections from reflective hardware or conops maneuvers (known as "glints") can result in pixels that exceed the threshold for detection resulting in bogus transient object alerts.

4. Discussion

To avoid bogus galaxy detections and bogus transient object alerts, we recommend the following requirements for trail detection algorithms and trail mask width. These algorithms must be adaptive and determined on a "streak by streak basis", as well as be allowed to vary for a given trail along the trail. The mask width must be determined by the streak profile signal relative to the image noise, and be sufficiently large to include significant flux for all possible variations in coadd assembly and source detection threshold. In general this is less then 100% efficient due to the high dimensional space. Simulations show that these mitigations will be difficult and only partially effective. The impact on LSST science will vary, depending on S/N required. A primary goal of LSST – discovering the unexpected – will be most impacted.

Acknowledgements

We acknowledge support from the US NSF (grant AST-2205095) and US DOE (grant DE-SC0009999).

Astronomy and Satellite Constellations: Pathways Forward
Proceedings IAU Symposium No. 385, 2026
P. Grimley, J. Barentine, D. Galadí-Enríquez & C. Walker
doi:10.1017/S174392132400084X

The satellite threat to observational astronomy

Priya Hasan

Department of Physics, Maulana Azad National Urdu University, Hyderabad, India
email: priya.hasan@gmail.com

Abstract. This paper will focus on the satellite threat to observational astronomy. As a member of the Satellite Constellation Working Group: Observatories Subgroup a recommendation was drafted for the Dark and Quiet Skies 2 Report, which shall be discussed. I shall also discuss SatHub activities and the education of observers about satellite constellations for regular citizens, amateur astronomers and professional astronomers and the outreach activities planned.

Keywords. Satellite constellations, observations, mitigation, SatHub

1. Introduction

Astronomy is an observational science which involves studying celestial objects and phenomena through direct observation. This has been fundamental to our understanding of the universe. Satellites have led to a revolution in communication, navigation, and Earth observations. However, their increasing numbers pose a significant challenge to observational astronomers.

Figure 1 shows an image of galaxy group NGC 5353/4 with satellite trails as an indicator of a possible dark future for observational astronomy. Within a few years, the number of satellites which is presently over 5000 will increase by one order of magnitude or two leading to very serious threats to observations (Walker, Hall, Allen et al. 2020; Walker et al. 2020). The effect of this is also on our common heritage as ancestral global commons (Venkatesan, Lowenthal, Prem et al. 2020).

2. The satellite threat

A satellite constellation is a group of artificial satellites working together as a system. In the present model, to provide high internet speeds, it is required to have satellite constellations made up of thousands of satellites at low altitudes to provide high-speed internet anywhere on the globe. In 2018, the U.S. Federal Communications Commission (FCC) approved the deployment of a fleet of 4 425 Starlink satellites (4-5 m) with orbit altitudes ranging from 550 to 1325 km which led to the trails in Fig. 1. OneWeb and Amazon have also plans to place satellites in LEOs for a similar purpose. Startups like AST SpaceMobile and Swarm plan smaller constellations of satellites aimed at specific markets such as mobile communications and the Internet of Things. Companies in China and various places have similar plans. The total number of proposed satellites in LEOs is of the order of 100 000 or even more.

Satellites have been in orbit since the launch of Sputnik 1 in 1957. At present there are a few thousands of them present in the sky, but mostly unobservable by telescopes. The new satellites that are being launched are brighter than 99% of the objects in orbit now (McDowell 2020). SpaceX will multiply this number by a factor of ten or more!!!

Figure 1. An image of NGC 5353/4 galaxy group made with a telescope at Lowell Observatory in Arizona, USA, on the night of Saturday 25 May 2019. The image shows the trails left by more than 25 of the 60 Starlink satellites launched, as they passed through the telescope field of view. Credit: Victoria Girgis/Lowell Observatory

Satellite launch has three phases: initial mission, operational and the de-orbit phase. In the initial phase, satelllites are at low-Earth orbits and with a non-standard orientation. Satellites reflect the light of the Sun and the near earth satellites are visible from the night-time side of the Earth at the begining ≈ 18 hours and end (dawn) of the night. Higher altitudes make satellites fainter, but they stay visible for a longer time. Lower orbits lead to shorter visibility times, but they appear brighter. Visibility also depends on the latitude and time of the year. The satellite has to be in sunlight or the penumbra, not in the Earth's shadow to be seen. Satellite apparent brightness depends on orbital inclination, altitude, attitude, orbital position, observation wavelength and time of the year.

3. Mitgation

SATCON1 recommended a brightness limit of 7th V mag at an orbital height of 550 km. The obvious choice for satellites is making them smaller, to reduce the effect of a single satellite as well as keeping their position predictable so that observations can be scheduled. Possible solutions to mitigate the problem were explored (Hainaut & Williams 2020). The first possible solution was 'DarkSat' to reduce reflectivity by painting them 'black'. However, it led to an increase of brightness in infrared. The next was 'VisorSat' to reflect the light elsewhere and to reduce its luminosity by a factor of 3. The visor was omitted on post-VisorSat spacecraft with laser communication which followed, but the company added a reflective layer which resulted in an intermediate brightness between original and VisorSat. SpaceX is applying advanced brightness mitigation techniques to their Generation 2 Starlink satellites which are larger. The first of these, called Minis, are dimmer than Gen 1 Starlinks despite their greater size.

Photometric observations need to verify that brightness mitigation efforts done by SpaceX substantially reduce spacecraft luminosity. However, satellites still have some negative impact on astronomical observations and the very large satellites planned for Gen 2 may interfere more seriously. There is an urgent need for support of an immediate coordinated effort for optical observations of LEO satellite constellation members, to characterize both slowly and rapidly varying reflectivity (Bassa, Hainaut & Galadí-Enríquez 2022).

4. Recommendations and conclusions

The IAU has set up a Centre for the Protection of the Dark and Quiet Sky from Satellite Constellation Interference (CPS). The CPS has started working on different fronts. There is the requirement of a comprehensive satellite constellation observing network with uniform observing and data reduction protocols for feedback to operators and astronomical programs. Mature constellations will have the added complexity of deorbiting the units and on-orbit aging, requiring ongoing monitoring. It has been recommended to determine the cadence and quality of updated positional information or processed telemetry, distribution, and predictive modeling required to achieve substantial improvement (by a factor of about 10) in publicly available cross-track positional determination.

It is also required to adopt a new standard format for publicly available ephemerides beyond two-line-elements (TLEs) in order to include covariances and other useful information. The application noted in the above recommendation should be compatible with this format and include the appropriate uncertainty estimation.

Urgent work is needed at two levels:

• Constellation operators should work towards reducing reflection through optimal satellite body orientation, Sun shielding, and surface darkening. It is also required to provide accurate and timely ephemerides, and publish information about the satellites (brightness model, transmission bandpasses, etc.) and alert the community about changes in orbits (after an avoidance maneuver, for example).

• Astronomers should perform observations to provide feedback to LEO satellite operators, compile accurate brightness and timing information on these objects, perform simulations to predict visibility, brightness, and timing of satellites, develop software and hardware tools to mitigate the impact of satellite trails in science images. The SatHub will be used as a platform to train and advise astronomers to prepare for this upcoming challenge. This training needs to be provided to regular citizens, amateur astronomers and professional astronomers and outreach activities have to be also developed.

There is an urgent need for astronomers to address this problem and work on the possible ways in which they can get ready for this challenge.

References

Bassa, C.G., Hainaut, O.R., & Galadí-Enríquez, D. 2020, A&A, 657, A75

Hainaut, O., & Williams, P. 2020, A&A, 632, A121

McDowell, J. C. 2020, AJ, 892(2), L36

Venkatesan, A., Lowenthal, J., Prem, P., et al. 2020, Nat Astro, 4, 1043

Walker, C., Hall, J., Allen, L. et al. 2020, BAAS, 52(2)

Walker, C., et al. 2020, Impact of Satellite Constellations on Optical Astronomy and Recommendations Toward Mitigations, NOIRLab

Journal abbreviations used in the references

A&A Astronomy & Astrophysics
A&AR The Astronomy and Astrophysics Review
A&AS Astronomy & Astrophysics Supplements
Acta Astron Acta Astronautica
Adv Astron Sci Advances in the Astronautical Sciences
Adv Space Res Advances in Space Research
AJ The Astronomical Journal
Ap&SS Astrophysics and Space Science
ApJ The Astrophysical Journal
ApJL The Astrophysical Journal Letters
ApJS The Astrophysical Journal Supplement Series
ARAA Annual Reviews of Astronomy and Astrophysics
ASP Conf Series Astronomical Society of the Pacific Conference Series
Ast Nach Astronomische Nachrichten
Astroph Space Sci Astrophysics and Space Science
Astroph Space Sci Proc Astrophysics and Space Science Proceedings
BAAS Bulletin of the American Astronomical Society
Bulg Astron J Bulgarian Astronomical Journal
Commun ACM Communications of the ACM
Comp Struct Biotech J Computational and Structural Biotechnology Journal
Exper Astro Experimental Astronomy
Front Astro Space Sci Frontiers in Astronomy and Space Sciences
Front Ecol Environ Frontiers in Ecology and the Environment
IEEE Commun Mag IEEE Communications Magazine
IEEE Trans Antennas Propag IEEE Transactions on Antennas and Propagation
J Astro Space Sci Journal of Astronomy and Space Sciences
J Astron Inst Journal of Astronomical Instrumentation
J Astronaut Sci Journal of the Astronautical Sciences
J Env Impact Manage Pol Journal of environmental Impact and Management Policy
JGR Journal of Geophysical Research
JOSS Journal of Open Source Software
JPCC Journal of Physics: Conference Series
JSSE Journal of Space Safety Engineering
MNRAS Monthly Notices of the Royal Astronomical Society
Nat Astro Nature Astronomy
New Astro New Astronomy
New Astron Rev New Astronomy Reviews
Opt Expr Optics Express
PASP Publications of the Astronomical Society of the Pacific
PNAS Proceedings of the National Academy of Sciences
Proc Amer Phil Soc Proceedings of the American Philosophical Society
Proc IEEE Proceedings of the Institute of Electrical and Electronics Engineers
Proc IISL Proceedings of the INternational Institute of Space Law
Proc SPIE Proceedings of SPIE
QJRAS Quarterly Journal of the Royal Astronomical Society
RAA Research in Astronomy and Astrophysics
Rem Sens Remote Sensing
Res Astro Astroph Research in Astronomy and Astrophysics
Sci Adv Science Advances
Sci Amer Scientific American
Sci Rep Scientific Reports
Sci Total Environ Science of the Total Environment

Author index